ADVANCES IN RADIATION CHEMISTRY

Volume 3

ADVANCES IN RADIATION CHEMISTRY

MILTON BURTON AND JOHN L. MAGEE, EDITORS

ADVISORY BOARD

Advances in

RADIATION CHEMISTRY

EDITED BY

MILTON BURTON and JOHN L. MAGEE

Radiation Laboratory
and Department of Chemistry
University of Notre Dame
Notre Dame, Indiana

VOLUME 3

WILEY–INTERSCIENCE

A DIVISION OF JOHN WILEY & SONS, INC.

NEW YORK · LONDON · SYDNEY · TORONTO

Introduction to the Series

Radiation chemistry was born in 1895 with the discovery of the X-ray. Progress at first was slow and the field really came alive as a separately needed discipline in 1942 at which time it received its present name. Since then there has been an ever-increasing flood of research publications, reviews, annual reports, theoretical presentations, revisions of theory, etc., at a rate and in various languages (including the separate recondite languages of the various research groups) which has challenged the capacity of the expert to keep abreast.

Some effort has been made by a few noble and adventurous souls to help the novice by publication of complete books which unveil the mysteries of the subject as they were known at the time of authorship. Also, there are continuing annual attempts to interrelate progress in broad general areas—or, at least, to inform the already submerged reader that such areas exist. What then can be the objective of what purports to be a new series? What general purpose can it serve?

A review series helps a scientist only if it decreases the amount of material he must read if he is to remain abreast of current activities. It is the intention of the Editors to gather together, with whatever frequency the subjects demand, developments in the various specialized areas fundamental to theory and experiment in radiation chemistry. For some subjects only brief, occasional reviews may be required. Other subjects in which contributions are scattered throughout the literature and are not outlined in a manner evidently useful to radiation chemists should be occasionally consolidated and amplified for their more efficient reading. It is the objective of this series to perform these functions, and such as may be intermediate between these extremes, with the frequency and the detail that the particular subject matter may require.

MILTON BURTON
JOHN L. MAGEE

June 1969

v

Preface

In a series such as this, the order of presentation of material from volume to volume is almost whimsical. What actually appears in print seems to depend not so much on the planning of the editors as on the writing speed of the authors. Fortunately, because the needs of radiation chemistry are eclectic but the field is broad, the melange of any one volume manifests an inherent unity.

The chapter by Ludwig is not the first in this series concerned with initial phenomena. However, it is the first concerned with observational elements of such study and it does point a direction which it is planned to develop in later volumes. Examination of the variety of excitation processes as well as of the immediately subsequent processes of energy transfer and decay becomes daily more and more penetrating, demands new orders of speed and precision and requires increasingly careful consideration of pertinent theory. Ludwig's chapter is, in such sense, in the tradition of the subject matter introduced in the earlier volumes both by Mozumder and by Fessenden and Schuler and serves, in its turn, as a preface to future material.

The chapter by Vereshchinskii of the U.S.S.R. continues to extend a theme introduced by Wagner of the United States in the first volume of this series, namely the "usefulness" of radiation chemistry. Wagner presented the criteria by which potential for industry may be judged; Vereshchinskii summarizes whole classes of syntheses opened up by the techniques of the field.

Radiation biology, in the sense of these volumes, is a more turgid branch of radiation chemistry, in which the processes are more elusive and ultimately more complicated. They embrace the initial acts, the subsequent immediate physical and chemical processes, the more remote, still elementary but nevertheless palpable, effects that occur in living systems, and finally the delayed effects that may become evident only after many generations and which are the source of a very human preoccupation. The specific nature of the early processes has been a matter of lively interest in the Gray Laboratory at Northwood over many years. This interest has been continued, stimulated, and refreshened by Adams, who himself has made so many of the important, recent contributions. The chapter by Adams does not exhaust the potential subject matter; it is addressed to the radiation chemist as an introduction to a very active field, a field that we hope to develop from a similar viewpoint in later volumes.

Corollary to the picture presented by Adams, a chapter by Emmerson—for those who would like a brief view of selected processes of radiation biology—reflects an enduring, major interest in the effect of radiation on DNA, its mechanism, and its consequences.

And so we discover that we continue to present subjects that naturally engross the interest of the radiation chemist. He learns from related areas and contributes to them. Thus we must—and intend to—have appropriate chapters addressed primarily toward broadening of horizons, so that the radiation chemist may better appreciate his potential for extension of knowledge of this radiant universe.

Finally, volumes like this do not simply "occur" after they have left the authors' hand. We have the happy privilege of expressing our thanks for the varieties of assistance given to us by the Misses Sharon Akers and Mary Ann Skubis, Mrs. Ruth Wiltrout, and Mr. John Lute, as well as the Wiley-Interscience staff.

MILTON BURTON
JOHN L. MAGEE

Windingbrook Park, Indiana
New Year's Day 1972

Contents

ADVANCES IN RADIATION CHEMISTRY

Volume 3

Short-Lived Transients

PETER K. LUDWIG, *Radiation Laboratory* and Department of Chemistry,
University of Notre Dame, Notre Dame, Indiana*

Contents

  * The Radiation Laboratory of the University of Notre Dame is operated under
contract with the U.S. Atomic Energy Commission.

1. INTRODUCTION

Flash photolysis and pulse radiolysis have become common experimental tools for the study of photoinduced or radiation chemically induced processes under nonequilibrium conditions. In their standard forms they provide useful information on reactions occurring during times as short as a few microseconds.

However, numerous reactions of interest occur on a time scale several orders of magnitude smaller, as for instance, relaxation of electronically excited states, geminate ion recombination, excitation energy transfer, or molecular reorientation. For brevity of expression a process is here referred to as fast when its decay time lies between 10^{-9} and 10^{-6} sec, and as ultrafast when below 10^{-9} sec. The experimental methods that have been developed to investigate such processes are in general based on the same principle as that of standard flash photolysis and pulse radiolysis. That is, the system to be studied is briefly perturbed by a pulse of light or high-energy radiation, and its relaxation to equilibrium is followed by observing the variation with time of optical phenomena associated with the relaxation processes. The phenomena commonly observed are luminescence and light absorption.

Despite this similarity design requirements for experimental equipment to be used even in the region of a few nanoseconds differ considerably from those at longer times. This situation results largely from the requirement that the apparatus possess a frequency response sufficiently broad to pass the high-frequency components of fast electrical and optical signals without undue attenuation and phase shift.[1, 2] For example, in order for an electric pulse with a rise time of 1 nsec to be passed through an electrical network without undue distortion, the frequency response must be as large as approximately 450 MHz.[2]

Although modern electrooptical devices such as photodiodes have time resolution as high as $\sim 10^{-10}$ sec, other considerations such as those relating to amplification and noise often prevent their direct use for the short-time work under discussion. More often, therefore, the time dependence of the optical signal of interest is studied by methods that employ some form of sampling technique with the photomultiplier assuming the function of a counter or integrator.

Progress in experimental techniques for studies of fast and ultrafast processes depended also on the availability of pulsed excitation sources delivering light or high-energy radiation of sufficiently short duration and intensity.

In the case of high-energy radiation, these requirements can now be satisfied with suitably designed Van de Graaff generators which permit pulse radiolysis studies of transients with lifetimes of a few nanoseconds. Additionally, linear accelerators now make feasible the investigation of processes as much as two orders of magnitude faster. In the case of excitation by nonionizing radiation, fast pulsed sources based on electric discharge in gases are useful for luminescence work. For flash photolysis the intensity of light from such sources is much too low. However, pulsed lasers have made such studies more promising. Such coherent light sources are of additional interest because pulses as short as 10^{-12} sec can be obtained with mode-locking techniques.

A variety of excitation sources and methods of detection for fast processes has been developed during the last five decades. Some of these techniques are discussed in detail in this chapter even though others may have provided, and still may provide, useful information. The methods discussed in more detail are those that are considered to be more commonly used or which, although not common, appear to represent major recent advances in the state of the art.

This series is concerned with radiation chemistry and thus predominantly with the effects of ionizing radiation. Nevertheless, this chapter presents the effects of nonionizing photoexcitation extensively. Inclusion of such material is appropriate not only because the instrumentation is frequently similar for both modes of excitation but also because experimental data and their interpretation are so often interrelated.

2. OPTICAL TECHNIQUES FOR STUDY OF FAST AND ULTRAFAST PROCESSES

2.1. Survey of Techniques

The first accurate method of measurement of fast luminescence phenomena was developed in 1926 by Gaviola[3, 4] on the basis of earlier work by

Abraham and Lemoine[5] and by Gottling.[6] This technique is the phase-shift method, at present commonly used for measurement of short luminescence-decay times. A more detailed description is given in Section 2.2. Improvements in the phase-shift method were made by Tumerman[7] and by Bailey and Rollefson[8] particularly. Practically all other methods employed to measure fast variations in emission or absorption signals are based on pulse excitation techniques.

In the early work pulsed high-energy radiation sources[9–11] only were employed for excitation because light sources yielding sufficiently short light pulses were then not available. For example, the X-ray tube described by Phillips and Swank[9] in 1953 delivered pulses of about 2 nsec duration and was used for measurement of fluorescence decay times. Fluorescence decay measurements with light-pulse excitation became possible with the pulsed lamp designed by Malmberg[12] in 1957. This lamp provided pulses of several nanoseconds duration and high repetition rates and was the basis for a variety of other designs.[13–17]

With comparatively fast excitation sources thus available, experimental efforts were directed toward improvement of the time resolution of the detection systems. One of the first attempts in this direction was made by Brody[18] in 1957, utilizing the fact that the transit time spread and sensitivity of a photomultiplier can be improved by operation at rather high voltages. In order to avoid damage to the tube when operating at excessive voltage, he employed a pulsed mode of operation. That is, high-voltage pulses were applied to the phototube, for the duration of several decay times of the luminescence under study, in synchrony with the repetition rate of the pulsed excitation source.

Bennett[19] made a skillful modification of the pulsed photomultiplier operation. His method represented a first example of optical sampling. The dynodes of the photomultiplier were directly interconnected by coaxial cable, without the usual voltage-dividing resistance chain. The length of the cable between adjacent dynodes was such that a negative ramp voltage applied to the photomultiplier appeared at the dynodes synchronously with the electron avalanche within the tube. Conditions for electron multiplication at the dynodes were thus established only for the width of the leading edge of the ramp voltage. If this width Δt was less than the decay time of the luminescence pulse, the photomultiplier output corresponded to the light intensity during Δt at some particular time t. The timing of ramp and luminescence pulses could be varied by use of a variable delay line. The luminescence intensity could thus be scanned over the time interval of interest. Although this pulsed mode of operation of photomultipliers is in principle very promising, it seems that experimental difficulties have restricted exploration of its potential.

Perhaps the most remarkable improvement in the time resolution and performance of instrumentation for measuring fast luminescence processes was achieved by Dreeskamp and Burton[11, 20] in 1959 by employment of a modified image converter tube as the sampling device. In the original version the instrument was operated in conjunction with a pulsed X-ray tube and later improved and adapted for use with fast flash lamps.[16, 17] The sampling of the optical signal by the image converter was accomplished in the following way. The luminescence excited (e.g., by the X-ray pulse) in the sample under study falls on a small central exposed slit of the photocathode of the image converter, producing photoelectron current proportional to the light intensity. These photoelectrons comprise a flat beam which passes through a field between deflection plates. The field is adjusted to change abruptly with time so that at a particular selected time the beam sweeps across a slit portion of the phosphorescent screen. The slit is thus exposed to the electron beam during a time interval Δt (ca. 10^{-10} sec). The phosphor emission at the slit is viewed with a photomultiplier for a suitable time interval. The amount of this *phosphorescence* emission depends on the intensity of the electron beam at the time of passage over the slit. That intensity in turn is related to the intensity of the *luminescence* pulse at some time t. The time width Δt is determined by the sweep rate and the width of the screen slit. The precise time at which the luminescence is sampled (over the interval Δt) is fixed by external adjustment of the initiation time of the sweep. For this purpose the electrical signal of the excitation source is passed through a variable delay line and used to trigger an electronic circuit which provides the deflection voltage. In this manner a high degree of synchronism between excitation pulse and the timing of the beam deflection is achieved. Summation of the photomultiplier output over many pulses and variation of the deflection timing yields the average shape of the luminescence pulse.

The time resolution of the detection system, estimated to be about 10^{-10} sec, is considerably better than that of the excitation sources employed.

Today, the most popular optical sampling method is the "monophoton technique" described by Bollinger and Thomas[21] and by Koechlin[22] in 1961. Because of its comparatively widespread application, a more detailed discussion is given in Section 2.3.

Several other methods of measuring fast luminescence decay described in the literature[23] are based on one or the other of the techniques mentioned. In general, with the instrumentation predominantly in use today, time resolution of the order of 10^{-10}–10^{-9} sec can be achieved. In practice, however, this resolution is not fully exploited because the time resolution of commonly used noncoherent excitation sources is of the order of one to several nanoseconds. For evaluation of decay times in this time range from experimental luminescence intensity–time curves, it is in general necessary to apply mathematical corrections which involve solutions of convolution integrals. Even when such corrections are applied, it is at times quite difficult to establish accurate error limits on the final results. The difficulty lies in setting up reliable criteria for what constitutes acceptable agreement between computed and experimental curves. For an excellent discussion of the mathematical procedures involved in extracting decay time data from experiments, the reader is referred to an article by Helman.[24]

Progress in fast flash photolysis and radiolysis techniques began much later than the fast luminescence work. The main problem was the development of excitation sources of sufficient intensity for production of photolytic or radiolytic transients detectable by absorption spectroscopy.

Fast pulse radiolysis first became possible in 1967 after modification of the 3-MeV Van de Graaff generator of the Argonne National Laboratory by Ramler, Johnson, and Klippert.[25] The instrument delivered repetitive electron pulses of several amperes for periods as short as 1 nsec. A year later, Bronskill and Hunt[26] demonstrated the potential of the linear accelerator as a source of electron pulses with duration of a few 10^{-11} sec.

Flash photolysis in the nanosecond region became feasible with the development of Q-switched lasers,[27] which typically deliver output pulses of about 10–30 nsec duration. The apparatus of Novak and Windsor,[28] which utilizes a device of this kind as the source of excitation and indirectly provides the analyzing light, appears to be the first (1967) fast flash photolysis setup. Since then several similar systems have been described in the literature. They are discussed in Section 2.4.

One of the most exciting developments in flash photolysis has resulted from application of mode-locked pulsed lasers[29] (see Section 3.5) to photochemical processes.[30–32] Excitation pulses of only a few picoseconds duration allow measurements of ultrafast relaxation processes, such as excited molecular states in the condensed media. A more detailed description is given in Section 2.5.

2.2. Phase and Modulation Fluorometry

2.2.1. *Operating Principle*

Phase and modulation fluorometry methods employ a continuous light beam, which is intensity modulated at high frequency, to excite the luminescent sample of interest. The intensity modulation of the resultant luminescence is shifted in phase with respect to that of the exciting light, and the degree of modulation is reduced. The shift in phase is a consequence of the comparative slowness of the luminescence process, which is typically of the order of 10^{-9}–10^{-7} sec.

The time-dependent, observable luminescence intensity $I(t)$ is related to the exciting light intensity $I_e(t)$ and the luminescence–decay function $h(t)$ by the convolution integral

$$I(t) = \int_0^t I_e(t') \cdot h(t - t')\, dt' \tag{2.1}$$

A particularly simple solution of this integral is obtained for the frequently occurrent case in which the luminescence decay is singly exponential and the modulation of the exciting light is sinusoidal. Then, the following two fundamental expressions are obtained

$$\tan \phi = \omega\tau \tag{2.2}$$

$$m_e/m_f = [(\omega\tau)^2 + 1]^{1/2} \tag{2.3}$$

where ϕ is the phase shift of the fluorescence signal with respect to the exciting light, ω is the angular frequency of modulation, and m_e and m_f denote, respectively, the degrees of modulation[1] of exciting light and of luminescence.

Measurement of the quantities ϕ or m_e/m_f thus permits determination of the mean luminescence decay time τ. In many applications of the phase-shift technique, ϕ alone is measured. Determination of both ϕ and m_e/m_f has served mainly as a check of the assumption of exponential decay, because only then do the two measurements give consistent results. In addition, Birks and co-workers[33, 34] showed that, with simultaneous measurements of ϕ and m_e/m_f, one can study luminescence decay phenomena that involve the sum or difference of two exponentials—for example, as in the case of monomer plus excimer emission of certain aromatic hydrocarbons. Although in principle phase and modulation fluorometry are not limited to studies of such simple decay characteristics,[35] it appears that luminescence decay with more complex time dependence has not been studied experimentally by this method.

2.2.2. Instrumentation

Many instruments have been developed particularly for phase fluorometry. Birks and Munro[23] have reviewed the various designs so that a brief description of their operating principle suffices here. A very detailed description of a modern instrument and a discussion of the experimental problems are presented by Berlow.[36]

Figure 1 shows schematically the main components of a phase fluorometer. The exciting light beam of a strong continuous source L is modulated with a frequency of several megahertz by the modulator M and impinges on the sample S. Modulation may be achieved using electrooptical devices,[37–39] such as Kerr or Pockels cells, or ultrasonic diffraction gratings.[7, 40–43]

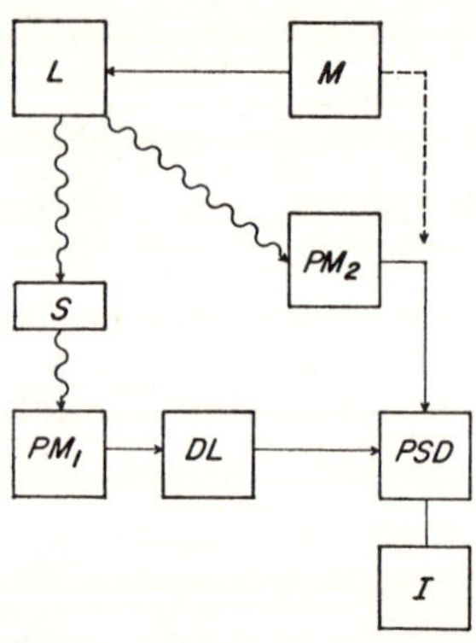

Fig. 1. Block diagram of phase-shift apparatus. In this diagram the wavy connecting lines represent light signals; the straight ones represent electric signals. Light originating from a source L, and with sinusoidal intensity modulation provided by the modulator M, excites the sample S. The resultant luminescence fluctuation stimulates an alternating electric current via the photomultiplier PM_1. The electric signal passes through the variable delay DL and is fed to the phase-sensitive detector PSD. An electric reference signal can be obtained via a second photomultiplier PM_2 (viewing the excited light) or directly via the modulator M. Comparison of the phases of the two signals is accomplished by variation in the delay of the PM_1 signal and monitoring of the PSD output with an indicating instrument I.

The modulated luminescence is converted into a corresponding electrical signal by the photomultiplier PM_1. A reference signal may be derived either from a second photomultiplier PM_2, which views part of the exciting light, or directly from the modulation power supply. Both signals, after equalization, are fed into a phase-sensitive detector PSD, the output of which can be monitored by a suitable indicating or recording instrument I. Several phase detection methods have been employed. For example, the two signals may be fed to the x and y inputs of an oscilloscope,[44] thus displaying the Lissajous pattern. In other instruments either balanced input circuits[8] or vector-summing amplifiers[36, 39] are used. The phase difference ϕ of interest is obtained by comparing the phase shift of the luminescence signal with respect to the reference signal. The phase shift is actually measured by adjusting the variable delay line DL until a "null reading" is obtained at the phase-sensitive detector output. With modern components and careful design, the accuracy of lifetime measurements in the range from 5 to 10 nsec is estimated to be 0.1 nsec. Various factors that determine the range and limitation of applications of the phase-shift technique are discussed by Berlow.[36]

2.3. Single-Photon Counting Technique

Among the methods for measuring fast luminescence processes that employ pulsed excitation sources, the single-photon counting technique[21, 22] is the most popular and versatile one.

The method is an adaptation to the optical wavelength region of the delayed coincidence techniques common in nuclear physics. Its operating principle rests on the probability relation between detection of a single photon emitted from the light source to be studied during a small time interval Δt at t and the intensity of that source.

Under certain conditions, discussed in detail in Section 2.3.1, a particularly simple one-to-one correspondence exists for detection of a photon during Δt at t and the photon flux in the same time interval. In other words, the intensity variation of the light source can be obtained from measurement of the time distribution of single-photon events as observed, for instance, with a photomultiplier. The method is particularly well suited for studies in which the luminescence time dependence is not known in advance or is known to be more complex than exponential.

In practice, one does not fully benefit from this advantage for processes faster than about 5 nsec because of the limited time resolution of commonly used excitation sources. In such and similar cases, mathematical procedures such as described by Isenberg and Dyson[45] are required to extract the relevant data from the experimental curves.

2.3.1. *Theory*

Consider a pulsed light source such as a fluorescent sample repetitively excited by a fast flash lamp or pulsed high-energy source. The average luminescence–pulse shape may be described by the function $I(t)$; see Eq. (2.1).

A fraction of the light with an average intensity of $\iota(t)$ photons per second is admitted to the photocathode of a photomultiplier. The time range during which the temporal variation in the light intensity is to be measured encompasses several decay times and is divided into n_c equal intervals, or channels, of width Δt. Because the light intensities employed are comparatively weak and because Δt is usually of the order of a few tenths of a nanosecond, the appearance of a photon during Δt at some time t is subject to large fluctuations and must be treated as a statistical phenomenon. In particular, it is necessary to explore the relation between the probability of appearance of a photomultiplier pulse and the light intensity at time t during the interval Δt.

In the following development the time difference between detection of a photon and its emission from the source is ignored because the primary concern is the shape of the light intensity–time curve and not its absolute position on the time axis. Furthermore, the separation between light source and photodetector is small and the decay processes are in the nanosecond region, so that effects of optical dispersion can be safely neglected.

The probability that n photons reach the photocathode in the small time interval Δt at time t is given by a Poisson distribution[46]

$$P_i(n, \Delta t, t) = \frac{\mu^n e^{-\mu}}{n!} \tag{2.4}$$

where $\mu = \int_t^{t+\Delta t} \iota(t')\,\mathrm{d}t'$ and $\iota(t')$ is the number of photons per second per pulse reaching the cathode at time t'.

Because the photodetector efficiency a is less than unity, conversion of photons to photoelectrons is a statistical process, and the probability that n photons produce z photoelectrons is given by the binominal distribution

$$P(z, n) = \frac{n!}{z!(n-z)!} \cdot a^z (1-a)^{n-z}$$

The probability of occurrence of a photodetector output pulse during the time interval Δt at time t can be obtained[183] by summing Eq. (2.4) over all values of n with appropriate consideration of the binominal expression. Thus one obtains

$$P_o(\Delta t, t) = 1 - \exp(-a\mu) \tag{2.5}$$

Consideration must next be given to the dead time of the photon detection instrumentation. During the time of processing and storing of a photomultiplier pulse by the associated electronic circuits and their recovery, the instrumentation is insensitive to any further pulses. This time is typically of the order of several microseconds. The instrument therefore sees only the first pulse that appears during the time range in which measurement of the variation in the luminescence intensity is desired (typically in the nanosecond region).

The dead-time effect can be taken into account by considering the probability that a photomultiplier pulse occurs during Δt at time t with no pulse having arrived up to time t. Stated differently, it is the probability that the pulse observed by the instrumentation is indeed a first pulse and not a second, and that the instrumentation is really responding to such a photon. That conditional probability is identically $P_r(\Delta t, t)$ and is given by

$$P_r(\Delta t, t) = P_o(\Delta t, t) \cdot \overline{P(n, t)} \tag{2.6}$$

where

$$\overline{P(n, t)} = 1 - \sum_{n=1}^{\infty} \frac{\mu^{*n} e^{-\mu^*}}{n!} = e^{-\mu^*} \tag{2.7}$$

with

$$\mu^* = \int_0^t \iota(t')\,\mathrm{d}t' \tag{2.8}$$

Combination of Eqs. (2.5), (2.6), and (2.7) yields

$$P_r(\Delta t, t) = (1 - e^{-a\mu})e^{-a\mu^*} \tag{2.9}$$

Equation (2.9) represents the probability of occurrence of the first photomultiplier pulse in the interval Δt at time t.

It is immediately apparent that for the case $a(\mu + \mu^*) \ll 1$ the probability of occurrence of a photomultiplier pulse is proportional to the light flux $\iota(t)$ so that

$$P_r(\Delta t, t) \propto a\mu \tag{2.10}$$

Thus in the case of an experiment with sufficiently low intensity, the measured temporal photomultiplier pulse distribution represents directly the shape of the light pulse under study. In practice, a counting efficiency of up to about 3 percent (3 photomultiplier pulses recorded for every 100 exciting light pulses) is commonly found to satisfy Eq. (2.10). The advantage of a faithful representation of the light intensity–time relation for sufficiently low counting efficiency suffers from the comparatively long time required for data collection. Use of a higher counting efficiency is therefore very desirable. In this case, as seen by inspection of Eq. (2.9), the linear relation between observed pulse distribution and light intensity is lost. However, mathematical corrections based on Eq. (2.9) can be applied to the experimental data and the light intensity–time curve can be recovered. Several practical expressions for this correction have been given in the literature.[21, 47, 48, 182] A particularly detailed theoretical and experimental investigation of this problem has recently been made by Gregory.[47] From this work it is apparent that much higher counting efficiencies can be employed with resultant much shorter data collection times and correspondingly greater accuracy of the results. A maximum signal-to-noise ratio can be obtained at 60-percent counting efficiency according to this study. The work also shows how the required data correction can be automatically applied by incorporation of a computer into the experimental apparatus.

2.3.2. *Instrumentation*

Because of the ready availability of commercial components, a monophoton apparatus can be easily assembled. Several such setups are described in the literature.[21, 22, 48–52, 147]

Figure 2 shows a typical assembly scheme of a monophoton apparatus involving optical excitation. *Light pulses* from a lamp L pass through a monochromator M which selects the desired wavelength region for excitation of the sample S. Depending on the nature of the sample, its luminescence pulses are viewed by a photomultiplier PM at a 90°, or much smaller, angle to the direction of the exciting light. A continuously variable neutral-density filter F permits adjustment of the luminescence intensity viewed by the photomultiplier to satisfy the single-photon condition. With the filter combinations FC, scattered exciting light is suppressed and the luminescence wavelength region of interest is selected. The discriminator and pulse-shaper

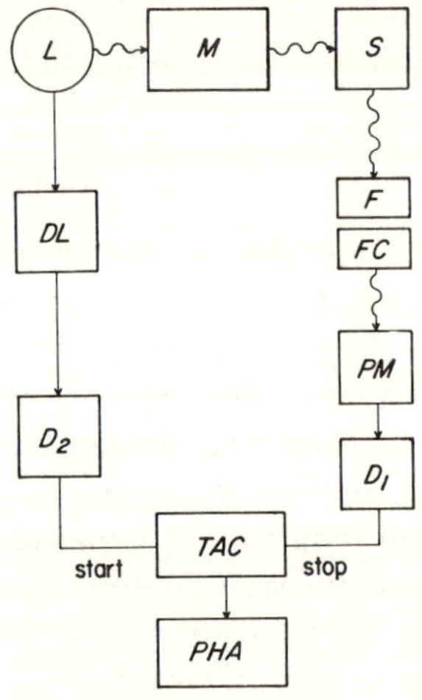

Fig. 2. Photon-counting apparatus for measurement of time dependence of luminescence intensity (see text). In this diagram the wavy connecting lines represent light signals; the straight ones represent electric signals. L, Pulsed light source; M, monochromator; S, sample; F, variable neutral-density filter; FC, filter combination for suppression of scattered, exciting light and selection of wavelength region of luminescence of interest; PM, photomultiplier; D_1 and D_2, discriminators and pulse shapers; TAC, time-to-amplitude converter; PHA, pulse-height analyzer; DL, delay line.

unit D_1, set to reject undesirable signals, delivers output pulses of uniform shape which enter the "stop" input of the time-to-amplitude converter TAC.

The *electrical pulses* associated with the light pulses of the lamp are very short and provide the time reference. They are passed through the variable delay line DL before entering a "discriminator/shaper" D_2, which provides the same uniform output pulses as appear at the output of D_1. The output pulses from D_2, when they reach the "start" terminal of the time-to-amplitude converter TAC, provide the reference time signal. Thus a pulse from D_2 activates for a preset time interval T the "internal clock" of the TAC, which then is stopped by the first pulse arriving at the "stop" terminal. For any such pair of electric pulses occurring within the interval T, an output pulse is generated with an amplitude proportional to the *difference of arrival times* of the two pulses. By storage of the TAC output in a pulse-height analyzer PHA and measurement for a sufficiently long time, one thus records the time distribution of photons from a large number of luminescence pulses.

Decay times may be obtained graphically by plotting the distribution with an x-y recorder or, better, by storing the data and subsequent computer evaluation.

Application of the monophoton technique is not limited to use with a pulsed light source, as described here. The source may be replaced by a pulsed X-ray tube, thus permitting investigation of luminescence–time characteristics for ionizing and nonionizing excitation. Moreover, even random excitation pulses may be employed, for example, as from radioactive materials.[48, 51, 52] In the case of even a moderately strong α-particle source, each such pulse or ray produces a large number of excitations per unit distance in the medium through which it passes. Then, the leading edge of the light signal thereafter emitted is *sufficiently* steep so that the output pulse of a (second) photomultiplier wide open to the system provides a zero-time reference signal; that is, a signal that occurs during the time resolution of the apparatus. The first photomultiplier (as represented in Fig. 2) observes only a very small fraction of the irradiated system, so that the probability of its seeing a single photon is directly related to the decay time of the excited species in the system under observation. Laustriat[51] used such a device primarily for α-particle studies, while d'Alessio and his co-workers[52] extended the application even to γ-ray studies.

2.3.3. *Comments*

The time resolution obtainable with the monophoton technique utilizing high-gain modern photomultipliers is estimated to be $\sim$1 nsec. In practice, the resolution is limited more by the time characteristics of the pulsed exciting source. Typically, half-widths of one to a few nanoseconds of lamp or X-ray pulses may be achieved. Frequently, however, a weak but much longer tail in the exciting pulse requires mathematical correction of the experimental luminescence curve for short decay-time studies.

In the case of random excitation with individual particles from a (weak) radioactive source, the duration of excitation is very short. Nevertheless, in this case much uncertainty may be introduced in the reference time signal if the number of photons emitted per particle is *small*, with the result that fluctuations in the output of the reference photomultiplier may occur.

2.4. Laser Flash Photolysis

The production of light pulses for nanosecond flash photolysis by conventional discharge lamps suffers from the requirements of simultaneous high intensity and short pulse duration. These requirements oppose each other; reduction of the pulse duration demands decrease in electrical discharge capacitance, which in turn reduces discharge (and light) intensity; increase in light intensity by increase in capacitance increases pulse duration.

Light pulses useful for fast flash photolysis became available with the advent of Q-switched lasers and, more specifically, solid-state lasers. Q-switching provided the means to convert the relatively long-lasting and fluctuating laser output to a very intense and rather well-defined pulse of typically 10–30 nsec duration. Even then these pulses are of limited photochemical interest because the wavelengths for the frequently used ruby and neodymium-glass lasers are 694 and 1060 nm, respectively. The second factor, in addition to Q-switching, which made such lasers useful for fast flash photolysis, is the nonlinear optical effect of frequency multiplication.

By frequency doubling or even quadrupling, light can be generated at wavelengths of 530, 347, and 265 nm, thus permitting fast flash photolysis studies of a variety of compounds of photochemical and radiation chemical interest.

The detection system of a fast laser flash photolysis apparatus is in principle similar to that of the more conventional technique. Observation of the rapidly changing absorption signal does, however, place some different requirements on the analyzing light source.

2.4.1. Q-Switching

Generation of short light pulses[27] is accomplished by inserting (in the resonator cavity) a component the light transmission of which is variable and controllable (a Q-switch). During the early part of the pumping pulse, the transmission of that component is kept low, so that optical amplification of the laser oscillator remains below threshold. Consequently, a large population inversion (i.e., in excited states compared to ground states) is built up in the laser material. If the element is then made to transmit light, the amplification suddenly lies above the threshold value and may result in a short light pulse of high power in the case of a large overpopulation of the inversion. Under suitable conditions the whole inversion may thus be removed in a single pulse.

A variety of Q-switches has been employed,[27, 53–55] but for fast switching, and thus short output pulses, devices based on either the Kerr effect or the Pockels effect (active Q-switches) and saturable absorbers (passive Q-switches) are most useful.

The Kerr effect is an electrooptic phenomenon displayed by optically isotropic liquids; they become birefringent on application of an electric field. Nitrobenzene and carbon disulfide are particularly notable in the magnitude of response. A similar phenomenon in solids is called the Pockels effect.

An *active switch* consists essentially of a polarizer (the function of which may also be assumed by the ruby, appropriately cut) and a Kerr cell or a Pockels cell; see Fig. 3.

As shown in Fig. 3, the light originating from the laser-active material L, after passage through the polarizer P, arrives at the Kerr cell K, plane-polarized. The angle between the direction of the plane of polarization of the light and of the electric field E applied to the cell is set at 45°. Because of the birefringent properties of the Kerr cell, when an electric field is applied the light beam splits into two components upon entering the cell. One of the components is polarized in a direction perpendicular, and the other parallel, to the applied electric field. As in the case of birefringence in a uniaxial crystal, the two components propagate with different velocities in the cell,

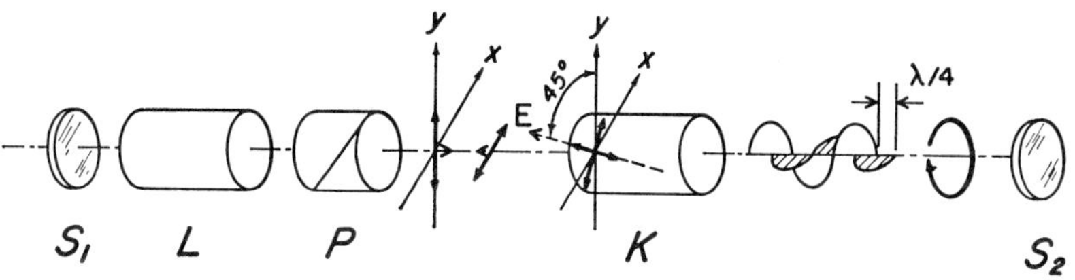

Fig. 3. Active Q-switching of a solid-state laser. The totally reflecting mirror S_2 and the partially reflecting mirror S_1, parallel to each other, form the optical cavity of the laser. L represents the laser-active material in which population inversion is effected by an intense pulsed (helical) light source (not shown). The polarizer P and the Kerr cell K form the active Q-switch. (For passive Q-switching a dye solution is put in their place.) The function of the switch is illustrated by showing the polarization of the light (heavy arrows) at several locations between P and S_2. The double-headed arrow in the y direction indicates the plane of polarization of the light that has passed the polarizer P from the left. Upon entering the Kerr cell, the light splits into two equal components; one is parallel to the static electric field E of the cell, the direction of which is set at an angle of 45° to the direction of the plane of polarization of the entering light; the second component is perpendicular to the direction of E. After passage through the Kerr cell, the two components emerge with a $\lambda/4$ phase shift, resulting in circular polarization of the light. As the consequence of reflection at S_2 and second passage through K, an additional $\lambda/4$ phase shift causes the returning light again to be plane-polarized. However, the plane of polarization is now rotated by 90° (double-headed arrow in the x direction) with respect to that of the original direction of the plane of polarization. The returning light is thus prevented from passage through P; the switch is closed. Interruption of the electric field applied to the Kerr cell destroys the effect on the plane of polarization of the light which consequently can thereupon pass through the polarizer back to L; the switch is open.

leading to a field-dependent phase shift between them. With appropriate adjustment of the field, the light can thus be made to emerge from the cell with circular polarization, corresponding to a phase shift of $\lambda/4$ between the two light components. Upon reflection from the end mirror and traversal of the cell in the opposite direction, the two components are subject to an additional phase shift of identical magnitude. The total phase shift is thus $\lambda/2$. The light *emerging* at the cell *entrance* window is therefore again plane-polarized. The plane of polarization, however, is now rotated by 90° (i.e., in a "round trip") with respect to the initial direction of polarization. The light is thus prevented from passage through the polarizer; that is, the shutter is closed. At a time at which optimal inversion has occurred in the laser-active material, the electric field is suddenly (and temporarily) removed. Because of the associated disappearance of the birefringent properties of the cell, the light now passes through the polarizer and the shutter is suddenly opened and the properties of the laser cavity required for light amplification are restored. The effect is to induce stimulated emission (laser action) which continues (in this case briefly) until the upper state is exhausted.

Commonly used *passive switches* consist of dye solutions with intensity-dependent absorption characteristics in the wavelength region of the laser emission. The transmission of the solution for low light intensity is adjusted in such a manner that the laser resonator reaches the threshold for light amplification at the time of maximum inversion t_s; that is

$$R \cdot V(t_s) \cdot T_0 \approx 1$$

where $R \equiv (R_1 \cdot R_2)^{1/2}$, with R_1 and R_2 being the reflectivities of the resonator mirrors, $V(t_s)$ the amplification of the laser material at time t_s, and T_0 the transmission of the switch at low light level. For several dyes the transmission T approaches unity within nanoseconds so that then

$$R \cdot V(t_s) \cdot 1 \gg 1$$

and a fast laser pulse is generated. In addition to being very fast, such passive switches operate at low loss and therefore yield shorter and more powerful output pulses than do other devices.

2.4.2. *Frequency Multiplication (Doubling)*

Generation of harmonics of the fundamental laser frequency when the light interacts with certain dielectric media is formally a consequence of the nonlinear relation between polarization and electric field strength.[56, 57] Let **P** denote the polarization of the medium, **E** be the electric field strength, and $\bar{\varepsilon}$ terms denote electric susceptibilities which in the most general case (anisotropic media) are tensors. Then, while over the whole range of classical optics the relation

$$\mathbf{P} = \bar{\varepsilon}\mathbf{E} \tag{2.11}$$

is valid, the relation must be extended at high field strengths to include higher-order terms

$$\mathbf{P} = \bar{\varepsilon}_1\mathbf{E} + \bar{\varepsilon}_2(\mathbf{E})^2 + \bar{\varepsilon}_3(\mathbf{E})^3 \cdots \tag{2.12}$$

It may be qualitatively shown from the form of the second right-hand term of Eq. (2.12) that a harmonic can be generated with twice the frequency of the fundamental. Detailed analysis shows that such effect occurs in a first approximation only with crystals that lack a center of inversion.

In practice, the laser beam is made to pass through double refracting crystals of KH_2PO_4 (KDP) or $(NH_4)H_2PO_4$ (ADP) for frequency doubling. The efficiency of conversion of the fundamental to the second harmonic depends on the power of the light pulse and on the angle between the incident light and the crystal axis. The angular dependence is a consequence of dispersion, which results effectively in cancellation of the second harmonic in a thick crystal. For a particularly chosen angle of incidence, however, the speed of propagation of the ordinary beam of the fundamental and that of the extraordinary beam of the second harmonic are equal, and maximum conversion of up to about 30 percent can be achieved (phase matching).[58, 59] For ruby light and KDP, this angle between crystal axis and incident light is 52°.

In order to obtain light at even shorter wavelength, the second harmonic may be passed through a second KDP crystal.

2.4.3. *Instrumentation*

In the flash photolysis apparatus, the frequency-doubled, or -quadrupled, laser output pulse is directed into the sample under study. As in standard flash photolysis, two methods of monitoring the resultant transient, absorbing photoproducts are employed; that is, spectroscopic and kinetic monitoring. Spectroscopic monitoring requires passage of a second light pulse through the sample at predetermined times after the excitation pulse and recording of the spectrum with a spectrograph. For kinetic monitoring a constant light beam is passed through the sample at a wavelength corresponding to the known absorption maximum of the transient, and the intensity variation with time is recorded. In both cases the experimental requirements differ somewhat from those of ordinary flash photolysis. For spectroscopic monitoring an analyzing light pulse is required, with time characteristics comparable with those of the exciting pulse. For kinetic monitoring a very intense beam is necessary to achieve a sufficiently high signal-to-noise ratio.

2.4.4. *Spectroscopic Monitoring*

The generation of a fast analyzing light pulse was achieved in a very elegant manner by Novak and Windsor,[28] who developed the first successful laser flash photolysis apparatus. By use of a Q-switched ruby laser, the fundamental and second harmonic were produced in the manner described in Section 2.4.2 and spatially separated by a prism. The ultraviolet pulse served for excitation of the sample, while the red light was focused onto a gas-filled quartz cell. Because of the high intensity of the red light in the focal region, electric breakdown occurs in the gas resulting in a spark with light emission over a wide spectral range. The choice of gas determines the duration of the emission. Oxygen, for example, gives a light pulse that coincides with the 30-nsec laser output pulse, while with xenon the emission persists for several hundred nanoseconds. The analyzing light thus available is passed through the sample which has been excited by the frequency-doubled laser pulse. With a fast analyzing light pulse (such as is obtainable with oxygen), an instantaneous absorption spectrum can be recorded by a spectrograph in a single exposure. For time-resolved spectra, the long-lasting emission of a spark in xenon is employed, and the variation in the absorption spectrum with time is recorded by an image converter camera.

Porter and Topp[60] have described an apparatus in which a brief analyzing light pulse is obtained from the fluorescence of a fast scintillator excited with part of the frequency-doubled laser light. By making the optical path of the analyzing light variable, absorption spectra of the excited sample can be recorded on a spectrograph at different times after excitation, and their change can be determined as function of time.

2.4.5. *Kinetic Monitoring*

The main requirement for fast kinetic monitoring, in order to obtain a good signal-to-noise ratio, is high analyzing light intensity. This goal is achieved by synchronization of light, from flash discharges, with the exciting laser pulses.[61, 62] At the peak of the emission of such a flash, the light intensity is practically constant for a period of time that is long compared to the fast decay time of the transients. In this manner the variation in the transient absorption as function of time can be followed electrooptically, just as in the case of ordinary flash photolysis. In addition, employment of pulsed analyzing light sources reduces the adverse effects of continuous intense light on the detecting photomultiplier.

2.5. Mode-Locked Laser Techniques

The most dramatic advance in optical pulse techniques during the last few years involves use of mode-locked solid-state lasers. As of 1971, such devices provided light pulses of duration time of the order of 10^{-12} sec (1 psec). Although mode-locked lasers are not yet commonly used in photochemistry, their potential for study of ultrafast molecular relaxation processes has been convincingly demonstrated.

2.5.1. *Generation of Ultrashort Laser Pulses: Theory*[29]

A formal description of mode locking can best be given for a continuous wave (CW) laser. According to the Fourier theorem,[1] a train of pulses with time-dependent amplitude $E(t)$ and a repetition rate $1/\tau$ can be represented by a series of individual sinusoidal functions

$$E(t) = \frac{E_0}{2} + \sum_{m=1}^{\infty} \left[E_m \cos\left(\frac{2\pi mt}{\tau}\right) + E_m' \sin\left(\frac{2\pi mt}{\tau}\right) \right] \qquad (2.13)$$

where m is an integer. In Eq. (2.13) all the frequencies are multiples of $1/\tau$ and have fixed phases.

As may be seen from a summary of its relevant features, a laser resonator has physical properties such as can provide the means for synthesis of a pulse train as described by Eq. (2.13). The axial-mode resonance frequencies v_m of a laser oscillator are given by

$$v_m = mc/2Ln_0 \qquad (2.14)$$

where L is the separation of the end reflectors, c is the velocity of light in vacuum, and n_0 is the refractive index of the medium.

The frequency separation δv (DeMaria, Stetser, and Glenn[63] use Δf) between two neighboring resonances v_m and v_{m+1} is

$$\delta v = c/2Ln_0 \qquad (2.15)$$

and the time for a light pulse to traverse the cavity in both directions is

$$\tau = 1/\delta v \qquad (2.16)$$

The number of axial modes activated when the laser oscillates depends on the laser line-gain profile and thus on the active medium. Only those Fabry–Perot resonances can oscillate that fall within the laser line profile for which the gain lies above the threshold for stimulated emission. Assuming a separation of the end reflectors of 1.5 m and a laser line centered at 1.06 μ with a width of $2 \times 10^{-2}\ \mu$, one finds that approximately 5×10^4 axial-mode resonances fall within this width.

A large number of frequencies is thus available but, as yet, without fixed phase relationship to render them suitable for pulse generation as required by Eq. (2.13). This second requirement, a fixed phase relation, can be satisfied, according to DeMaria, Stetser, and Glenn,[63] by inserting, into the feedback path of the laser, a component (of any one of a variety of kinds) that permits amplitude modulation of the optical frequencies. The phase- or *mode-locking* effect of modulation may be qualitatively understood in terms of the following considerations.

The laser oscillation begins (Fig. 4) at a resonance v_m close to the maximum of the laser gain profile. If the modulation frequency f is chosen such that $f = \delta v$, where δv is the frequency separation of neighboring axial-mode resonances, then side bands at $v_m \pm \delta v$ are generated. The two side bands and v_m are thus coupled with well-defined phase and amplitude. In passage through the modulator, these side bands generate their own side bands and the process is repeated until all the axial modes falling within the line width are coupled.

In analogy to the synthesis of a train of pulses by interference of the Fourier components in Eq. (2.13), the constructive and destructive interferences of all these mode- (or phase-) locked laser oscillations likewise result in the formation of a train of pulses. If the number of frequencies

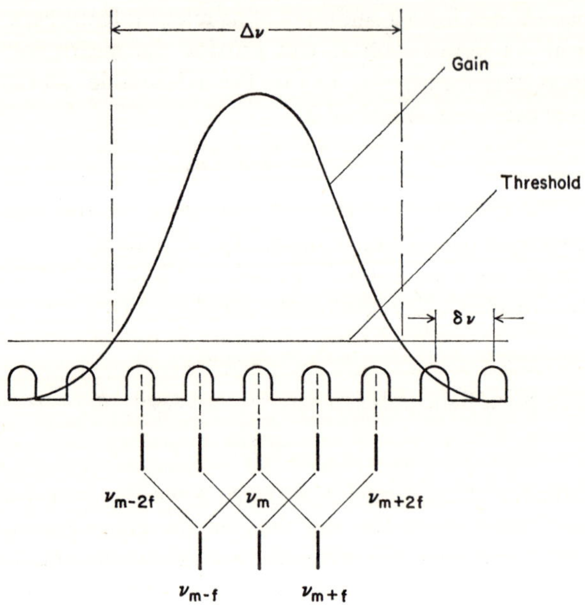

Fig. 4. Mode locking of a CW laser; schematic representation of a laser line gain profile superimposed on the Fabry–Perot resonances of the laser cavity. The laser frequency ν_m closest to the gain maximum is first to appear in stimulated emission and, when amplitude-modulated with a frequency $f = \delta\nu$, has side bands coinciding with the neighboring cavity resonances $\nu_m \pm f$. Upon repeated reflection between the laser end mirrors (S_1 and S_2 of Fig. 3), each of these side bands is in turn modulated and yields its own side bands so that side bands also appear at frequencies $\nu_m \pm 2f$, $\nu_m \pm 3f$, and so on. The axial modes that fall within $\Delta\nu$ are thus forced into fixed phase relationship. The horizontal line represents the threshold for light amplification (i.e., laser action). (See DeMaria, Stetser, and Glenn.)[63]

involved is large, the pulse width is given by

$$\Delta t \simeq 1/\Delta\nu \tag{2.17}$$

where the pulse width is defined as the width at half-amplitude, and $\Delta\nu$ is the similar width of the laser line profile.

The pulse repetition rate $1/\tau$ is given by Eq. (2.16) and is the reciprocal of the time required for the light to pass through the laser cavity in both directions.

It follows that the characteristics of the pulse train are determined by the spectral properties of the active laser material and the length of the laser cavity. The pulse separation can be varied by changing the spacing of the end mirrors of the laser, while the pulse width depends on the laser line width.

Mode locking[53, 64] of CW lasers has been achieved with a variety of active modulators such as Pockels cells and ultrasonic standing-wave diffraction gratings. It seems, however, that these methods require rather critical adjustment of components and that they are therefore not convenient for photochemical work with pulsed solid-state lasers. For this purpose mode locking with passive modulators[29, 65, 66] appears preferable. This method is particularly simple because the dye solution acting as a Q-switch may also serve as (a passive) modulator. In this case no additional element is required and mode locking occurs automatically. The mode-locking effect of such an absorber in conjunction with a pulsed solid-state laser may be described in the following way.[63]

A dye solution with an absorption band in the wavelength region of the laser light is inserted in the laser cavity close to one of the reflecting end mirrors (for example, S_2 in Fig. 3). The relaxation time of the excited-dye state, reached as the result of absorption of the laser light, is short compared to the time required for a light signal to traverse the laser cavity in both directions.

During the early stages of optical pumping, the spontaneous emission from the laser material is sufficiently weak so as to be completely absorbed by the dye. As in the case of passive Q-switching, laser action is thus prevented and a high population inversion can be achieved in the laser material. However, as the excited laser state is more and more populated, the intensity of spontaneous emission can become large enough to cause optical saturation of the dye, provided the pumping rate of the laser material is sufficiently large. In this situation the rate of excitation of the dye is comparable to the rate of relaxation of the excited states reached; the dye thus becomes transparent. As in the case of Q-switching, stimulated emission now becomes the predominant process and results in a very intense light pulse bouncing back and forth between the end mirrors of the laser *cavity*. Because of its very short relaxation time, the excited-dye state can, to a large extent, return to the ground state within the time interval required for the light pulse to traverse the *entire* cavity in both directions. Each time the pulse reaches the dye, absorption occurs predominantly at the leading and trailing edges (for such absorptions correspond to *time intervals* of the light pulse during which optical saturation is *not* achieved in the dye). The overall effect is a progressive steepening of the rise and decay and a time-narrowing of the pulse. A limit is reached when the harmonic composition of the pulse corresponds to the bandwidth of the laser system.[63] The effect of the absorber may be described as a pulse-shaped varying loss which has a fundamental frequency equal to the reciprocal of the round-trip transit time within the laser cavity.

2.5.2. *Instrumentation*

Lasers particularly useful for studies of the temporal properties of molecular states are the neodymium-glass and ruby lasers. Because of their wide line widths, as compared to gas lasers, pulses of a few picoseconds duration may be generated by mode locking [Eq. (2.17)]. Moreover, the larger power permits frequency multiplication so that several wavelengths (related in the ratios 1, $\frac{1}{2}$, $\frac{1}{4}$, and so on) are available.

For mode-locked laser operation a suitable modulator, such as outlined in Section 2.5.1, must be added. Although several devices have been employed for this purpose, the passive modulator (such as a suitable dye solution referred to in the previous section) is commonly employed for the applications of interest here.

2.5.3. *Measurement of Ultrashort Light Pulses*

The use of electrooptical devices for measurement of optical pulse shapes is limited by their time resolution, which is of the order of 10^{-10} sec. For study of shorter pulses such as are obtainable from mode-locked solid-state lasers, new experimental techniques are required. Two of these techniques based on nonlinear optical effects are outlined in the following.

2.5.3.1. Two-Photon Excitation of Fluorescence. A method of measuring picosecond light pulses by utilization of liquids that exhibit two-photon fluorescence was first described by Giordmaine and his co-workers[30] and refined by Rentzepis and Duguay.[31]

Two light pulses of different wavelengths (fundamental and doubled laser frequencies) are passed in overlapping paths but in opposite direction through a suitable solution of a scintillator. The absorption properties of the scintillator are such that the wavelength of one of the pulses with an intensity I_1 does not produce fluorescence. The wavelength of the other with intensity I_2 may produce weak fluorescence. The fluorescence intensity in this case is kept low. The summed optical frequencies of the two light pulses I_1 and I_2 are large enough to induce strong fluorescence by two-photon absorption. Such fluorescence occurs, along that region of the light path where the two pulses overlap, with an intensity proportional to $I_1 \cdot I_2$. With a knowledge of the rate of propagation of the two light pulses in the scintillator solution and from measurement of the spatial extent of the fluorescence intensity along the path, the pulse width can be deduced.

In the actual experiment of Rentzepis and Duguay,[31] the mode-locked pulses at the fundamental and doubled frequency of a neodymium-glass laser were studied utilizing a 5×10^{-2} M solution of diphenylcyclopentadiene (DPCPD) in tetrahydrofuran. Such a solution is not excited to fluorescence by strong pulses (≈ 1 GW/cm²) of the 1060-nm laser light. The fluorescence resulting from interaction with the frequency-doubled light (I_2) is weak when the pulse power is about 1 MW/cm².

By passing the composite beam through 9 cm of bromobenzene, normal dispersion caused a separation of the two pulses by 25 psec. The pulses then entered the DPCPD solution in a cell 2 cm long impinging at normal incidence on the end wall formed by a dielectric mirror. The reflectivities of this mirror were 75 percent at 1060 nm and less than 5 percent at 530 nm.

A fluorescence spot in the solution indicated the position where oncoming 530-nm pulses coincided with 1060-nm pulses reflected from the end mirror. By variation in the intensity of the light of shorter wavelength (I_2), the background fluorescence could be held at a low level.

From the width of the luminescence spot in the cell as recorded on film, the full width at half-height of the laser pulses was established to be about 3 psec. (It is in such work conveniently assumed that the pulse shape is Gaussian.)

A magnified display of the luminescence spot can be obtained by removing the bromobenzene so that the two light pulses of different optical frequencies enter the DPCPD solution simultaneously. In the scintillator solution the 1060-nm pulse moves progressively ahead of the shorter wavelength pulse, because of dispersion, and the fluorescence intensity diminishes correspondingly with distance from the entrance window. On the assumption of a Gaussian pulse shape, a pulse width of 2×10^{-12} sec was derived from the fluorescence intensity variation.

2.5.3.2. Second-Harmonic Generation by Reflection. Bloembergen and his co-workers,[67, 68] in their extensive studies of nonlinear optical effects, showed that for laser light falling on the surface of certain crystals there exist several orientations of polarization with respect to the crystal axis such that no reflected harmonic of a particular polarization is produced.

This phenomenon has been employed by Armstrong[69] to measure the shape of ultrashort light pulses. In his method the pulse train of a Q-switched, phase-locked neodymium-glass laser is divided into two parts with polarizations vertical to one another. The beams impinge on the surface of a single crystal of GaAs. The orientation of the crystal axes is such that neither beam alone produces second harmonics in reflection polarized parallel to the plane of incidence. However, if pulses of the two beams overlap at the crystal surface, the net polarization is such as to produce strong reflected second harmonic light with polarization parallel to the plane of incidence. The intensity of the reflected second harmonic is then measured as a function of overlap of the two pulses by use of a variable optical delay in the path of one of the beams.

Laser pulse widths of 4×10^{-12} sec have been measured in this manner and it is expected that a time resolution of 10^{-13} sec is feasible.

2.6. Pulse Radiolysis

Fast processes induced by ionizing radiation have been studied for many years with some of the techniques based on detection of luminescence. Although such methods as the monophoton technique are sufficiently sensitive to permit observation of feeble luminescence, often in fine detail, their application is obviously not too general. Consequently, studies of many processes of radiation chemical interest are precluded.

Such limitations have been largely reduced with the extension of pulse radiolysis techniques to the nanosecond and even subnanosecond region. In principle, therefore, these methods promise to provide insight into the dynamics of radiation chemical events on a very short time scale.

2.6.1. *Van de Graaff Generator*

Many of the nanosecond pulse radiolysis data up to the present time have been provided by a 3-MeV Van de Graaff generator of the Argonne National Laboratories.[25] The machine differs from conventional ones in its gridded electron gun, which is designed for the high-frequency requirements associated with the fast electrical signals that are handled.

The principle of generating the electron pulses is similar to that employed in the operation of a triode in which the electron current is controlled by the grid voltage. Correspondingly, the voltage supplied to the grid of the electron gun permits control of the electron beam. In the "off" periods a negative potential at the grid with respect to the cathode prevents electrons from reaching the accelerating tube. Electron pulses are admitted to the accelerating system by brief, periodic reversal of the polarity of the grid potential.

In the actual design the pulsing components are placed within the high-voltage terminal of the Van de Graaff. The components include a variable line discharger providing the gating pulses for the grid, an electromagnetically activated switch with an associated driving circuit to activate the line discharger, and a photomultiplier which, upon receiving external light signals, activates the switch and thus the pulse sequence.

A Lucite light pipe couples externally generated light pulses to the photomultiplier in the high-voltage terminal and permits control of the pulse repetition rate.

The machine is capable of delivering 5-A, 3-MeV electron pulses with a duration as short as about 1 nsec and a repetition rate variable from a single shot up to 100 pulses per second.

2.6.1.1. Detection System. The method of monitoring the change in the absorption signal of the transient radiolytic products is quite similar to that in ordinary pulse radiolysis. However, in the short-time range of interest the contributions to noise from fluorescence, Čerenkov radiation, and random electrical disturbance become more troublesome. To obtain a high signal-to-noise ratio, the intensity of the analyzing light is made very high. Such objective is accomplished (as mentioned in Section 2.4.5) by operating the analyzing light source in a pulsed mode,[70] thus permitting a significant increase in the lamp power. The output of a 450-W xenon lamp used in conjunction with the Argonne machine, for example, was increased 50-fold over the steady-state power level when the lamp was operated to give 1-msec pulses.[71]

2.6.2. *Linear Accelerator (Linac)*

Linear accelerators furnish another source of high-energy radiation for fast pulse radiolysis studies. In conventional use commercially available devices can be operated to provide pulses as short as a few nanoseconds with a repetition rate of several hundred pulses per second. Thus the pulse characteristics seem similar to those of a Van de Graaff machine. Comparison between the two types of accelerators for use in pulse radiolysis is, however, difficult for it involves performance requirements such as energy range, energy spread, and stability, and other factors such as cost.

An advantage of a linac for pulse radiolysis applications is its potential to provide experimentally useful electron pulses of a few times 10^{-11} sec duration. Such potential has been recognized and demonstrated by Bronskill and Hunt[26] and his collaborators at the University of Toronto. Although at present limited to the study of carefully selected systems, their development represents a significant advance in experimental technique for study of early radiation chemical processes. An outline of the principles underlying this method is therefore appropriate.

2.6.2.1. Subnanosecond Pulse Radiolysis: Radiation Source. The electron beam pulse emerging from a continuously operated linac is not itself actually continuous but consists of a sequence of narrowly spaced "fine-structure" pulses. This fine structure is the consequence of the microwave field employed for electron acceleration and related to the principle of phase stability.[72] Only those electrons injected into the accelerator tube within a certain phase interval of the microwave field experience synchronous acceleration between the accelerating electrodes. The fine-structure pulse width is related to this phase spread. For the microwave frequency $2.86 \times 10^9 \text{ sec}^{-1}$ (of the "S-band") and a phase spread of $20°$ this fine-structure pulse width is equal to $(20/360)/(2.86 \times 10^9) = 2 \times 10^{-11}$ sec. The fine-structure pulse spacing equals the reciprocal of the microwave frequency, in this case 35×10^{-11} sec.

Table I summarizes the parameters of the linac at the University of Toronto, which is the first one that has been regularly used for subnanosecond pulse radiolysis.

TABLE I

Parameters of the University of Toronto Linac,
as Used for Stroboscopic Pulse Radiolysis[a]

Manufacturer	Vickers-Armstrong Ltd., Swindsen, England
Beam energy	>30 MeV
Beam current	1 A per 30-nsec pulse; $\sim$30 A per fine-structure pulse
Accelerating microwave frequency	2.86×10^9 Hz
Fine-structure pulse spacing	0.35 nsec
Minimum beam diameter	<4 mm

[a] See Bronskill and Hunt.[26]

2.6.2.2. Detection System. Direct observation of concentration changes of transient radiation products with lifetimes of about 10^{-10} sec or less is not possible with conventionally available electrooptical devices. Consequently, Bronskill and Hunt[26] devised a stroboscopic detection system in which the Čerenkov light associated with the passage of the fine-structure pulse through air is utilized as the analyzing light. Figure 5 indicates the principle of operation of this method of detection. Figure 6 illustrates the nature of the sequence of electron and Čerenkov pulses and the rapidly decaying signals of the associated absorbing products.

Čerenkov light pulses and product pulses are synchronous. However, the light pulses can be delayed (by appropriate adjustment of the mirror system) to pass through the irradiated sample in such a manner as to fall on any preselected point of a (later) product pulse. An absorption signal corresponding to this point summed over a series of fine-structure pulses can thus be obtained with a photomultiplier which views the transmitted light. Variation of the delay of the light pulses with respect to the successive product pulses permits variation of the "time point" of overlap of the two pulses. Thus absorption in a series of time regions (determined by the location of the mirrors) is systematically measured. The change of absorption with time delay is recorded by scanning across the time interval between the fine-structure pulses. Hunt's method in principle represents a form of optical sampling akin to that employed by Dreeskamp and Burton[11] in their X-ray-induced luminescence studies.

2.6.2.3. Design Considerations. For application of the operating concepts outlined to the design of a feasible experimental system, Hunt and his co-workers list three major areas of consideration: design of a suitable,

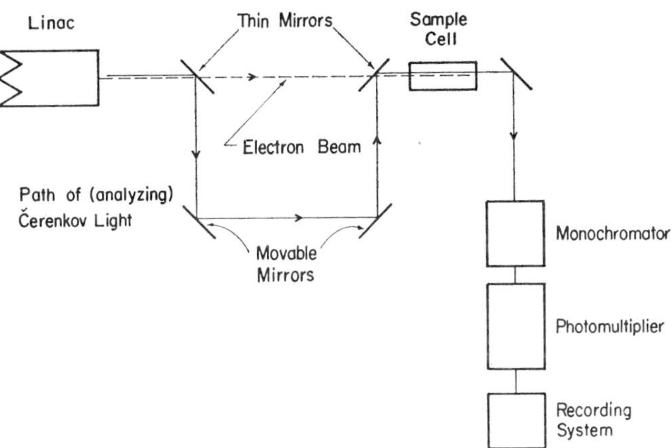

Fig. 5. Scheme of detection system for ultrafast pulse radiolysis (technique of Hunt and co-workers). The analyzing light, (thin line) derives from the Čerenkov radiation associated with a "fine-structure" electron pulse (broken line), from the linac passing through air. Unlike the electron beam, the light pulse is made to traverse a variable path between a set of mirrors before entrance into the sample cell. Light and electron pulses traverse the sample cell along the same path (for clearness of presentation they are drawn separated), but their *temporal* spacing when entering the cell depends on the length of the *adjustable* optical path length between the mirrors. The optical density of radiolytic intermediates produced by the electrons can be probed at variable time intervals with respect to the electron pulse by measurement of light intensity as a function of light path length.

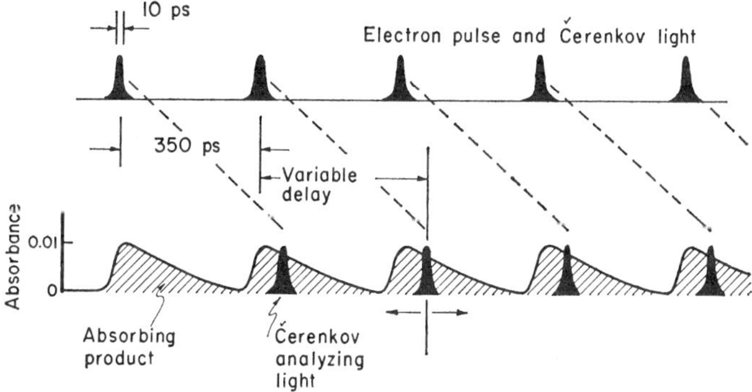

Fig. 6. Stroboscopic analysis of ultrafast pulse radiolysis signals. The upper drawing shows a sequence of fine-structure electron, and associated Čerenkov light, pulses after exiting from the accelerator. The lower drawing illustrates the transient absorption induced by the electron pulses in the sample and their temporal relation to the probing light pulses for a particular optical delay setting. Note that the first probing light pulse coincides with the absorption signal of the second electron pulse. (See Bronskill and Hunt[26].)

variable method of delay for the Čerenkov pulses; establishment of product concentrations, in individual fine-structure pulses, sufficient for detection; signal-to-noise ratio.

Time Delay.   An overall time delay of at least two fine-structure pulse intervals is required for experimental verification of synchrony of fine-structure and product pulses. The required delay of up to 0.6 nsec is achieved by variation of the analyzing light path in air over a range of 18 cm (by use of the pair of reflecting mirrors indicated in Fig. 4). Note that the velocity of electrons in this case is βc, where β is almost unity. For maximum time *resolution* consideration must be given to the difference in the rate of propagation of the electron beam and of the light in the sample. For passage through a medium with refractive index n and of length L cm, this difference is given by

$$\Delta t = \frac{nL}{c} - \frac{L}{\beta c}$$

and is equal to 0.01 L nsec in the case of water. Thus to keep Δt small a short cell is desirable. However, a large value of L is required for high light absorption (which occurs *while* the light and electron pulses are traveling together in the sample). Hunt chose a sample of 2 cm length as a compromise between high absorption and short time resolution, yielding a theoretical time resolution of 2×10^{-11} sec.

Detectable Concentration of Transient.   For a 30-MeV, 1-A electron beam of 5 mm diameter, the average dose per fine-structure pulse is about 250 rads. This corresponds to a solvated electron concentration of approximately 6×10^{-7} M or to an absorption signal of 4 percent, which is considered easily detectable.

Noise.   There are two major contributions to the noise of the system; shot noise resultant from statistical fluctuations of the number of photoelectrons produced at the photocathode of the detecting photomultiplier, and noise resulting from fluctuations of fine-structure pulses.

Shot Noise.   According to a criterion set by Bronskill and Hunt,[26] to obtain a 100:1 signal-to-noise ratio in the detection of a 4 percent absorption signal, about 4×10^{9} photons per Čerenkov pulse are required on the assumption of photocathode efficiency and light-collection efficiency of 10 percent each. As Bronskill and Hunt show, this requirement can be fulfilled for light at 700 nm and 10-nm bandwidth when a 30-nsec, 30-MeV, 0.5-A electron beam passes through 10 cm of air at 1 atm.

Pulse Intensity Fluctuations.   The actual pulse *seen* by the detecting photomultiplier *when the sample is being irradiated* may be denoted as A. The quantity A consists of the analyzing (i.e., Čerenkov) light B entering the cell, *less the absorption by the products B'*, plus the additional Čerenkov emission C produced by the electrons in the sample cell. The absorption signal B' is therefore $A - (B + C)$. Because A, B, and C may fluctuate from one 30-nsec pulse to another, they must be averaged over many such pulses to obtain meaningful data. In the experimental setup the three light pulses A, B, and C are measured separately by recording the photomultiplier output under appropriate conditions. To obtain A the electron beam together with the analyzing light is admitted to the sample cell. B is obtained by blocking the electron beam and admitting the analyzing light only, and C by passing the electron beam through the cell but excluding the analyzing light. The required averaging over these three signals is accomplished by a rotating chopper wheel synchronized with the 30-nsec pulses from the linac and designed in such a manner as to produce repetitively the signal sequence $A, B, C, A \ldots$. These signals are then electronically processed to give the desired absorption signal $A - (B + C) = B'$.

Suitable Chemical Systems. The apparatus is applied successfully to studies of transients which essentially decay within the time interval between fine-structure pulses, that is, within 0.3 nsec. Slower-decaying radiolysis products may be studied also, but the analysis of the time dependence of the absorption signal is more complicated because of the buildup of products.

3. EXPERIMENTAL RESULTS

An essential requirement for the understanding of radiation chemical processes is a knowledge of the energy flow as function of time, as schematically outlined by Mozumder in Volume 1 of this series, Chapter I, Section 3.6. For the experimentalist the acquisition of such knowledge translates into the problem of identification, and possibly quantitative determination, of the intermediates as function of time after the energy loss by the very energetic particle interacting with the system under study. Much of the information of this nature can and has been obtained by steady-state experiments such as scavenging studies, or by moderately fast dynamic studies such as conventional pulse radiolysis. In principle, pulse radiolysis on a nanosecond scale, as described in Section 2.6, can give more-direct information on short-lived intermediates such as excited states or electrically charged species. In addition, it should permit observation of the temporal behavior of such species. In this sense, nanosecond pulse radiolysis may provide information complementary to that obtained from more conventional experimental studies.

Most of the fast pulse radiolysis work has centered on studies of formation and decay of excited states and ionic species in organic liquids and solids, and on electron trapping in organic glasses. A critical survey of the papers on the subject shows that the data and their interpretation are not always consistent and that at times quantitative results may be questionable. Nonetheless, patterns of observations do emerge, in a variety of cases, consistent with models developed from the vast amounts of experimental data that have been accumulated in the past.

3.1. Pulse Radiolysis of Liquid Organic Systems

3.1.1. *Excited States*

The paths by which excited states may be formed as the result of high-energy irradiation of organic liquids are more numerous and more complex than in the case of optical excitation. Nevertheless, just as in the case of nonionizing excitation, the only excited states observable on the nanosecond scale are those of the lowest excited singlet and triplet states. This observation is consistent with the expectation that in a condensed medium upper excited states convert very rapidly to lowest excited levels.

Identification of the lowest excited states is based on comparison of their absorption or emission spectra (or both), as well as their lifetimes, with the corresponding properties obtained in optical excitation, such as by flash photolysis or luminescence-decay studies. Additional guidance in the identification of the excited species involved can also be derived from studies of the effects of quenching additives.

For the radiation chemist mere identification of the lowest excited states appearing in pulse radiolysis of organic solutions is possibly of minor interest because, usually, formation of such states has already been established by much simpler steady-state and other techniques. Of larger interest is the mode of formation of these states, for such knowledge may provide direct insight into the modes of energy flow (see Mozumder in Volume 1 of this series). Studies of pure aromatic and aliphatic liquids, as well as of their solutions of suitable additives, have been particularly interesting in this respect.

3.1.2. *Pure Liquids*

3.1.2.1. Cyclohexane. According to Hunt and Thomas,[73] fast pulse radiolysis of pure cyclohexane yields a transient absorption spectrum, at wavelengths below 300 nm only, which is assigned to the cyclohexyl radical C_6H_{11}.

3.1.2.2. Benzene. In the case of electron-irradiated benzene, two major absorption bands were observed by Thomas and Mani.[74] One of these exhibits a maximum at 320 nm and decays by a first-order process with a half-life $t_{1/2} = 112$ nsec. The other band is located at 515 nm and has a half-life $t_{1/2} = 18.5$ nsec (or $\tau = 1/k = 26.7$ nsec).

The decay time for the *absorption at 515 nm* is close to that of the known benzene fluorescence, and it was therefore initially assigned[75] to the transition from the first excited singlet state of benzene, $^1E_{2g} \leftarrow ^1B_{2u}$. According to the calculations of Pariser,[76] however, that transition is expected at $\lambda = 350$ nm. Moreover, the effects of temperature and of benzene concentration[75] on the absorption are inconsistent with such an assignment. More recently, there-fore, the absorption has been reassigned to a $^1E_{1u} \leftarrow ^1B_{2u}$ transition of the benzene excimer, the existence of which has been well established from fluorescence work; see Birks, Braga, and Lumb[77] and Ludwig and Amata.[78] The position of the observed band is consistent with the one calculated by Vala et al.[79] for the excimer. No excitation transition has ever been observed, even at low benzene concentrations, that can be properly attributed to the first excited state of benzene monomer.

The *absorption at 320 nm* cannot be associated with an excited state of benzene. Rather it appears to be related to the *reaction product* of an excited state; that is, presumably of the lowest excited triplet state of benzene. This conclusion is based on scavenging experiments with biacetyl, piperylene, naphthalene, and oxygen. The first three additives lower the intensity of the absorption signal at 320 nm but do not affect its decay rate. Oxygen, however, both decreases this absorption intensity and increases its decay rate. These characteristics seem to indicate that a triplet state is implicated. The effect of benzene concentration on the rate of growth of the 320-nm absorption signal is qualitatively in agreement with the assumption of a triplet state (assumed by Thomas and Mani[74]) as a precursor to the observable entity; in 10 percent solutions of benzene in cyclohexane, this rate is lower than in pure benzene.

Finally, the relative yield of the 320-nm absorption, as a function of benzene concentration, parallels the corresponding relative yield of the known benzanthracene triplet formed in benzene–cyclohexane solutions of benzanthracene. On the assumption that this triplet state is formed by triplet energy transfer from benzene, the conclusion is that benzene triplet is the precursor to the species exhibiting absorption at 320 nm. The specific nature of the species is, however, uncertain. It has been suggested by Thomas and Mani[74] that the product species in question may be the biradical formed from benzene triplet, as postulated in another connection by Bryce-Smith and Longuet-Higgins.[80]

3.1.2.3. Toluene. Cooper and Thomas[75] report an absorption with a maximum at 550 nm in irradiated toluene. The absorption signal decays with a *half-life* of 21.6 nsec, close to that calculated from the fluorescence *decay time* measured for this compound.[81, 82] As in the case of benzene, the initial assignment of the absorption to the lowest excited singlet state of toluene monomer has been given up in favor of an assignment to the excimer.

3.1.3. *Ionic Intermediates in Nonpolar Solutions*

Pulse radiolysis data obtained with pure organic liquids reveal little detail regarding the mode of formation of the observable excited states. On a nanosecond scale the rate of their formation is so fast that little information can be directly derived as to possible differences between the processes of their formation and the corresponding processes in optical excitation. However, as in the techniques employed in steady-state work, addition of suitable additives to the organic solvents frequently provides a means of obtaining some indication as to the possible precursors of these excited states. Ions and higher excited electronic states may be considered as particularly likely precursors.

3.1.4. *Solutions in Aliphatic Solvents*

Direct observation of ionic species of the liquid aliphatic solvents has not been successfully accomplished in fast pulse radiolysis, presumably because of the lack of suitable and well-characterized absorption spectra and also because overlap with radical spectra would make identification difficult. Observation of ions has therefore been limited (by Thomas et al.[83]) to solutions of suitable additives such as biphenyl or anthracene. Identification of the ions is based on the similarity of the transient fast pulse radiolysis spectra observed with those of the same compounds in irradiated glassy solutions. An additional means of identification is the effect of a second additive (acting as electron or positive-charge scavenger) on the absorption spectrum.

Figure 7 represents an example of the transient absorption spectra obtained in fast pulse radiolysis of cyclohexane *solutions of biphenyl* under a variety of experimental conditions. In a solution containing 10^{-3} M biphenyl, the spectrum peaking at 660 nm is similar to that given in the literature for the biphenyl anion. The lowering effect of N_2O on this absorption is consistent with assignment to this anion. However, there is also a change in the shape of the spectrum such that the peak absorption is shifted toward the red. This shift suggests the presence of a second absorbing species. Conclusions about the nature of this species have been derived from the effect of adding a positive-charge scavenger (aniline) to the biphenyl solution. Figure 7 shows that in this case the initial absorption is again lowered but the peak absorption is shifted significantly toward the blue. It is noteworthy that this absorption spectrum corresponds closely to the "difference spectrum"

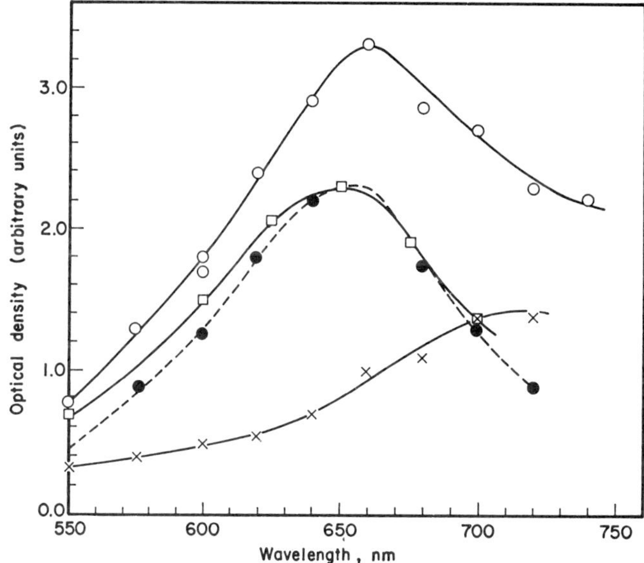

Fig. 7. Transient absorption spectra obtained during fast pulse radiolysis of $10^{-3}\,M$ solutions of biphenyl in cyclohexane. $\bigcirc$, $10^{-3}\,M$ biphenyl only; $\times$, saturated with N_2O; $\square$, also containing $2 \times 10^{-2}\,M$ aniline; $\bullet$, difference between the spectra of solution of $10^{-3}\,M$ biphenyl only and solution of $10^{-3}\,M$ biphenyl saturated with N_2O. (Thomas et al.[83].)

obtained by subtracting the absorption for the biphenyl plus N_2O solution from that of the solution containing biphenyl only.

On the basis of the similarity of the spectra obtained to those given in the literature for the biphenyl anion and cation,[84-86] plus the effects reasonably to be expected of electron and positive-charge scavengers, Thomas and his co-workers[83] assigned the spectra to the two ion species.

The convenience of differentiation between the two types of biphenyl ions plus the fact that the extinction coefficient of the anion at $\lambda = 600$ nm is known[84] to be 10,600 make possible the measurement of the yield of the negative species as a function both of biphenyl concentration and of time. Figure 8a shows the yield of ϕ_2^- as a function of concentration at the end of a 12-nsec accelerator pulse and also about 1 μsec thereafter. For low biphenyl concentrations ($\sim 10^{-3}\,M$), decay of the ion absorption signal is observable over many microseconds. At higher concentrations the decay in the first 50–100 nsec after the pulse is fast; a subsequent slower rate of decay is similar to that for the more dilute solution.

In addition to the spectra attributed to the biphenyl ions, there also appears an absorption at $\lambda = 370$ nm. This absorption is characterized by

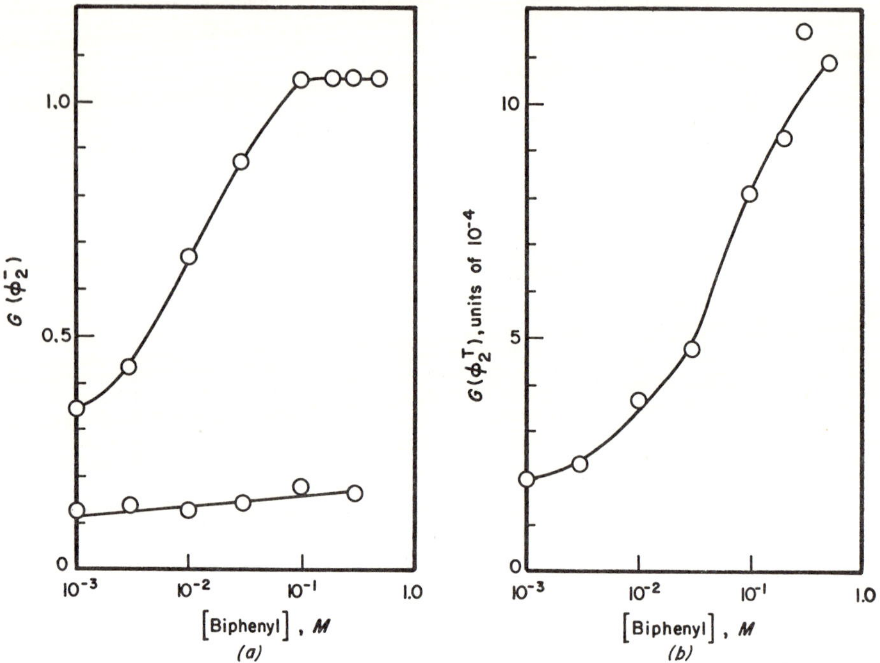

Fig. 8. Dependences of biphenyl anion yield and triplet yield on biphenyl concentration in cyclohexane. (a) Biphenyl anion. The upper curve refers to the observed yield at the end of a 12-nsec accelerator pulse; the lower line represents the yield about 1 μsec thereafter. (b) Biphenyl triplet, presumably at the end of a 12-nsec pulse. (Thomas et al.[83].)

an initial increase during the first 100 nsec after the pulse and a subsequent slow decay. Because the absorption spectrum of the triplet state of biphenyl, as determined by Porter and Windsor,[87] has its maximum in this region, the 370-nm signal is assigned to this state.

The results obtained for *anthracene in cyclohexane* are analogous to those obtained with biphenyl. As indicated by the effect of charge scavengers, both anion and cation are formed. The yield of the anthracene anion has likewise been measured, for (as in the case of the biphenyl anion) the corresponding extinction coefficient, $\varepsilon = 10{,}500$ at $\lambda = 725$ nm, is known.[84]

Again, as in the case of biphenyl, an additional absorption, attributed to the triplet state, has been observed. The signal for the triplet immediately after the pulse increases during the following 100 nsec, finally reaching a plateau.

Figure 9 shows the growth of the triplet absorption A^T, as represented by a plot of $A_\infty^T - A^T$ versus time. A_∞^T refers to the triplet absorption signal at the plateau level. The interesting aspect of this plot is the close correlation between rate of triplet formation and the decay of the anion signal. Addition of low concentrations of N_2O to the anthracene–cyclohexane system largely affects only the growing portion of the triplet. Only at higher N_2O concentrations (as in saturated solution) is the instantaneous triplet absorption signal also reduced.

The pulse radiolysis data of Thomas[88] for solutions of 1,2-benzanthracene and naphthalene in aliphatic solvents are similar to those for benzene and anthracene. However, analysis of the data is complicated by the fact that the absorption spectra of singlet states, triplet states, and ionic species overlap.

3.1.5. *Interpretation of Experimental Data*

The results outlined in the previous paragraphs are, on the whole, consistent with current concepts regarding the interaction of low LET high-energy radiation with liquids of low dielectric constant; see Volume 1 of this series, p. 83. According to such concepts a predominant early process is electron ejection from the solvent molecules, with the majority of electrons reaching thermal energy within about 80 Å from their origin. A large fraction of these is assumed to recombine very rapidly because of strong coulomb interaction. On the assumption that the electron motion may be described by diffusion, some fraction of the electrons escape immediate recombination and may be trapped by impurities in the case of the "pure" solvent or by scavenger molecules in the case of solutions. Depending on the form of the

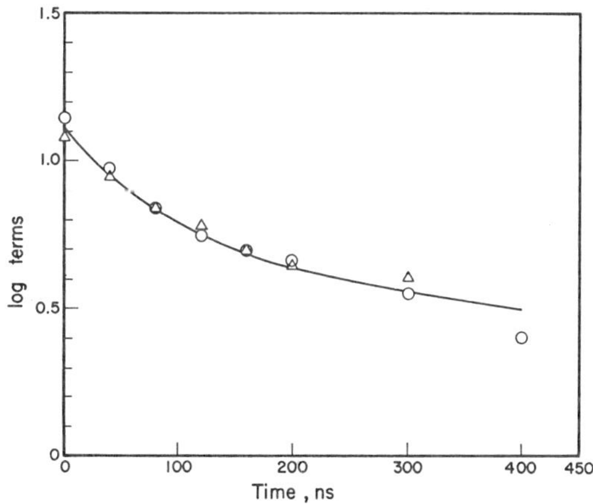

Fig. 9. Decay of anion and growth of triplet of anthracene A, in cyclohexane. ○ refers to log (A⁻). The growth of the triplet is expressed as $\log (A_\infty{}^T - A_T) + 0.3$, and the measured points are denoted by △. (Cf. Thomas et al.,[83] in which the symbolism is slightly different but the meaning is equivalent.)

initial cation-anion (or electron) distribution function, a wide spectrum of neutralization times is to be expected.

In pure aliphatic solvents no direct observation of neutralization processes has as yet been made. As indicated in Section 3.1.4, such failure may be related to the possibility that light absorption by aliphatic ionic species either is nonexistent or is inadequate for detection with present techniques. In the case of the solutions, however, ionic species of the aromatic solutes are observed. The pulse radiolysis data indicate that their formation must occur on a time scale faster than the time resolution of present instrumentation.

In the case of biphenyl solutions, for which the results are less complicated by overlap of spectra of different species, the decay characteristics of the anion and its dependence on solute concentration give perhaps the most direct evidence for the existence of a nonhomogeneous ion-pair distribution. At low biphenyl concentrations the anion decay is slow, with only a small indication of an initial, fast-decay component. Evidently, under these conditions there is possibility of scavenging only those electrons that escape geminate recombination and thus reach relatively large initial separation from the solvent cation. At sufficiently long times after the exciting pulse or for very dilute solutions, second-order decay may be expected. However, no such time dependence of the ion decay has been observed. The signal-to-noise ratio at such low ion concentrations apparently does not permit reliable determination of the shape of the decay curve.

Qualitatively, time dependence of ion absorption signals, or, more specifically, those of solute anions, can be accounted for within the framework of present models of interaction of high-energy radiation with aliphatic solvents. However, the results have so far contributed little to the development of more quantitative theoretical models—largely because of the qualitative nature of the experiments themselves. Even in cases in which more quantitative work has been attempted, results are not always consistent with data from steady-state experiments. Nevertheless, it should be noted that Rzad et al.[89] have applied a semiempirical description of ion-pair recombination to the biphenyl–anion decay curve observed by Thomas et al.[83] Their description is based on studies of steady-state charge-scavenging effects and leads, at least for biphenyl solution, to a very satisfactory correlation between their results and the fast pulse radiolysis data.

In addition to ions, presence and formation of solute excited states are also observed. Both excited singlet and lowest triplet appear during the excitation pulse. However, in some cases (e.g., anthracene), a significant yield of triplet can also be observed to form after the pulse. The observation of formation of anthracene triplet after the pulse is of particular interest because in this case the rate of triplet formation is much the same as that of ion decay. The observation may be considered as direct evidence that triplet-state species result from ion recombination. The interpretation is also consistent with the effect of the addition of N_2O. With increasing electron scavenger concentration, the growing-in part of the triplet is progressively suppressed, in correspondence with a reduced yield of anions.

That ion recombination is the source also of the solute triplets observed *immediately after the pulse* is less clear. Other processes may also yield excited states. However, the nature of such processes is quite speculative because present experimental data do not permit distinction between several possibilities. For example, it has been suggested by Hunt and Thomas[73] that excitation by slow electrons may result in "immediate" formation of solute triplets.

The observation by Hirayama and Lipsky[90] of comparatively long-lived fluorescence from ultraviolet and high-energy excited liquid aliphatic compounds requires reconsideration of solvent-solute energy transfer as a possible path for the "immediate" formation of excited solute states. Such a process, predominantly cited in the case of aromatic solvents, had been largely discarded for aliphatics because of the lack of convincing evidence for sufficiently long-lived excited states of aliphatic compounds.[91, 92] Of course, the adequacy of the yield of such excited states for production of a significant yield of solute excited states remains to be established.

3.1.6. *Solutions in Aromatic Solvents*

Benzene and toluene have been employed as solvents in studies of fast pulse radiolysis of naphthalene,[75, 88] anthracene,[75] biphenyl,[75] 1,2-benzanthracene,[88] and biacetyl[75] by Thomas and his co-workers.

The main feature of the results with such systems as compared with those described for aliphatic-solvent systems is the absence of large ion yields. In benzene and toluene solutions of biphenyl, for example, G values for ion production are only ca. 0.1. However, yields of solvent, as well as of solute, excited states are high. The G values for triplet-state formation in solutions of naphthalene, anthracene, and biacetyl are given as about 3 to 4. For pure benzene $G(^3B_{1u}) = 1.85$ and $G(^1B_{2u}) = 1.62$ have been claimed. (Although these numbers suggest agreement with steady-state data, the method of extraction of these values from the data published is not clear.)

The fact that the G value for ion production as measured with biphenyl–benzene solution is ca. 0.1, within 1 nsec after the pulse, suggests that ion recombination in benzene (and presumably in other) aromatic solvents is very fast. In order to explain this observation, Cooper and Thomas[75] speculate that in aromatic solvents the thermalization distance of the secondary electrons is much shorter than in liquid aliphatic hydrocarbons. They attribute the effect to the availability of low-lying states in aromatics, which could conceivably lead to a more effective energy loss by the electrons. However, consideration of the availability of lower electronic states in aromatics as compared to aliphatic hydrocarbons as the *sole* cause for a shorter thermalization distance is probably too simplistic. According to theoretical models of thermalization of electrons in liquids, a major factor contributing to the thermalization distance is the energy loss mechanism for subexcitation electrons; (see Mozumder in Volume 1 of this series, p. 45). It seems that attention should also be addressed to differences in the liquid structures of aromatics and aliphatics. Furthermore, according to Mozumder, virtual negative-ion formation likewise known to occur in aromatics can conceivably affect the recombination time.[93]

Clearly, the difference between ion yields in aromatic and aliphatic liquids, as observed with fast pulse radiolysis, is not thoroughly understood.

It appears that ionic processes are of minor significance in formation of excited solute states in aromatic solvents because of the apparently short lifetimes of charged species. Their formation can be understood in part by assuming excitation energy transfer between excited solvent and the solutes. Such processes have been well established under conditions of photoexcitation of the solvent.[108, 171] Thus the formation of solute singlet states in benzene can be accounted for by singlet energy transfer from the $^1B_{2u}$ state of benzene while solute triplets may be derived in part from the analogous triplet-state interaction and in part from intersystem crossing from the solute singlet.

The latter process should be conveniently observable in those cases in which the solute singlet lifetime is sufficiently long. Thus for solutions of naphthalene an increase of the triplet absorption signal at 430 nm, during about 100 nsec following excitation, would be expected.

In this case, however, such an increase of the triplet absorption signal has not been observed. The lack of the expected effect was initially taken as an indication that intersystem crossing in naphthalene solutions may not be an important process.[88] A more likely interpretation, supported by laser flash photolysis studies of naphthalene solutions,[88] is that the signals for triplet and singlet absorption overlap at the wavelength of measurement. As a result, the decaying singlet signal is largely compensated by the growing triplet signal leading effectively to a fairly constant observable absorption signal. It may be mentioned here that Theard, Peterson, and Holroyd,[173] who studied pulse radiolysis of liquid naphthalene at 140°C, do not come to the same conclusion. However, in their case the singlet lifetime is much shortened as compared to that in solutions at room temperature, so that processes competing with intersystem crossing may be prevalent.

3.1.7. *Energy Transfer from Higher Excited States*

The suggestion has been made that energy transfer from higher excited states of benzene may contribute to solute triplet formation.[75] The conclusion is based on the observation that yields of triplet excited anthracene are consistently higher than those for naphthalene in their respective solutions at the same concentrations. Addition of xenon to these solutions leads to an increase in naphthalene triplet yields but has a smaller effect on those of anthracene triplet. These two observations have been used as a basis of speculation that anthracene singlets derive energy from an upper excited singlet state of benzene which is not accessible to naphthalene.[75] In explanation of the xenon effect, it was suggested that in its presence intersystem crossing occurs from an upper state to a benzene triplet state from which energy transfer to naphthalene is then possible.[75]

Both experimental observations might seem thus to be accounted for. It should be emphasized, however, that the data do not represent satisfactory evidence for the occurrence of energy transfer from upper electronic states of benzene. There is, for example, no a priori reason that the triplet yields of naphthalene and anthracene, at least for low concentrations, are required to be equal. For example, such difference might actually be expected if the rate constants for excitation energy transfer from triplet benzene differ in the two cases studied *and* if the measured benzene triplet lifetime is short (as was suggested by Hentz et al.[152b] in comment on the early results of Dubois[150] and of Cundall[151]) because of reaction with benzene itself.

A more serious objection may be raised in regard to the interpretation of the effect of xenon on the yield of triplet naphthalene. Assuming a diffusion-controlled rate for the proposed interaction of xenon with a *higher* excited state of benzene (resulting in intersystem crossing), it is necessary that, at the xenon concentration employed, the lifetime of that higher excited state be several nanoseconds in order that such quenching be observed. Such a long lifetime, however, is in contradiction to the general experience that in condensed phases excited electronic states higher than the lowest one are, except in the case of azulene, not observed.

To provide a more satisfactory interpretation of the difference in the xenon effect on anthracene and naphthalene triplet yields, it is here suggested that the interaction of xenon is with the first excited singlet states of these solutes rather than with the solvent. The difference in triplet yields can then be attributed to the difference in rates for the process

$$^1S + Xe \xrightarrow{k} {}^3S + Xe$$

where ^1S and ^3S represent, respectively, the singlet and triplet states of the solute and k is the specific rate of xenon-induced intersystem crossing. Consider that in alcoholic solutions k has been measured to be 1.1×10^9 dm^3 $mole^{-1}$ sec^{-1} for anthracene and 0.69×10^9 dm^3 $mole^{-1}$ sec^{-1} for naphthalene,[94] and that moreover the singlet lifetimes are 4×10^{-9} sec and about 70×10^{-9} sec for anthracene and naphthalene, respectively.[82] One easily calculates that xenon at the concentration of the experiment should quench the singlet state of naphthalene almost 100 percent as compared to only about 35 percent in the case of anthracene. It appears therefore that the differences in triplet yields in the presence of xenon can be accounted for by such quite conventional considerations.

The difference in triplet yields for the two solutes in the absence of xenon and at the higher concentrations remains puzzling because energy transfer from benzene to anthracene and naphthalene may be expected to be complete at high solute concentration. Thomas, in a private communication, indicates, however, that the extinction coefficients used to evaluate the triplet yields may be uncertain. In view of this uncertainty and in view of the interpretation of the xenon effect given here, it appears that the experimental data do not provide conclusive evidence for energy transfer from higher excited electronic states in the condensed phase.

General Comment. Fast pulse radiolysis studies on the nanosecond time scale in liquid organic systems reflect to a large extent the results and conclusion derived from steady-state work, conventional pulse radiolysis, and luminescence studies. Much of the work reported here is more of an exploratory rather than of a quantitative nature. Results are therefore frequently not in good quantitative agreement with data obtained from other experimental investigations, nor is their interpretation unambiguous. More systematic studies may reduce or eliminate some of these deficiencies.

A more fundamental limitation for obtaining new information on early radiation chemical processes seems to be in the inherent time resolution associated with fast pulse radiolysis techniques. While the fate of the lowest excited states of many of the organic compounds can be readily observed on a nanosecond time scale, the preceding processes, such as the major fraction of ionic reactions or relaxation processes of higher excited states, appear to occur in much shorter times than are resolvable with nanosecond techniques.

3.2. Fast Pulse Radiolysis of Glasses

Steady-state studies of high-energy irradiated organic glasses have provided some insight into radiation chemical processes, among which behavior and fate of secondary electrons have been of particular interest.[95] Because of high viscosity, many processes are slowed down in such glasses. Consequently, application of fast pulse radiolysis techniques presents fewer problems of time resolution than those encountered in studies of liquid systems.

Richards and Thomas[96] applied fast pulse radiolysis to studies of electron trapping in low-temperature glasses of 3-methylpentane, 3-methylhexane, 2-methyltetrahydrofuran (MTHF), cumene, ethanol, and alkaline ice. In all cases trapped electrons were observed.

3.2.1. *Aliphatic Hydrocarbon Glasses*

The results for aliphatic hydrocarbon glasses at 77°K are characterized by the appearance of a broad spectrum at the end of a 12-nsec electron pulse which extends from about 400 nm (the short wavelength limit of observation) toward the red, with a sharp cutoff at $\lambda = 1600$ nm. Within 5 μsec after the excitation pulse, the absorption at $\lambda > 1400$ nm increases and extends to at least 1700 nm (the long-wavelength limit of observation). During this time the spectrum at $\lambda < 1400$ nm remains essentially unaltered. The spectrum observed after the longer time interval resembles that obtained for trapped electrons when hydrocarbon glasses are irradiated with ^{60}Co γ-radiation. The effects of addition of 2 percent hexene-1 or MTHF (both considered positive-hole traps) to the glass seem to be inconsistent. Addition of hexene has little effect on the early and late spectra. However, MTHF enhances the spectrum below about 1600 nm but does not affect the growing-in portion of the spectrum at longer wavelength.

In the presence of electron scavengers such as biphenyl, naphthalene, or carbon tetrachloride, the intensities of the immediate and growing-in spectra are reduced. A weak absorption observed in the visible at the same time is attributed to the cation and the anion of the additive. When naphthalene-containing glass is irradiated at 100°K, fluorescence of naphthalene and a triplet absorption spectrum appear. The rate of growth of the naphthalene triplet increases with increasing temperature of the glass.

Absorption spectra observed in pulse radiolysis of aliphatic glasses have been attributed to the electron trapped in the matrix. The forms of the spectra, at least at long times after excitation, and the effect of charge scavengers appear to be consistent with this interpretation. The observation of fluorescence and triplet formation of naphthalene at higher temperatures, at which viscosity is reduced, agrees with the model of thermal release of electrons from their traps and their subsequent neutralization by the naphthalene cation. The most interesting result obtained was, however, a reported variation in the absorption spectrum with time. The most satisfying interpretation of this observation seemed to have been given by Hamill.[97] From bleaching experiments at different wavelengths with ^{60}Co-γ-irradiated aliphatic glasses, he concluded that the trapped electron absorption spectrum corresponds to transitions to bound states at wavelengths $\gtrsim 1600$ nm, while at shorter wavelengths the spectrum is attributable to photoionization. The remarkable coincidence of the photoionization limit at about 1600 nm with the sharp cutoff at this wavelength in the immediate absorption spectrum obtained from fast pulse radiolysis suggests that the initial trapping sites are such as to allow photoionizing transitions only. According to Hamill's model, such sites correspond to shallow traps. He suggests that, as time proceeds and the medium surrounding the trap relaxes, the trap deepens so that the first excited state drops below the unbound level and thus permits low-energy transitions to bound states. Such transitions correspond to longer wavelengths (>1600 nm) and account for the part of the spectrum that develops during at least several hundred nanoseconds after the irradiation pulse.

Klassen, Gillis, and Walker[175] suggest that the time variation of the absorption spectra may be the result of an instrumental effect. They found the response time of the solid-state detectors they employed in the wavelength region studied to be itself a function of wavelength. However, according to Richards and Thomas,[71] their apparatus was not afflicted with this difficulty.

A statement by Richards and Thomas[96] that in general positive-charge scavengers enhance the trapped electron yield seems inconsistent with their data. Careful examination of these data shows that the statement in this general form is inaccurate. Only two positive-charge scavengers, hexene-1 and MTHF, were utilized. Hexene-1 had very little, if any, effect. MTHF, however, increased only the growing portion of the spectrum (i.e., 1250–1750 nm). However, in this case effects different from positive-charge scavenging may play a role. MTHF by itself exhibits a much larger trapped electron yield than do the aliphatic hydrocarbon glasses. The observable predominant effect of MTHF on the growing-in portion of the absorption spectrum of irradiated aliphatic hydrocarbon glass may therefore be associated with a contribution of MTHF-trapped electrons rather than with positive-hole trapping.

3.2.2. *Aromatic Hydrocarbon Glasses*

Cumene (isopropyl benzene) glass is the only aromatic hydrocarbon glass of which a study has been reported[96] up to 1971. At $-160°C$ only a small growing-in absorption at $\lambda > 1600$ nm is observed. The yield of trapped electrons appears to be only a few percent of that characteristic of aliphatic hydrocarbon glasses. An additional, short-duration absorption at 500 nm, also present, has been attributed to an excimer. The experimentally obtained data on this system are limited. Nonetheless, the comparatively low yield of trapped electrons in this aromatic glass seems to reflect the differences in yields of charged species reported for liquid aromatic and aliphatic compounds (cf. Sections 3.1.5 and 3.1.6).

3.2.3. *Glasses of Polar Compounds*

3.2.3.1. 2-Methyltetrahydrofuran Glass. The results obtained with MTHF glass are similar to those obtained with aliphatic glasses.[96] Immediately after the pulse, a spectrum with long-wavelength cutoff at $\lambda \sim 1600$ nm is observed. At longer times the absorption spectrum develops beyond this wavelength but simultaneously there is also considerable growth over the whole spectrum at shorter wavelengths. The yield of trapped electrons is large in this system as compared with the aliphatic hydrocarbon glasses.

The interpretation given by Hamill for hydrocarbon glasses may also apply in this case. According to Hamill,[97] at least two different trapping sites must be assumed, each with its own photodetachment energy. Deepening of the traps, as time proceeds and the system relaxes (see Section 3.2.2), promotes the probability of absorption to bound states and thus accounts for the development of the growing-in part of the spectrum toward its red end (i.e., in the range around 1600 nm and longer). The simultaneous increase of absorption in the remaining observable part of the spectrum may be accounted for by the expected blue shift for the transitions corresponding to photodetachment.

3.2.3.2. Ethanol Glass. Difficulties in development of a unified model for the trapping of electrons in organic glasses are enhanced by the results obtained in fast pulse radiolysis studies of ethanol glass. Figure 10 shows that in the case of this rather polar matrix the absorption immediately after the

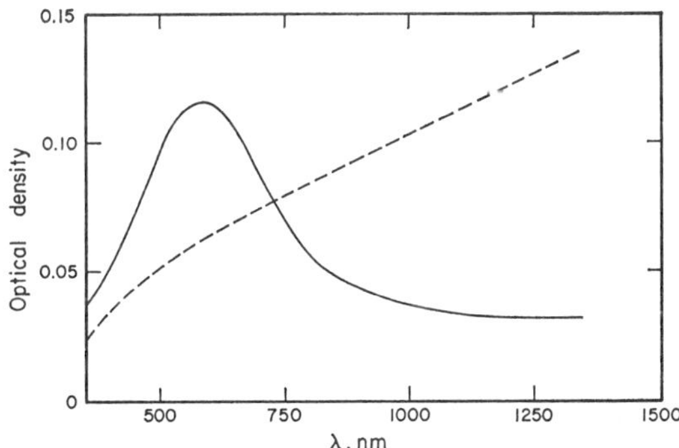

Fig. 10. Transient absorption spectra observed in fast pulse radiolysis of ethanol glass at 77°K. – – –, Spectrum at the end of 12-nsec electron pulse; ———, spectrum 4 μsec thereafter. (Richards and Thomas.[96])

pulse increases monotonically toward the red in the region of observation. After 4 μsec, however, the region at the blue end of the spectrum becomes markedly enhanced. This increase in absorption around 600 nm is accompanied by a decrease in the absorption in the red region.

In a very general way, such change in the spectrum with time has been associated with relaxation and deepening of the electron traps. However, both the nature of the transitions involved and the contrast with the spectral changes in the other glasses remain at present unclear. Pulse radiolysis studies of electron trapping of other aliphatic alcohols, as a function of chain length, seem not to have been made. Such investigation may possibly provide a clue as to the reasons for the differences in the temporal changes in the absorption spectra related to matrix polarity.

3.2.3.3. Alkaline Ice. Glass of alkaline ice (6 M KOH) exhibits no transient absorption spectra when submitted to fast pulse radiolysis at 77°K. The absorption spectrum obtained is the same as that observed under irradiation with ^{60}Co γ rays.

Summary. The main observations made in the study of trapping of electrons in organic glasses by fast pulse radiolysis relate to temporal variation in their absorption spectra. Qualitatively, this result can be related to relaxation of the trapped electrons during the first several hundred nanoseconds after their generation; after that time the spectra tend to assume the form observed under conditions of ^{60}Co γ-irradiation. The nature of the transitions giving rise to absorption spectra and to their changes with time is at present not clearly understood. In the cases of aliphatic hydrocarbons and MTHF, the temporal changes of the spectra appear to provide some additional evidence for a model that assumes that the spectra at long times involve transitions between bound states as well as transitions leading to photodetachment of the trapped electron.

3.3. Ultrafast Pulse Radiolysis

Application by Hunt and his co-workers[98, 99] of the ultrafast pulse radiolysis technique outlined in Section 2.6.4 has so far been largely concentrated on studies of solvated electrons in aqueous and alcoholic solutions. Practical considerations, as well as scientific interest, seem to make studies of such systems particularly useful. However, the optical and chemical properties of the solvated electrons are well documented in an extensive amount of work; cf. Thomas in Volume 1 of this series, p. 103. Their lifetimes can be varied over a wide range so that they may be adjusted to be compatible with the 350-psec spacing of the substructure electron pulses of the linac. However, there is much interest in the early stages of the history of solvated electrons as they relate to spur theory, dielectric relaxation, and excited states of the polar solvents.

3.3.1. *Solvated Electron in Water*

In order to reduce the problem of analyzing the decay data resulting from buildup of electron concentration between substructure pulses, Hunt and his co-workers[99] used acidic solutions for their ultrafast pulse radiolysis studies. The decay time of the electrons was thus made less than 350 psec, the time spacing of the substructure pulses. Figure 11 shows the time dependences of the solvated electron absorption signal for various concentrations of $HClO_4$. Analysis of these curves reveals a pseudo-first-order decay with a specific rate $k_2 = (1.2 \pm 0.2) \times 10^{10}$ dm³ mole⁻¹ sec⁻¹, where k_2 relates to the reaction

$$e_{aq}^- + H_{aq}^+ \xrightarrow{k_2} H$$

In derivation of this value of k_2, consideration was given to a shift in baseline resulting from incomplete signal decay between substructure pulses at low acid concentrations. Without such correction an artificial variation in k_2 with acid concentration was obtained. While k_2 is thus found to be constant for acid concentrations ranging from about 0.5 to 5 M, an increase at lower concentrations seems to be real.

The absorption spectrum at the peak of the absorption signal is similar to that obtained from microsecond pulse radiolysis, as shown in Fig. 12.

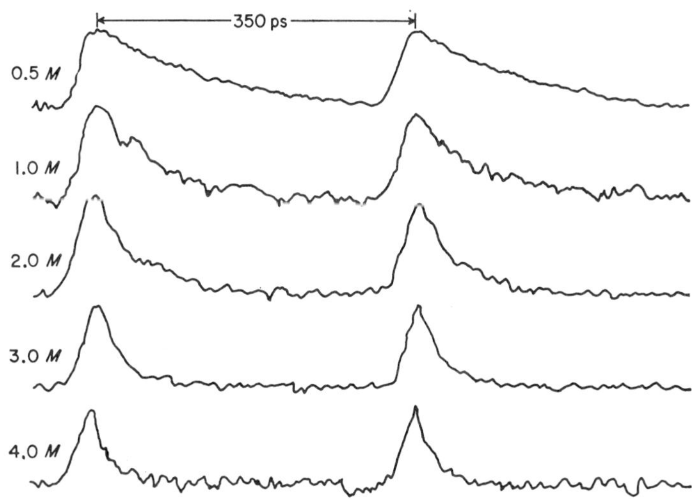

Fig. 11. Picosecond pulse radiolysis of aqueous $HClO_4$ solutions. Time dependence of the e_{aq}^- absorption signal for various acid concentrations. (Bronskill, Wolff, and Hunt.[98])

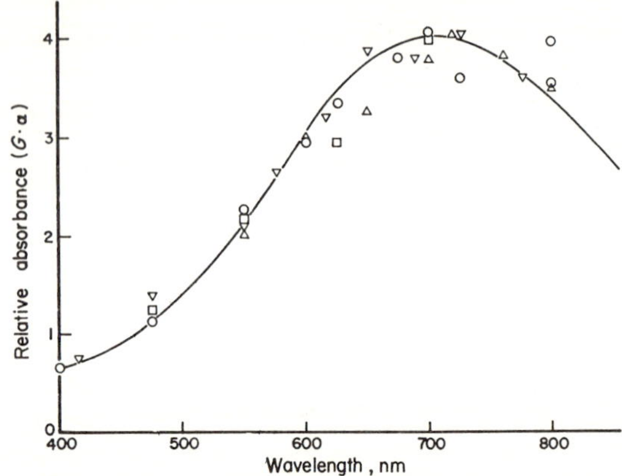

Fig. 12. Absorption signal observed in picosecond pulse radiolysis of aqueous solutions. $\triangledown$, 1.0 M $HClO_4$; $\bigcirc$, 0.5 M $HClO_4$; $\square$, 0.175 M Cd^{2+}; $\triangle$, 1.0 M $HClO_4$. The spectra were obtained at the times of maximum absorption and normalized to equality at 700 nm. The solid curve represents the absorption spectrum of e_{aq}^- obtained in microsecond pulse radiolysis. (Bronskill, Wolff, and Hunt.[98])

3.3.2. *Solvated Electron in Alcohols*

Hunt and his co-workers[99] also observed the spectrum and the decay of the solvated electron in methanol, ethanol, 1-propanol, 2-propanol, and ethylene glycol. Sulfuric acid served as electron-scavenging agent in these cases. As of early 1971, the decay kinetics have not been published. The absorption spectrum, at least in the case of methanol, is reported to be identical with that obtained in microsecond pulse radiolysis work. Table II compares the *relative* yields of solvated electrons for the alcohols in picosecond work with the corresponding actual values from microsecond pulse radiolysis. There is good agreement between corresponding data for the two time domains. The conclusion is that the electrons are fully solvated at the time of earliest observation.

3.3.3. *Formation Time of Solvated Electrons*

In most pulse radiolysis studies, the decaying portion of the pulse is analyzed. However, much information on the formation of observable intermediates may in principle also be derived from the rising part of the pulse. In practice, such evaluation is frequently limited by the fact that overall instrument response is considerably slower than the physical process of interest. In the case of ultrafast pulse radiolysis such as discussed by Hunt and his co-workers,[99] the major contributions to instrumental time resolution are desynchronization between the fine-structure linac pulses and the analyzing

TABLE II

Relative Yields of e_{solv}^- in Alcohols as Calculated from Absorption Maxima in Picosecond Work[a]

	Relative $G(e_{solv}^-) \cdot \varepsilon_{max}$[b] (measured)	Microsecond pulse radiolysis data		Relative $G \cdot \varepsilon_{max}$ (calculated)
		$G(e_{solv}^-)$	ε_{max}	
H_2O	1.00	2.8	18,500	1.00
MeOH	0.37	1.1	17,000	0.36
EtOH	0.29	1.0	15,000	0.29
1-Propanol	0.29	1.0	13,000	0.25
2-Propanol	0.25	1.0	14,000	0.27
Ethylene glycol	0.36	1.2	14,000	0.32

[a] According to Wolff, Bronskill, and Hunt.[99]

[b] ε_{max} is the extinction coefficient of e_{solv}^- at the wavelength of maximum absorption.

light pulse within the sample cell, the width of the electron pulse, and the variation in dose between front and back of the sample cell. The contributions of other instrumental factors to the response time are comparatively small.

Evaluation of the effects of the factors listed permits an estimate of the upper limit for the time of formation τ_f of the solvated electron.

The results are summarized in Table III, in which observed rise times are

TABLE III

Dielectric Relaxation Times and e_{aq}^- Formation Times (in picoseconds)[a]

Solvent	Relaxation time		Observed rise time	Theoretical minimum rise time (τ_{min})	$2\sigma + \tau_f$[d]
	t_d[b]	t_d'[c]			
H_2O (2-cm cell)	10	0.2	24 ± 2	18	6 ± 2
H_2O (1-cm cell)			17 ± 2	9	8 ± 2
Methanol	69	3.8	26 ± 2	18	8 ± 2
Ethanol	143	10.6	24 ± 2	19	5 ± 2
1-Propanol	530	49	23 ± 2	20	3 ± 2
2-Propanol			31 ± 5	20	11 ± 5
1-Butanol	650	72	25 ± 3	21	4 ± 3
Ethylene glycol			26 ± 5	23	3 ± 5

[a] According to Hunt and co-workers.[99]

[b] From C. P. Smyth, *Dielectric Behaviour and Structure*, McGraw-Hill, New York, 1955.

[c] t_d' is the relaxation time at constant charge $t_d' = t_d(\varepsilon_\infty/\varepsilon_s)$ (see Mozumder[101]), where ε_∞ is the high-frequency dielectric constant and ε_s is the static dielectric constant.

[d] See text for interpretation of $2\sigma + \tau_f$.

compared to the theoretical minimum rise times (of desynchronization) for the different solvents studied. The differences are attributed to the combined effect of fine-structure pulse width 2σ and e_{solv}^- formation time. The value given for $2\sigma + \tau_f$ thus represents an upper limit for τ_f.

3.3.4. *Reaction Rate of e_{solv}^- with Scavengers*

The observed interactions of H^+ with e_{aq}^- on a picosecond time scale are quite similar to those observed with conventional pulse radiolysis.[98, 100] It appears, however, that other electron-scavenging additives reduce the immediate e_{aq}^- yield dramatically, while at the same time the decay rate of the e_{aq}^- signals is not affected. Figure 11 and Table IV summarize the results obtained with H_2O_2, NO_2^-, acetone, NO_3^-, and Cd^{2+}.

Figure 13 gives a plot of the immediate relative yield of e_{aq}^- as function of scavenger concentration. In Table IV the effect of the scavengers on the "initial yield" (i.e., at first observation) is expressed by C_{50}, the scavenger concentration required for reduction of this yield by 50 percent. For comparison, the specific rates for the reaction e_{aq}^- plus scavenger are listed together with the effects of scavenger on the hydrogen yields as measured by conventional methods in earlier work.

TABLE IV

Specific Rates of Reaction between e_{aq}^- and Various
Scavengers and the Effectiveness of the Latter
in Reduction of "Initial Yields" of e_{aq}^- and H_2

Scavenger	$k,$ $dm^3\,mole^{-1}\,sec^{-1}$ [a]	e_{aq}^- C_{50}, M [b]	Molecular H_2 C_{50}, M [c]
H_{aq}^+	2.06×10^{10}	>9	
H_2O_2	1.23×10^{10}	1.0	0.39[d]
NO_2^-	$4.6\ \times 10^9$	1.2	0.394[e]
Acetone	$7.6\ \times 10^{9}$[f]	1.0	0.5[g]
NO_3^-	$1.1\ \times 10^{10}$	0.35	0.2[h]
Cd^{2+}	$1.2\ \times 10^{10}$	0.31	

[a] Anbar and Neta.[176]

[b] Concentration of scavenger required for 50-percent reduction of yield "initially observed."

[c] Concentration of scavenger required for 50-percent reduction of yield by conventional methods.

[d] Ghormley and Hochanadel.[177]

[e] Schwarz[178].

[f] Bronskill (unpublished results).

[g] Appleby.[179]

[h] Mahlman and Sworski.[180]

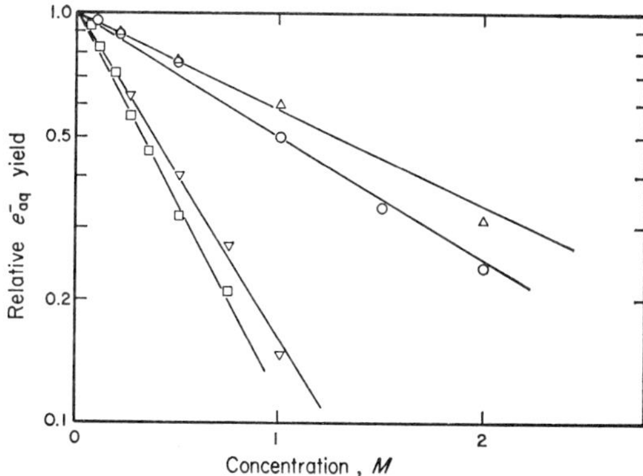

Fig. 13. Effect of additives on the "immediate" yield of e_{aq}^- in picosecond pulse radiolysis. $\bigcirc$, Acetone; $\triangle$, H_2O_2; $\bigtriangledown$, NO_3^-; $\square$, Cd^{3+}. (Wolff, Bronskill, and Hunt.[99])

For clarification, it should be noted that the relative e_{aq}^- yields have been corrected for the reduction of the peak signal with decreasing decay times. Such decrease in peak intensity occurs quite generally when the decay time of the signal approaches the time resolution of the apparatus. This problem has been discussed in detail by Bronskill, Wolff, and Hunt[98] for picosecond pulse radiolysis measurements.

3.3.5. *Comments*

The results of ultrafast pulse radiolysis have direct bearing on several of the unsettled aspects of radiation chemistry of polar liquids in general and of water in particular. Among these are the question of solvation time of electrons in polar liquids, the role of spurs, and the more recently discussed possibility of chemical reaction of the electron prior to solvation.

The rate of solvation of electrons is commonly taken to be related to the dielectric relaxation of the medium in question.[101] In other words, it is assumed that the time required for the alignment of solvent molecules in the neighborhood of an electron is comparable to the macroscopic dielectric relaxation time τ_d of the solvent. τ_d refers to the relaxation time for *constant electric field*. According to Mozumder,[101] the dielectric relaxation time for *constant charge* τ_d' is more appropriately related to the solvation time. τ_d' is connected to τ_d by $\tau_d' = (\varepsilon_\infty/\varepsilon_s)\tau_d$, where ε_∞ and ε_s are, respectively, the high-frequency and the static dielectric constants of the solvent. The relaxation times are considerably longer than the pertinent values of τ_f, the e_{solv}^- formation time. It has been suggested by Mozumder[101] that the macroscopic relaxation times given are possibly too large. Even if this is the case the question remains open whether or not solvation of electrons can be described in so relatively simple a manner. The relaxation times given are derived under conditions of comparatively weak electric field strength. One might speculate that the electric field in the neighborhood of an electron may have a much more drastic effect on the matrix molecules than is anticipated from the value of the macroscopic relaxation time. Hamill, as well as Hentz (see Mozumder[101]), has suggested that the solvation time of electrons is possibly more appropriately connected to a microscopic dielectric relaxation process. Such a process accordingly may require rotational reorientation of solvent molecules in the immediate vicinity of an electron. The corresponding "relaxation time" may thus be much shorter than either τ_d or τ_d'. The likelihood of such a process seems to be borne out in experiments by Taub and Eiben,[181] who observed the immediate presence of the solvated electron spectrum in the pulse radiolysis of ice. The macroscopic relaxation time for ice[101] at 273°K is 2×10^6 sec, in complete disagreement with the experimental observation if either τ_d or τ_d' be presumed to be the pertinent quantity.

It might be expected that employment of the time scale possibilities of ultrafast pulse radiolysis studies would make a direct test of the spur model possible. A model suggested by Hunt and his colleagues[98] (following an idea originally suggested by Schwarz) envisions within the spur an initially rapid decay of concentration of solvated electrons (in a diffusion-controlled process approximately first order in those electrons alone) accompanied by a slower process

$$e_{aq}^- + H^+ \rightarrow H + aq$$

which is dependent both on solvated electron and (solvated) H^+ ion concentrations. If the rate of the approximately first-order reaction be written $k'[e_{aq}^-]$ and that of the second-order process be written $k''[e_{aq}^-][H^+]$, it follows that the actually observed specific rate k observed in the very-fast pulse work should have the form

$$k = (k' + k''[H^+])/[H^+] \tag{3.1}$$

Equation (3.1) corresponds approximately to the results of Bronskill, Hunt, and Wolff,[98] shown in Fig. 14, in their high-velocity linac pulse work. It fails, of course, at very low H^+ ion concentration, for k does not go to infinity. The results of Bronskill et al. were, of course, preliminary. Indeed, Schwarz[185] showed in a later paper that their data could be adequately fitted at low H^+ ion concentration by application of an improved time-dependent rate theory, which would however predict a minimum in k at $[H^+] = 1\ M$ and some significant increase at higher H^+ ion concentration—all within the limits of error of the work reported.

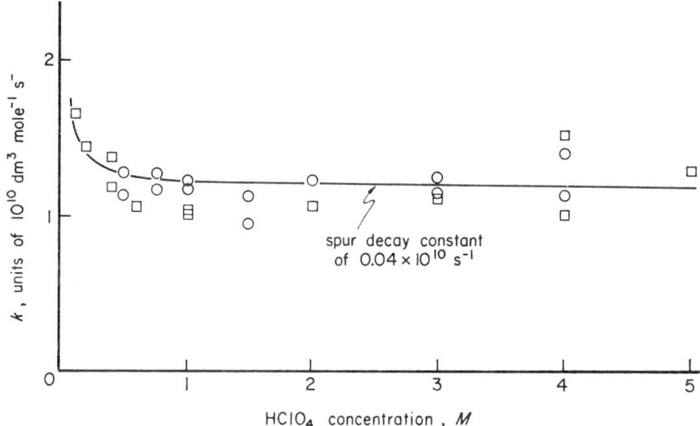

Fig. 14. Specific rate k for the reaction $e_{aq}^- + H_{aq}^+$ as a function of HClO$_4$ concentration. The curve was calculated with a spur decay constant $k_1 = 4 \times 10^8 \text{ sec}^{-1}$; the two sets of points represent two different sets of experiments. (Bronskill, Wolff, and Hunt.[98])

The adequate representation of the experimental values of k in Fig. 14 by Eq. (3.1) is by no means conclusive evidence for a spur model. Other interpretations of the increase of k at lower H$^+$ concentrations can be given. These include ionic strength effects and the effect of buildup of radiolysis products (H$_2$O$_2$) between linac pulses, as also considered by Bronskill, Hunt, and Wolff.[98]

The specific rates of reaction of the hydrated electron are of the same order of magnitude for all the scavengers studied by Hunt and his co-workers (see Table IV). Nonetheless, the effect on the immediate yields of e_{aq}^- varies dramatically depending on the scavenger. The effect is shown in Table IV in terms of C_{50}; that is, in terms of the scavenger concentration required to reduce the initial e_{aq}^- yield by 50 percent.

This effect has the appearance of a reaction of a precursor to the solvated electron. Suggestions have been made that this precursor might be either a highly excited state of water or the electron prior to solvation. Hunt and his collaborators[99] favor the second possibility, although involvement of an excited state of water cannot be ruled out completely. The interaction of the scavengers with e^- prior to solvation appears consistent with the conclusions derived by Hamill[184] based on studies of the reduction of the H$_2$ yield as a function of scavenger concentration. [*Editors' Comment:* In a paper published after this chapter was committed to press, Schwarz[185] commented (as a consequence of his own theoretical treatment) that a simpler mechanism, "in which the hydrated electron is the only species which reacts with the solutes," is adequate to explain the data.]

The Hamill model[84] envisions a competition between solvation and scavenging of e^-

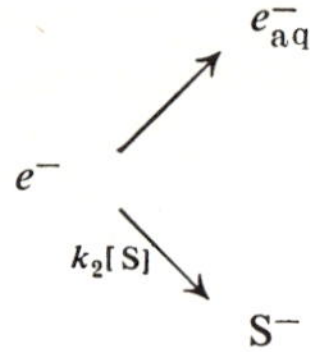

Hunt and his collaborators note that in such a case the ratio of concentrations of solvated electrons obtained in the absence and presence of scavenger should be

$$[e^-_{aq}]_0/[e^-_{aq}] = 1 + \frac{k_2[S]}{k_1} \qquad (3.2)$$

A plot of $1/[e^-_{aq}]$ versus [S] for acetone as scavenger, however, does not reveal the linear relationship to be expected from Eq. (3.2). Hunt emphasized therefore that the mechanism of e^- scavenging does not constitute a simple competition between the two processes.

Such a conclusion may be somewhat unjustified for it assumes applicability of the conventional formalism of diffusion-controlled reactions. In view of the high concentrations of reactants, H_2O, and scavengers, such an assumption is not likely to be justifiable. It is, however, interesting to note that the trend of the observed $1/[e^-_{aq}]$ versus [S] curve can be correctly simulated by including the time dependence of bimolecular specific rates in the derivation of Eq. (3.2). This modification is justified because on the time scale of Hunt's experiment the time-dependent term in Smoluchowski's expression for bimolecular specific rates[103]

$$k_2 = 4\pi N'RD\left[1 + \frac{R}{(Dt)^{1/2}}\right] \qquad (3.3)$$

(where R is the radius of interaction and D the sum of the diffusion coefficients of the two reactants) can not be ignored; [cf. Schwarz,[185] *Ed.*].

While such procedure may serve as a reminder that on the time scale of ultrafast experimental techniques nonstationary distributions of reactants may have to be considered, it remains questionable whether ordinary diffusion kinetics can be applied under Hunt's experimental conditions. It is in particular uncertain how to describe the motion of an electron in the condensed media in question.

Another interesting aspect of the results of Hunt and his co-workers is the exponential dependence of the hydrated electron yield on scavenger concentration. Hunt explains such dependence on the assumption that electrons can react with the scavenger while still "hot."

If we denote the probability of reaction of a hot electron with a scavenger molecule S by P, the ratio of scavenger molecules to water molecules by ρ, and the number of collisions of the electron prior to thermalization by n, the ratio of solvated electron concentrations in the presence and absence of scavenger is obtained as

$$[e_{aq}^-]/[e_{aq}^-]_0 = (1 - \rho P)^n \tag{3.4}$$

If the number of collisions $n \gg 1$ and $\rho P < 1$, the right-hand side of Eq. (3.4) takes the form $\exp(-\rho P n)$, a form consistent with the experimentally observed exponential dependence of the hydrated electron yield on scavenger concentration.

One may object to such specific description on the basis that the dynamics of the scavenging process are not revealed even on a picosecond time scale nor are the concepts of description of the behavior of an electron in the condensed phase established.

It is not difficult to derive the exponential relationship between solvated electron yield and scavenger concentration from more general considerations. A clue to such an approach is the close resemblance of Hunt's experimental data, under consideration, to those encountered for static fluorescence quenching.[103] In such work addition of quencher reduces the luminescence yield without affecting the luminescence decay time. Just as in Hunt's observation regarding the dependence of e_{aq}^- yield on scavenger concentration, luminescence yield is often found to depend exponentially on quencher concentration.

Following a suggestion by Hentz,[105] the exponential dependence can be derived in an elementary fashion from a model that assumes an effective quenching radius around the quenching molecule. Whenever an excited molecule happens to be within this radius, its emission is quenched by an unspecified mechanism. Analogously, in the case of electron scavenging, one may assign an effective scavenging radius to the scavenger molecule, such that the electron is scavenged when within this radius and becomes hydrated when outside.

Denoting the number of scavenging molecules in the volume V by n, each having an effective scavenging volume v, the probability for the electron *not* to be scavenged is given (cf. Hertz[104] for a related model) by

$$w = \left(1 - \frac{v}{V}\right)^n$$

which for *large* values of n becomes

$$w \simeq \exp\left(-nv/V\right)$$

According to the assumptions initially made, an electron not scavenged becomes hydrated. Thus, noting that $n/V = 6.02 \times 10^{20} \cdot [S]$, Eq. (3.4) assumes the form

$$[e_{aq}^-]/[e_{aq}^-]_0 = \exp\left(-N'v[S]\right) \qquad (3.5)$$

where $N' = 6.02 \times 10^{20}$.

Combination of Eq. (3.5) with the values for C_{50}, the quencher concentration required to reduce the initial e_{aq}^- yield by 50 percent, as given in Table IV, gives values of 6–10 Å for the effective scavenging radius for the more effective scavengers.

3.4. Luminescence-Decay Studies

Application of fast experimental techniques to time-dependent luminescence studies is of a variety far exceeding the scope of this chapter. Aspects of such studies have been extensively discussed in several books and review articles.[23, 82, 103, 106–108] Consequently, this review of the relation between observable lifetimes and the relevant relaxation rates is very brief and only a few recent results of more specific radiation chemical interest are described.

3.4.1. *Lifetimes and Relaxation Processes of Electronically Excited States*

The most common quantity determined experimentally by the fast techniques described in Sections 2.2 and 2.3 is the mean luminescence decay time τ. It is related to the luminescence intensity–time function $I(t)$ by

$$\tau = \frac{\displaystyle\int_0^\infty I(t) \cdot t \, dt}{\displaystyle\int_0^\infty I(t) \, dt} \qquad (3.6)$$

τ has unique significance when $I(t)$ represents a single exponential decay of one excited state. In this case τ, the mean luminescence decay time, is related to the specific rates of radiative and nonradiative relaxation processes of this state by

$$1/\tau = k_e + \sum k_{nr} \qquad (3.7)$$

where k_e is the specific rate of spontaneous emission of the excited state (and thus can in principle be evaluated from the emission and absorption spectra[103]) and k_{nr} describes the intermolecular nonradiative decay processes (which are essentially intersystem crossing and internal conversion). At least in condensed media, k_e and k_{nr} may be considered constant and referring only to the lowest excited singlet, and triplet, states. The reason for such a statement is that relaxation of higher electronic or vibronic states to the lowest excited singlet or triplet state is generally much faster than emission. It should be emphasized that Eq. (3.7) is not applicable in the case of very high intensity of exciting light (laser excitation), when the emitting state is directly excited, because stimulated emission may strongly modify the emission rate.[109]

When the emitting state is directly populated by absorption or, more generally, when all the excitation is channeled into this state, the mean decay time τ is related to the luminescence quantum yield η by

$$\eta = \tau/\tau_e \tag{3.8}$$

where $\tau_e = 1/k_e$.

For a complete description of the relaxation of an excited state, lifetime measurements must be supplemented by luminescence quantum yield measurements and, if multiple channels of nonradiative decay exist, by quantum yields of the corresponding final states. The specific relaxation rate $1/\tau$ is frequently modified by interaction of the excited molecule with the environment. In such a case additional terms must be introduced in Eq. (3.7). Lifetime and quantum yield data thus not only provide information on intramolecular processes but are fundamental for consideration of the kinetics of photochemical and radiation chemical processes involving excited states.

Collections of lifetime and quantum yield data for many organic molecules are given in several references,[23, 36, 81, 106, 107] and particularly in two very extensive surveys.[82, 110]

3.4.2. *Luminescence Decay in the Gas Phase*

In condensed media emission is usually observable from the lowest excited electronic states only. In the gas phase, at pressures where the mean molecular collision time becomes comparable to the excited state lifetimes, the situation is more complex. For that pressure region lifetimes and emission spectra depend on both pressure and excitation energy. This fact is readily understandable; collision-induced nonradiative processes compete with radiative and internal nonradiative relaxation processes of the initially excited state.

Much theoretical work[111–117, 172] and, to some extent, experimental luminescence-decay studies[118–124] have been concerned with relaxation of excited, comparatively large organic molecules, particularly at pressures sufficiently low that collisions may be ignored. Such activity was greatly stimulated by the observation of Kistiakowsky and Parmenter[125] that the fluorescence quantum yield of benzene even at very low pressures is much below unity. As a result of all this work, it is now generally recognized that nonradiative relaxation processes of excited states of comparatively large organic molecules, such as benzene, can occur even in the absence of inter-molecular interactions.

The processes considered in the theories of nonradiative molecular transitions are intersystem crossing and internal conversion, although, particularly for higher excited levels, chemical decomposition cannot be ignored. Under certain conditions (which appear to be satisfied for molecules such as benzene, naphthalene, and so on), theory leads to the following expression[111b, 117] for the specific rate of nonradiative transitions

$$k_{nr} = \frac{2\pi}{\hbar} \beta^2 \rho F \tag{3.9}$$

where β represents the matrix element for coupling of the two electronic states between which transition occurs, ρ denotes the density of vibronic states (per cm^{-1}) of the final electronic state at the level of the transferring singlet, and F is an average Franck–Condon factor for the initial and final levels. While theory provides a general understanding of the principles underlying inter-system crossing and internal conversion, its quantitative application to specific compounds suffers at present from the lack of knowledge of the (numerical values of the) various parameters such as are required for use of Eq. (3.9).

Luminescence decay and spectral properties have been studied most extensively for *benzene vapor*. In this case the specific rates k_e and k_{nr} have been evaluated for three specific vibrational levels of the $^1B_{2u}$ state (i.e., the first excited singlet state). These levels correspond to the vibrations $(v_6' + v_1')$, v_6', and the zero-point level of this singlet.[126] Table V summarizes the results.

TABLE V

Specific Rates of Emission and Nonradiative Decay from Vibronic
Levels of the $^1B_{2u}$ State of Benzene[a]

Vibrational level	Vibrational energy above zero point, cm^{-1}	k_e, sec^{-1}	k_{nr}, sec^{-1}
Zero point	0	2.4×10^6	8.7×10^6
v_6	523	3.4×10^6	9.1×10^6
$v_6 + v_1$	1446	3.7×10^6	11.0×10^6

[a] See Ware et al.[123]

In relating the observed variation of k_{nr} to Eq. (3.9), it is assumed that β is constant and that intersystem crossing is the nonradiative relaxation process. The density of final states ρ can be evaluated from knowledge of the normal modes of benzene.[127] The variation of k_{nr} with vibrational energy is then the result of the increasing density of states and the decrease in magnitude of the Franck–Condon factors. For the three levels under consideration, the experimental values of k_{nr}, when combined with Eq. (3.9), give an exponential decrease in the Franck–Condon factors.[123] This result appears consistent with the theoretical works of Byrne, McCoy, and Ross[114] and of Siebrand.[133]

Variation in the radiative rate k_e for benzene (shown in Table V) can be associated with the fact that the electronic transition $^1A_{1g} \leftarrow {}^1B_{2u}$ is vibronically induced. In this case the radiative transition probabilities are described by the Herzberg–Teller theory.[128–130] However, comparison of the ratio of the observed values of k_e with those calculated from this theory shows only partial agreement.[121, 123]

The results in respect to k_{nr} seem to settle the discussion as to whether the transitions $^1B_{2u} \rightarrow {}^1B_{3u}$ from the lowest vibrational levels of the $^1B_{2u}$ state of benzene correspond to the case of the statistical limit.[131, 132] It had been suggested that intersystem crossing from these low excited levels might be quite ineffective, and that Eq. (3.9) may not be applicable. However, the fact that the fluorescence quantum yield is found to be less than unity even at pressures as low as 10^{-5} torr indicates that a nonradiative process is predominant. Moreover, evaluation of fluorescence quantum yield data and decay times of these low levels using Eq. (3.9) seems to give results consistent with theory.

Other compounds whose luminescence decay has been studied as a function of excitation energy include β-naphthylamine,[118] hexafluoroacetone,[122] naphthalene,[124] and anthracene.[134] All cases, with the exception of anthracene, show an increase in decay rate with vibrational energy.

For β-naphthylamine, variation in k_{nr} with excitation energy is found to be exponential. Fischer and Schlag[116] account explicitly for this exponential dependence by a theory of nonradiative transitions based on the theory of rate constants developed by Kubo[135] and Yamamoto.[136]

The variation in k_{nr} for hexafluoroacetone is more complex but in this case, in addition to intersystem crossing, chemical decomposition also occurs.

Measurements of the luminescence lifetimes of naphthalene vapor by Laor and Ludwig[124, 134] reveal a nearly linear increase in $1/\tau$ with energy when the exciting frequency is within the range of the two lowest excited singlet states. For higher excitation energy the dependence is more closely exponential. These measurements have been refined and supplemented by relative fluorescence quantum yields.[134] The refined measurements show, additionally, that for excitation into the lowest vibrational levels of the first absorption band the values of k_{nr} are considerably higher than might be expected from extrapolation of the nearly linear curve to those energies.

The overall increase in k_{nr} with excitation energy appears qualitatively consistent with the trend of k_{nr} observed with other compounds. It is not clear at present whether the linear variation of k_{nr} within the range of excitation of the first and second absorption bands represents a fortuitous combination of the energy-dependent factors in Eq. (3.9), or whether it reflects a more general physical property of nonradiative processes.

It should be emphasized that the nearly linear variation in k_{nr} with excitation energy[124] is over an energy region covering two electronic excited states. There is no observable break in the curve at the point where excitation of the second excited singlet state occurs. This fact is presumably related to the observation by Watts and Strickler[137] that internal conversion from the second to the first state is very rapid. Therefore, emission corresponding to high vibrational levels of the first state is alone observed. The combined decay time and quantum yield measurements[134] suggest moreover that intersystem crossing from the second excited state is also negligible compared to internal conversion.

The much larger variation in k_{nr} with excitation energy on excitation to the third electronic state of naphthalene is similar to that observed for β-naphthylamine and hexafluoroacetone. It remains an open question whether in these cases photochemical decomposition or reaction may occur in addition to intersystem crossing.

For anthracene vapor, luminescence-decay times as a function of excitation energy are presently available, while fluorescence quantum yield measurements for vibronic states have apparently not been published. The lifetime measurements exhibit no dependence on excitation energy for any single electronic state excited. However, each of the electronic states has a different lifetime which decreases with its energy. Comparison of anthracene with naphthalene and benzene shows that its lower electronic transitions are rather strong. This strength is reflected in the fact that the radiative lifetime is about 12 nsec as compared to about 100 and 400 nsec for naphthalene and benzene, respectively.[82] The empirical correlation between (1) strength of radiative transition and (2) vibrational energy dependence of nonradiative processes may be more than accidental. However, study of a large number of molecules is required to generalize such a correlation.

Table VI gives the fluorescence lifetimes of naphthalene vapor at 0.07 torr for excitation within various vibronic levels. The bandwidth of excitation is 50 Å. Table VII contains corresponding data for anthracene vapor at 0.1 torr. Both sets of data are previously unpublished.

TABLE VI

Experimental Fluorescence Lifetimes of Naphthalene Vapor

λ_{ex}, nm	ν, cm^{-1}	τ, nsec	$\pm\Delta$[a]	$1/\tau$, units of 10^7 sec^{-1}	$\pm\Delta$[a]
308	32,468	179.1	4.7	0.558	0.014
304	32,895	133.0	4.2	0.752	0.024
300	33,333	116.7	1.7	0.857	0.012
295	33,898	93.7	3.9	1.067	0.044
289	34,602	77.0	2.2	1.30	0.040
279	35,842	58.9	0.3	1.70	0.01
275	36,364	53 4	0.5	1.87	0.02
269	37,175	47.8	0.4$_5$	2.09	0.02
265	37,736	43.5	0.4	2.30	0.02
258	38,760	37.6	0.3	2.66	0.02
255	39,216	36.6	0.3	2.73	0.02
250	40,000	32.5	0.3	3.08	0.02
240	41,667	25.3	0.3	3.95	0.04
235	42,553	20.1	0.3	4.97	0.07
230	43,478	10.6	0.4	9.4	0.4
225	44,444	7.2	0.3	13.9	0.6
220	45,455	5.8	0.3	17.2	0.9
215	46,512	2.2	0.1	45.5	2.1

[a] Δ represents the mean deviation of the experimental data for τ and $1/\tau$.

3.4.3. *High-Energy-Induced Luminescence*

There are two aspects of interest in dynamic studies of radioluminescence: development of fast particle detectors such as initiated by Kallmann[138] in 1947, and information on radiation chemical processes potentially available from the shapes of luminescence intensity–time curves. (The practical aspect, scintillation counting, is reviewed by Birks[139] and is not discussed here.)

The second interest in such studies is related to the observation that the luminescence decay curves of fast scintillator solutions frequently exhibit two "components," an initial decay largely characteristic of the scintillator itself, and a slower one the origin of which seems less clear. It is thought, however, that this slow decay is associated with processes of radiation chemical interest because, for ultraviolet excitation of scintillators in liquid solutions, the decay is frequently found to be exponential. Several interpretations have been advanced to account for the slow emission: slow ion–electron recombination,[140] triplet-singlet intramolecular transitions,[141] formation of metastable exciton pairs in organic crystals,[142, 143] triplet-triplet interaction,[144, 145] ion-ion combination.[92, 146]

TABLE VII

Experimental Fluorescence Decay Rates of Anthracene Vapor ($p \simeq 0.1$ torr; 5-nm bandwidth of exciting light)

Excitation λ_{ex}, nm	Decay rate $(1/\tau)$, units of 10^8 sec^{-1}	$\pm \Delta$[a]
362	16.1	0.3
358	15.9	0.3
353	17.2	0.3
345	16.9	0.3
337	17.2	0.3
330	17.5	0.3
322	16.7	0.3
315	16.4	0.5
300	16.7	0.6
290	17.5	0.9
275	17.2	1.2
265	18.2	2.0
260	18.5	1.5
255	22.7	1.0
250	22.7	0.5
246	22.7	0.5
239	21.7	0.9
233	22.7	1.0
226	25.0	1.3
220	25.0	2.5

[a] Δ represents the mean deviation of the experimental data.

In view of the fact that a slow decay component is observed in a wide variety of systems (such as pure aromatic crystals, dilute solutions of organic scintillators in liquid aromatic solvents, and also aliphatic solvents), it is questionable whether any single mechanism accounts for the observations in all these situations.

Two models have been considered in greater detail for liquid solutions.

Laustriat[51] and Voltz and King[148] conclude that triplet-triplet interaction leading to delayed emission is the main factor responsible for the slow component. They carried out detailed calculations of theoretical luminescence decay curves incorporating the spur concept.[149] Their results agree very well with observed luminescence decay both of high-energy-excited organic crystals and of solutions of organic scintillators in aromatic solvents.[51] They offer the plausible suggestion that triplet-triplet interaction accounts for "delayed" fluorescence in organic crystals. In this case motion of triplet excitons can be sufficiently fast to permit triplet-triplet interaction during the lifetime of such triplet states.[142, 143]

One may, however, properly question for aromatic solvents the possibility of occurrence of the corresponding process for the solvent triplets. One objection is inherent in the fact that in pure benzene and several of its derivatives the triplet lifetime is found to be of the order of 10 nsec.[150–152] Such a fact makes it difficult to correlate the delayed emission, which occurs over several hundred nanoseconds, with the triplet state of the solvent.

A somewhat obvious alternative is the assumption that the triplet state in question is not that of the solvent but that of the solute (scintillator). Birks[153] has taken up this possibility in more detail. He asserts that such interpretation of the model of Voltz and Laustriat does not affect the form of the slow luminescence decay. This interpretation may be correct but perhaps still requires an analysis of the original calculations of Voltz and King.

There is an even more serious inconsistency, however, between the model of triplet-triplet annihilation and the data on fast pulse radiolysis of organic solutes in aromatic solvents. As discussed in Section 3.1.6, the triplet *absorption* signals of several solute molecules observed in a time interval of $\sim$350 nsec approximately 50 nsec after the ionizing pulse show almost no decay. However, luminescence intensity in the experiments of Laustriat and co-workers in a time interval of only $\sim$150 nsec, approximately 50 nsec after the pulse, decreases by as much as 70–80 percent. This comparison seems to suggest that the long-luminescent component is not related to triplet-triplet combination.

Ludwig and Huque[92] studied the luminescence of nanosecond pulsed X-ray and ultraviolet excited solutions of *p*-terphenyl in aliphatic solvents such as cyclohexane, and also the effect of charge scavengers such as N_2O, CCl_4, and piperidine on the luminescence-decay curve. For ultraviolet excitation the fluorescence decay is exponential, and the addition of any of the scavengers has either no or negligible effect on the peak intensities and decay times. There is consequently no interaction with the emitting state of ultraviolet excited scintillators. However, as in the other cases mentioned, a long component is observed when the solutions are excited by pulsed ionizing radiation. In the presence of N_2O, the total luminescence–time curve is found to be lowered, while CCl_4, in addition, enhances the decay of the slow component. Piperidine noticeably suppresses only the slow component without observable effect on the peak intensity. On the basis of these observations and the scavenger characteristics, it is concluded that the emitting state of the scintillator is presumably produced by ion-ion neutralization.

The processes of significance may be written as

$$S \rightsquigarrow S^+ + e^- \tag{1}$$

$$e^- + S \longrightarrow \text{uncertain products or intermediates} \tag{2}$$

$$e^- + X \xrightarrow{\text{very fast}} X^- \tag{3}$$

$$X^- + S^+ \longrightarrow X^* \text{ and other excited states} \tag{4}$$

$$X^+ + X^- \longrightarrow X_2^*, X^* \text{ and other excited states} \tag{5}$$

Here S denotes the solvent, X the scintillator, and X^* the emitting scintillator state. The reactions of quenchers may be symbolized by

$$Q + e^- \rightarrow Q^- \qquad (Q = N_2O) \tag{6}$$

$$Q + X^- \rightarrow Q^- + X \qquad (Q = CCl_4) \tag{7}$$

$$Q + S^+ \rightarrow Q^+ + S \qquad (Q = \text{piperdine}) \tag{8}$$

In the simplest model the form of the luminescence decay curve can then be rationalized in terms of the initial spatial distribution function of the separation between the components of ion pairs ($X^- \cdots S^+$, $X^- \cdots X^+$). Thus the initial formation of X^* is fast (compared to the fluorescence lifetime of X^*) because of neutralization of ion pairs with small separation. The observable decay is therefore similar to that of the directly excited scintillator. As time advances, the rate of neutralization slows down because of the larger distances separating the members of the pairs. The decay time of an excited p-terphenyl molecule is 1.4 nsec.[81] The rate of emission at longer times is determined by the rate of formation of the excited scintillator state. At such times process (5) is expected to contribute predominantly to the formation of X^*. The different effects of the three scavengers on the luminescence decay curves observed are consistent with processes (6) to (8).

The interpretation of the high-energy-induced luminescence-decay curves in terms of ion recombination processes appears to be consistent with more recent data on fast pulse radiolysis of solutions of aromatic solutes in liquid aliphatic hydrocarbons.[74] As mentioned in Section 3.1.3, with this method the decay of the solute ions can be directly observed. It is found that in solutions most of the solute ions observed in absorption decay on a time scale short compared to the fluorescence lifetime of PTP and that there is a subsequent slow decay which stretches into the microsecond region.

It is not entirely clear whether or not the ionic mechanism suggested also accounts for the slow fluorescence component in aromatic solvents. In the pulse radiolysis experiments on these systems, the yield of ions is found to be small ($G \sim 0.2$), so that the disappearance of their absorption signals cannot be followed in detail. These data cannot therefore be conclusively employed as evidence for an ion-recombination model to account for a slow luminescence component in aromatic solutions.

3.5. Mode-Locked Laser Studies

The present literature on pulsed, mode-locked lasers is predominantly concerned with the development of experimental techniques. Only a few studies deal with applications of these techniques to processes of some radiation chemical interest. Among these are studies of lifetimes of excited vibronic states of organic molecules in solution, not accessible to conventional experimental methods, and a study of intermolecular energy transfer in liquid solution at times less than 10^{-9} sec.

3.5.1. *Radiative and Nonradiative Molecular Transitions in the Liquid Phase*

3.5.1.1. Intramolecular Processes. The rate of vibrational relaxation of excited levels of large organic molecules in the liquid phase is in general very fast and thus not directly observable by conventional means. However, by use of the mode-locked laser methods described in Section 2.5, Rentzepis[32] has directly observed vibrational relaxation processes for azulene in alcoholic solution. The procedure described in Section 2.5 served to measure the duration of the mode-locked laser pulses. The fluorescent compound employed in this case was such that a fluorescent state could be reached only by simultaneous interaction of two picosecond light pulses traversing the sample collinearly but in opposite direction. As a consequence, a higher-lying fluorescent level can be populated only in those regions of the solution where the two pulses encounter each other, since here the combined intensity of the two pulses is sufficient to reach the fluorescent state via *multiphoton* absorption.

The situation is different for such systems in which a molecular state, which can be populated by photon absorption from the faster pulse, has a finite lifetime. Here, as in the previously described case, a higher fluorescent state may be reached by multiphoton absorption in the overlap region of the two pulses traversing the sample in opposite but collinear directions. Fluorescence is therefore observed, as before, from the region where the two pulses spatially overlap. In addition, however, fluorescence may also occur along the path of the second pulse beyond the encounter region. This fluorescence is produced by the second pulse now encountering, along its path, molecules previously excited by the first pulse. The energy of the second pulse is such as to induce transitions from the short-lived excited states to a higher fluorescent state the emission of which is observed. Because the short-lived excited states decay rapidly, their population, as seen by the second pulse as it traverses the sample, decreases rapidly. The fluorescence intensity thus decreases along the path of the second pulse. From the spatial extent and decrease of fluorescence intensity, and knowledge of the rate of propagation of the pulses in the sample, the lifetime of the short-lived excited state can be derived. The method may be considered a sophisticated modification of flash photolysis in the sense that the first pulse produces the state to be studied and the second pulse serves to probe, in this case, the temporal behavior of that state. Unlike flash photolysis, however, information is not obtained from observation of the time dependence of an *absorption* signal associated with the transient state but rather from *fluorescence* related to the intermediate.

At present, the method has been applied to a study of various states of azulene,[32, 154] a compound that has properties particularly well suited for the type of measurements described. However, with appropriate modifications, the method should be applicable to a variety of other molecules. The following data for azulene (see Fig. 15) serve to illustrate the kind of data that can be obtained.

A picosecond pulse of the second harmonic ($\bar{\nu}_2 = 18863$ cm^{-1}) of a Nd^{3+}-glass laser excites the fourth vibrational level of the first excited singlet state of azulene. The second pulse (the fundamental at $\bar{\nu}_1 = 9431$ cm^{-1}) leads to transition to the second excited singlet $^1A_1(2)$ state, from which fluorescence is observed. As already outlined, the intensity of this fluorescence decreases as the second pulse probes the excited vibronic level at progressively later times after encountering the first. The nonradiative relaxation time τ'_{nr} is found to be 7.5×10^{-12} sec. In a similar manner, but utilizing the picosecond pulses of a mode-locked ruby laser, the relaxation time of the vibrationless level of the $^1B_1(1)$ state is found to be 8×10^{-12} sec.

The path of this relaxation process (i.e., whether it involves intersystem crossing or direct transition to the ground state) may be inferred from the

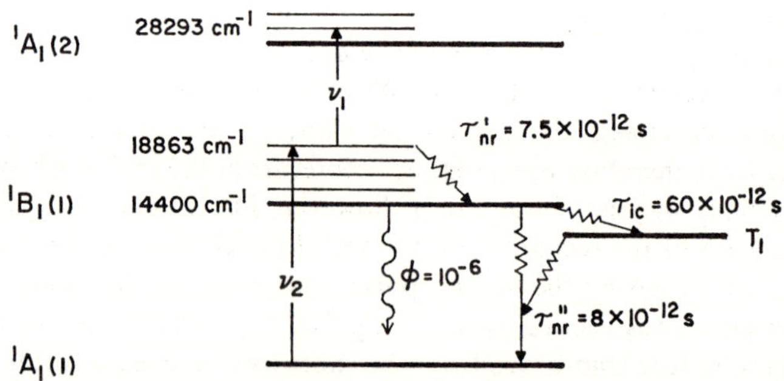

Fig. 15. Schematic representation of observed relaxation process involving the lowest excited singlet state of azulene; see text. (Rentzepis.[32, 102])

effect of heavy-atom solvents. Because such solvents generally enhance inter-system crossing, and because only a minor effect on the relaxation time of the vibrationless first excited state is observed, the conclusion is that direct coupling to the ground state is mainly responsible for relaxation of this state. This conclusion is supported by direct measurement of the buildup of the triplet-state population, utilizing a probing light pulse of an appropriately chosen wavelength. The time required for the intersystem crossing is 60 psec.

It is interesting, in view of the special characteristics of azulene, that direct laser excitation of its first excited singlet state permits observation of a hitherto unobservable fluorescence spectrum of that state. According to Rentzepis,[102] the fluorescence quantum yield is approximately 10^{-6}.

3.5.1.2. Intermolecular Processes. The application of lasers described in the foregoing section exploits the short-time characteristics and high intensity of mode-locked laser pulses. Eisenthal[155, 156] also made use of the *polarization* of laser light to study the kinetics of relaxation processes in a liquid system.

The applicability of his method to this type of investigation is exemplified by his study of intermolecular energy transfer between rhodamine-6G molecules in glycerol. The experiment is based on the following concept. A picosecond, intense, linearly polarized laser pulse is made to traverse an absorbing system and produces an anistropy in the orientational distribution of ground-state molecules. This effect results from the light absorption by predominantly those molecules with a component of the transition moment parallel to the direction of light polarization. After excitation the initial isotropic distribution tends to be restored by molecular rotation as well as by any process that leads to the return of excited molecules to the ground state. Such processes include emission, external quenching, and intermolecular excitation energy transfer. While in general all of these processes are effective, judicious choice of the experimental conditions may cause one or the other to predominate. Thus, in systems of high viscosity and low concentration of absorbing species, emission may be expected to be the most involved, while, at low viscosity, molecular rotation is dominant.[155] The return to isotropy can be monitored experimentally by measurement, as function of time, of the absorption of short, comparatively weak pulses of light with intensity components of polarization vertical $I_\perp$ and parallel $I_\parallel$ to that of the exciting pulses. The variation of the absorption ratio for both directions of polarization is related to the rate with which the isotropic distribution of the ground-state molecules is restored.

The technique has been applied to a direct measurement of the rate of resonance energy transfer between excited and ground-state molecules of the dye rhodamine-6G in glycerol solutions. Under the conditions of the experiment, it can be shown that energy transfer between excited and ground-state dye molecules and, to some extent, emission are the predominant processes that lead to restoration of orientational isotropy of the ground-state molecules.

The variation in the relative transmission $(I_\parallel - I_\perp)/(I_\parallel + 2I_\perp)$ is taken to be proportional to the average probability $\langle p(t) \rangle$ of finding the initially excited molecules still excited at time t. According to Förster[167, 168]

$$\langle p(t) \rangle = \exp \left[-t/\tau_0 - A(t/\tau_0)^{1/2} \right] \tag{3.10}$$

where A depends on the dye concentration and other parameters characteristic of a given solute and τ_0 is the excited-state lifetime in the absence of energy transfer.

The time dependence indicated by Eq. (3.10) is indeed found for the time interval 10^{-10} to about 10^{-9} sec, leading to a value of Förster's critical transfer distance $R_0 = 39$ Å, a quite acceptable value.

While similar results for this particular system have been obtained from measurements of fluorescence depolarization, it should be emphasized that the method described does *not* require luminescence in the system to be studied.

3.6. Laser Flash Photolysis

The elegant work of Novak and Windsor,[28, 157] who were the first to employ a Q-switched, pulsed laser for flash photolysis, has considerably widened the experimental means available for study of short-lived excited states and other photolytic transients. Modified versions of their technique[61, 88, 157–162] have since been used to study a variety of organic compounds. Among these studies those of Porter and Topp[158, 159] are perhaps the most extensive and systematic. Their data provide information on singlet- and triplet-state lifetimes and absorption spectra, on the energy of higher excited states, and on estimated specific rates of intersystem crossing. Table VIII, taken from the work of Porter and Topp, gives such data for a number of hydrocarbons, while Tables IX and X of the same investigators represent the triplet lifetimes of benzophenone in various solvents and of some quinone derivatives.

Laser studies of liquid systems studied also by fast pulse radiolysis (Section 3.1) include work on benzene,[163] solutions of naphthalene and 1:2 benzanthracene in cyclohexane and benzene,[88] and tetramethyl*para*-phenylenediamine (TMPD) in cyclohexane, ethanol, and water.[164]

Laser excitation of liquid benzene involves a two-photon process. The resultant absorption spectrum at 5200 Å has a decay time similar to that of the fluorescence of benzene. This absorption is assigned to the benzene

TABLE VII

Higher Singlet Levels and Correlation of Lifetimes with Fluorescence Data[a]

Molecule	Solvent[b]	Singlet S_1 absorption, nm	Level of higher S state, cm^{-1}	Triplet T_1^c absorption, nm	Singlet absorption decay time, nsec	Fluorescence decay time, nsec
Phenanthrene (d_{10})	Cx	(545), 515	⊂8200	520, 510, 481, 454	61.1	63.5
Phenanthrene (h_{10})	PMMA	(545), 515	⊂8200	520, 510, 481, 454	65	67.2
Triphenylene	Benzene	500, 465, 433	⊂8800	428	44.2	43.0
	PMMA	500, 465, 433	⊂8800	428	45.0	44.0
1,2-Benzanthracene	Cx	550, 495	⊂4200	540, 490, 461	52.7	49.4
	PMMA	550, 495	⊂4200	540, 490, 461	51.7	52.5
1,2,3,4-Dibenzanthracene	Cx	540, 500	45700	445	51.2	53.5
	PMMA	540, 500	45700	445	50.3	52.5
Pyrene	Cx	515, (480), 470	48000	520, 483, 416	296[d]	261[d]
	PMMA	515, (480), 470	48000	520, 483, 416	326	319
Coronene	Dx	(600, 570), 530, 495, (465)	42500	525, 480	319	307
	PMMA	(600, 570), 530, 495, (465)	42500	525, 480	390	380
3,4-Benzpyrene	Cx	(590), 535, 510	43700	480[e]	49.1	57.5
	PMMA	(590), 535, 510	43700	480[e]	39.4	40.5
3,4-Benzphenanthrene	Cx	595, 525	43200	517	68.5	70
	PMMA	595, 525	43200	517	76	81
3,4,9,10-Dibenzpyrene	Cx	575, 552	40400	595	140	143
1,2,5,6-Dibenzanthracene	PMMA	(570)	(42900)	532, 480	(40.0)[f]	37.5

[a] According to Porter and Topp.[159]
[b] Dx, dioxan; PMMA, polymethylmethacrylate: Cx, cyclohexane.
[c] See Porter and Windsor.[169]
[d] $3 \times 10^{-4} M$.
[e] Data from Craig and Ross.[170]
[f] Growth of triplet absorption.

TABLE IX

Decay of Benzophenone Triplet in Various Solvents[a]

Solvent	Decay time, nsec	Decay constant (k), sec^{-1}	D_0/D_∞ [b]
Isopropanol	46 ± 6	2.2×10^7	1.7 ± 0.2
Ethanol	104 ± 15	9.7×10^6	1.5 ± 0.2
Dioxane	200 ± 20	5.0×10^6	1.5 ± 0.2
Cyclohexane	300 ± 20	3.3×10^6	2.0 ± 0.3
C_6H_6	2500 ± 300	4.0×10^5	9.0 ± 0.5
C_6D_6	3450 ± 300	2.9×10^5	12.5 ± 1.0
C_6F_6	435 ± 40	2.3×10^6	15.0 ± 2.0

[a] According to Porter and Topp.[159]
[b] D_0/D_∞ Ratio of initial and final optical densities.

excimer[165] on the basis of the temperature dependence of the absorption intensity and the wavelength position of the spectrum. As in the case of fast pulse radiolysis, the spectrum is suppressed in the presence of 0.1 M naphthalene and replaced by absorption spectra attributed to the singlet and triplet states of the solute.

There is also one additional transient absorption at 300 nm and, after prolonged photolysis, a permanent one at 350 nm. While the origin of the absorption at 300 nm is tentatively assigned to a biradical,[80] the 350-nm absorption is possibly to be attributed to fulvene.

The laser photolysis data for benzene are thus quite analogous to those obtained with fast pulse radiolysis. This is likewise the situation for solutions of naphthalene and 1,2 benzanthracene in benzene and cyclohexane. An important practical aspect of the laser data is the observation of strong overlap of the absorption spectra of singlet and triplet states of the hydrocarbons studied. Caution is therefore required in interpretation of the time dependence of spectra such as discussed in Section 3.1.1.

TABLE X

Duroquinone and Chloranil Triplet Wavelength Maxima and Decay Constants[a]

Quinone	Solvent	λ_{max}, nm	k, sec^{-1}
Duroquinone	Benzene	490	3.2×10^5
Chloranil	Ethanol	500	8.4×10^5
Chloranil	Cyclohexane	500	5.0×10^5

[a] According to Porter and Topp.[159]

The combined flash and laser photolysis of TMPD in several solvents[164] yields a bewildering number of experimental data. In cyclohexane and ethanol solutions, fluorescence from the singlet state and an absorption spectrum of the TMPD triplet are observed. There is also delayed fluorescence in cyclohexane. From the latter it is concluded that triplet-triplet annihilation represents a major mode of triplet decay. However, while this process is possible, it is questionable that the experimental data and the mathematical description given are sufficient evidence to show that triplet-triplet interaction does play a major role. The seemingly exponential decay of the delayed fluorescence as well as the mathematical formalism by which this time dependence is described can be valid only when triplet-triplet annihilation is negligible compared to first-order triplet decay. Also, the requirement that, in this case, delayed fluorescence decay with twice the rate of triplet absorption is not in good agreement with the experimental decay data for triplet and delayed fluorescence; notably, $\tau_{(triplet)}$ is only $1.6 \times \tau_{(delayed\ fluorescence)}$.

Aqueous solutions of TMPD do not exhibit observable fluorescence or triplet absorption. Rather, spectra of TMPD$^+$ and of the hydrated electron are observed.

Application of the laser technique to gas-phase studies has been reported[166] in which vibrational relaxation of triplet states of anthracene and of 1,2 benzanthracene has been observed.

References

1. M. Schwartz, *Information Transmission, Modulation, and Noise*, McGraw-Hill, New York, 1959.
2. I. A. D. Lewis and F. H. Wells, *Millimicrosecond Pulse Techniques*, Pergamon Press, Oxford, 1959.
3. E. Gaviola, *Z. Physik*, **35**, 748 (1926).
4. E. Gaviola, *Ann. Physik*, **81**, 681 (1926).
5. II. Abraham and J. Lemoine, *Compt. Rend.*, **129**, 206 (1899).
6. P. F. Gottling, *Phys. Rev.*, **22**, 566 (1923).
7. L. A. Tumerman, *J. Phys. (Moscow)*, **4**, 151 (1941).
8. E. A. Bailey and G. K. Rollefson, *J. Chem. Phys.*, **21**, 1315 (1953).
9. H. B. Phillips and R. K. Swank, *Rev. Sci. Instr.*, **24**, 611 (1953).
10. R. K. Swank, H. B. Phillips, W. L. Buck, and L. J. Basile, *IEEE Trans. Nucl. Sci.*, **NS-5**, 183 (1958).
11. H. Dreeskamp and M. Burton, *Phys. Rev. Letters*, **2**, 45 (1959).
12. J. H. Malmberg, *Rev. Sci. Instr.*, **28**, 1027 (1957).
13. Q. A. Kerns, F. A. Kirsten, and G. C. Cox, *Rev. Sci. Instr.*, **30**, 31 (1959).
14. T. G. Innes and G. C. Cox, *Lawrence Radiation Lab., Berkeley, Rept.*, No. CC8-31, 1961.
15. J. B. Birks, T. A. King, and I. H. Munro, *Proc. Phys. Soc.*, **80**, 355 (1962).
16. J. T. D'Alessio, P. K. Ludwig, and M. Burton, *Rev. Sci. Instr.*, **35**, 1015 (1964).
17. J. T. D'Alessio and P. K. Ludwig, *IEEE Trans. Nucl. Sci.*, **NS-12**, 351 (1965).

18. S. S. Brody, *Rev. Sci. Instr.*, **28**, 1021 (1957).

19. R. G. Bennett, *Rev. Sci. Instr.*, **31**, 1275 (1960).

20. M. Burton and H. Dreeskamp, *Z. Elektrochem.*, **64**, 165 (1960).

21. L. M. Bollinger and G. E. Thomas, *Rev. Sci. Instr.*, **32**, 1044 (1961).

22. Y. Koechlin, Ph. D. Thesis, University of Paris, 1961.

23. J. B. Birks and I. H. Munro, in *Progress in Reaction Kinetics*, G. Porter, Ed., Vol. 4, Pergamon Press, Oxford, 1967, p. 257.

24. W. P. Helman, *Intern. J. Radiation Phys. Chem.*, in press.

25. W. J. Ramler, K. Johnson, and T. Klippert, *Nucl. Instr. Methods*, **47**, 23 (1967).

26. M. J. Bronskill and J. W. Hunt, *J. Phys. Chem.*, **72**, 3762 (1968).

27. F. J. McClung and R. W. Hellwarth, *J. Appl. Phys.*, **33**, 828 (1962).

28. J. R. Novak and M. W. Windsor, *J. Chem. Phys.*, **47**, 3075 (1967).

29. A. J. DeMaria, D. A. Stetser, and H. A. Heynan, *IEEE, J. Quantum Electron.*, **QE1**, 174 (1965).

30. J. A. Giordmaine, P. M. Rentzepis, S. L. Shapiro, and K. W. Wecht, *Appl. Phys. Letters*, **11**, 216 (1967).

31. P. M. Rentzepis and M. A. Duguay, *Appl. Phys. Letters*, **11**, 218 (1967).

32. P. M. Rentzepis, *Chem. Phys. Letters*, **2**, 117 (1968).

33. J. B. Birks and D. J. Dyson, *J. Sci. Instr.*, **38**, 282 (1961).

34. J. B. Birks, D. J. Dyson, and I. H. Munro, *Proc. Roy. Soc. (London)*, **A275**, 575 (1963).

35. E. W. Schlag, S. J. Yao and H. V. Weyssenhoff, *J. Chem. Phys.*, **50**, 732 (1969).

36. P. P. Berlow, Ph.D. Thesis, Cornell University, Ithaca, N.Y., 1969.

37. B. H. Billings, *J. Opt. Soc. Am.*, **39**, 797, 802 (1969); **42**, 12, (1952).

38. R. O. Carpenter, *J. Opt. Soc. Am.*, **40**, 225 (1950).

39. A. Muller, R. Lumry, and H. Kokubun, *Rev. Sci. Instr.*, **36**, 1214 (1965).

40. P. Debye and F. W. Sears, *Proc. Natl. Acad. Sci., U.S.*, **18**, 409 (1932).

41. H. E. R. Becker, W. Hanle, and O. Maerks, *Phys. Z.*, **11**, 951 (1936).

42. O. Maerks, *Z. Physik*, **109**, 685 (1938).

43. E. W. Schlag and H. V. Weyssenhoff, *J. Chem. Phys.*, **51**, 2508 (1969).

44. B. D. Venetta, *Rev. Sci. Instr.*, **30**, 450 (1959).

45. I. Isenberg and R. D. Dyson, *Biophys. J.*, **9**, 1337 (1969).

46. For a very instructive derivation of this distribution and the conditions of its applicability see, for instance, G. P. Wadsworth and J. G. Bryan, *Introduction to Probability and Random Variables*, McGraw-Hill, New York, 1969, p. 67.

47. T. Gregory, Ph.D. Thesis, University of Notre Dame, Notre Dame, Ind., 1971.

48. R. L. McGuire, E. C. Yates, D. G. Crandall, and C. R. Hatcher *IEEE Trans. Nucl. Sci.*, **NS-12**, 25 (1965).

49. H. Dupont, Ph.D. Thesis, University of Strasbourg, France, 1966.

50. C. D. Amata and P. K. Ludwig, *J. Chem. Phys.*, **47**, 3540 (1967).

51. G. Laustriat, *Mol. Cryst.*, **4**, 127 (1968).

52. H. Lami, J. T. D'Alessio, J. Zampach, R. Radicella, and J. M. Kesque, *Nucl. Instr. Methods*, **92**, 333 (1971).

53. P. Kofalas, J. Masters, and E. Murray, *J. Appl. Phys.*, **35**, 2349 (1964).

54. P. Sorokin, J. J. Luzzi, J. R. Lankard, and G. D. Pettiti, *IBM J. Res. Develop.*, **8**, 182 (1964).

55. L. M. Frantz and J. S. Nodvik, *J. Appl. Phys.*, **34**, 2346 (1963).

56. N. Bloembergen, *Nonlinear Optics*, W. A. Benjamin, New York, 1965.

57. P. Franken, A. E. Hill, C. W. Peters, and G. Weinreich, *Phys. Rev. Letters*, **7**, 118 (1961).

58. P. D. Maker, R. W. Terhune, M. Nisenhoff, and C. M. Savage, *Phys. Rev. Letters*, **8**, 21 (1962).
59. J. A. Giordmaine, *Phys. Rev. Letters*, **8**, 19 (1962).
60. G. Porter and M. R. Topp, *Nature*, **220**, 1228 (1968).
61. R. Bonneau, J. Faune, and J. Joussot-Dubien, *Chem. Phys. Letters*, **2**, 65 (1968).
62. R. McNeil, J. T. Richards, and J. K. Thomas, *J. Phys. Chem.*, **74**, 2290 (1970).
63. A. J. DeMaria, D. A. Stetser, and W. H. Glenn, Jr., *Science*, **156**, 1557 (1967).
64. A. J. DeMaria, *J. Appl. Phys.*, **34**, 2984 (1963).
65. D. A. Stetser and A. J. DeMaria, *Appl. Phys. Letters*, **9**, 118 (1966).
66. H. W. Mocker and R. J. Collins, *Appl. Phys. Letters*, **7**, 270 (1965).
67. J. Ducuing and N. Bloembergen, *Phys. Rev. Letters*, **10**, 474 (1963).
68. R. K. Chang and N. Bloembergen, *Phys. Rev.* **144**, 775 (1966).
69. J. A. Armstrong, *Appl. Phys. Letters*, **10**, 16 (1967).
70. M. W. Dowley, K. B. Eisenthal, and W. L. Peticolas, *Phys. Rev. Letters*, **18**, 531 (1967).
71. J. T. Richards and J. K. Thomas, *J. Chem. Phys.*, **55**, 3636 (1971).
72. M. S. Livingston, *High Energy Accelerators*, Interscience, New York, 1954, p. 83.
73. J. W. Hunt and J. K. Thomas, *J. Chem. Phys.*, **46**, 2954 (1967).
74. J. K. Thomas and I. Mani, *J. Chem. Phys.*, **51**, 1834 (1969).
75. R. Cooper and J. K. Thomas, *J. Chem. Phys.*, **48**, 5097 (1968).
76. A. Pariser, *J. Chem. Phys.*, **24**, 250 (1956).
77. J. B. Birks, C. L. Braga, and M. D. Lumb, *Proc. Roy. Soc. (London)*, **A283**, 83 (1965).
78. P. K. Ludwig and C. D. Amata, *J. Chem. Phys.*, **49**, 326 (1968).
79. M. T. Vala, Jr., I. H. Hillier, S. A. Rice, and J. Jortner, *J. Chem. Phys.*, **44**, 23 (1966).
80. D. Bryce-Smith and H. C. Longuet-Higgins, *Chem. Commun.*, 593 (1966).
81. M. Burton, P. K. Ludwig, M. S. Kennard, and R. J. Povinelli, *J. Chem. Phys.*, **41**, 2563 (1964).
82. I. Berlman, *Handbook of Fluorescence Spectra of Aromatic Molecules*, Academic Press, New York, 1965.
83. J. K. Thomas, K. Johnson, T. Klippert, and R. Lowers, *J. Chem. Phys.*, **48**, 1608 (1968).
84. J. Jagur-Grodzinski, M. Feld, S. L. Yang, and M. Swarc, *J. Phys. Chem.*, **69**, 628 (1965).
85. K. H. J. Buschow, J. Dieleman, and G. J. Hoijting, *Mol. Phys.*, **7**, 1 (1963).
86. T. Shida and W. H. Hamill, *J. Chem. Phys.*, **44**, 2375 (1966).
87. G. Porter and M. W. Windsor, *Proc. Roy. Soc. (London)*, **A245**, 238 (1958).
88. J. K. Thomas, *J. Chem. Phys.*, **51**, 770 (1969).
89. S. J. Rzad, P. P. Infelta, J. M. Warman, and R. Schuler, *J. Chem. Phys.*, **52**, 3971 (1970).
90. F. Hirayama and S. Lipsky, *J. Chem. Phys.*, **51**, 3616 (1969).
91. M. Burton, *Mol. Cryst.*, **4**, 61 (1968).
92. P. K. Ludwig and M. M. Huque, *J. Chem. Phys.*, **49**, 805 (1968).
93. A. Mozumder, Volume 1 of this series, p. 45.
94. A. R. Horrocks, A. Kearvell, K. Tickle, and F. Wilkinson, *Trans. Faraday Soc.*, **62**, 3393 (1966).
95. W. H. Hamill, in *Radical Ions*, E. T. Kaiser and L. Kevan, Ed., Interscience, New York, 1968, p. 321.
96. J. T. Richards and J. K. Thomas, *J. Chem. Phys.*, **53**, 218 (1970).
97. W. H. Hamill, *J. Chem. Phys.*, **53**, 473 (1970).
98. M. J. Bronskill, R. K. Wolff, and J. W. Hunt, *J. Chem. Phys.*, **53**, 4201 (1970).

99. R. K. Wolff, M. J. Bronskill, and J. W. Hunt, *J. Chem. Phys.*, **53**, 4211 (1970).

100. For a review, see J. K. Thomas, Volume 1 of this series, p. 103.

101. A. Mozumder, *J. Chem. Phys.*, **50**, 3153, 3162 (1969).

102. Cf. P. M. Rentzepis, *Chem. Phys. Letters*, **3**, 717 (1969).

103. T. Förster, *Fluoreszenz Organischer Verbindungen*, Vandenhoeck and Ruprecht, Göttingen, 1951, p. 205.

104. P. Hertz, *Math. Ann.*, **67**, 387 (1909).

105. R. R. Hentz, private communication.

106. J. B. Birks, *Photophysics of Aromatic Molecules*, Wiley, New York, 1970.

107. H. H. Jaffe and M. Orchin, *Theory and Applications of Ultraviolet Spectroscopy* Wiley, New York, 1962.

108. M. Burton, K. Funabashi, R. R. Hentz, P. K. Ludwig, J. L. Magee, and A. Mozumder, in *Transfer and Storage of Energy by Molecules*, Vol. 1, G. M. Burnett and A. M. North, Ed., Interscience, New York, 1969, p. 161.

109. H. E. Lessing, E. Lippert, and W. Rapp, *Chem. Phys. Letters*, **7**, 247 (1970).

110. A. Schmillen and R. Legler, Landolt-Bornstein, *Zahlenwerte und Funktionen aus Naturwissenschaften und Technik*, Neue Serie, Gruppe II: *Atom und Molekular-Physik*, Band 3, Springer Verlag, Berlin, 1967.

111. For reviews see (a) P. Seybold and M. Gouterman, *Chem. Rev.*, **65**, 413 (1965); (b) B. R. Henry and M. Kasha, *Ann. Rev. Phys. Chem.*, **19**, 161 (1968); (c) J. Jortner, S. A. Rice, and R. M. Hochstrasser, *Advan. Photochem.*, **7**, 149 (1969).

112. G. W. Robinson and R. P. Frosch, *J. Chem. Phys.*, **37**, 1962 (1962); **38**, 1187 (1963).

113. G. R. Hunt and I. G. Ross, *J. Mol. Spectry.*, **9**, 50 (1962).

114. J. P. Byrne, E. F. McCoy, and I. G. Ross, *Australian J. Chem.*, **18**, 1589 (1965).

115. S. H. Lin and R. Bersohn, *J. Chem. Phys.*, **48**, 2732 (1968).

116. S. Fischer and E. W. Schlag, *Chem. Phys. Letters*, **4**, 393 (1969).

117. K. F. Freed and J. Jortner, *J. Chem. Phys.*, **52**, 6272 (1970).

118. E. W. Schlag and H. von Weyssenhoff, *J. Chem. Phys.*, **51**, 2508 (1969).

119. E. W. Schlag, H. von Weyssenhoff, and M. Starzak, *J. Chem. Phys.*, **47**, 1860 (1967).

120. C. S. Parmenter and A. H. White, *J. Chem. Phys.*, **50**, 1631 (1969).

121. M. Nishikawa and P. K. Ludwig, *J. Chem. Phys.*, **52**, 107 (1970).

122. A. M. Halpern and W. R. Ware, *J. Chem. Phys.*, **53**, 1969 (1970).

123. W. R. Ware, B. K. Seliger, C. S. Parmenter, and M. W. Schuyler, *Chem. Phys. Letters*, **6**, 342 (1970).

124. U. Laor and P. K. Ludwig, *J. Chem. Phys.*, **54**, 1054 (1971).

125. G. B. Kistiakowsky and C. S. Parmenter, *J. Chem. Phys.*, **42**, 2942 (1965).

126. For normal mode numbering, see E. B. Wilson, Jr., *J. Chem. Phys.*, **3**, 276 (1934).

127. T. Chen and E. W. Schlag, in *Molecular Luminescence*, E. C. Lim, Ed., W. A. Benjamin, New York, 1969, p. 381.

128. G. Herzberg and E. Teller, *Z. Physik. Chem. (Leipzig)*, **321**, 410 (1933).

129. A. D. Liehr, *Z. Naturforsch.* **13a**, 311, 596 (1958).

130. A. C. Albrecht, *J. Chem. Phys.*, **33**, 156, (1959).

131. G. W. Robinson, *J. Chem. Phys.*, **47**, 1967 (1967).

132. J. Jortner, S. A. Rice and R. M. Hochstrasser, in *Advances in Photochemistry*, Vol. 7, J. N. Pitts, G. S. Hammond, and W. A. Noyes, Jr., Eds., Wiley, New York, 1969, p. 149.

133. W. Siebrand, *J. Chem. Phys.*, **46**, 440 (1967).

134. J. C. Hsieh, U. Laor, and P. K. Ludwig, *Chem. Phys. Letters*, **10**, 412 (1971).

135. R. Kubo, *J. Phys. Soc. Japan*, **12**, 570 (1957).

136. T. Yamamoto, *J. Chem. Phys.*, **33**, 281 (1960).

137. A. J. Watts and H. J. Strickler, *J. Chem. Phys.*, **44**, 2423 (1966).
138. H. Kallmann, see W. Bloch, *Natur und Technik*, July 1947, p. 15; H. Kallmann, *Research* (London), **2**, 62 (1949); *cf.* I. Broser and H. Kallmann, *Z. Naturforschung*, **2a**, 439 (1947).
139. J. B. Birks, *The Theory and Practice of Scintillation Counting*, Pergamon Press, Oxford, 1964, Chapter 6.
140. W. L. Buck, *IRE Trans. Nucl. Sci.*, **NS-7**, 11 (1960).
141. J. B. Birks, *IRE Trans. Nucl. Sci.*, **NS-7**, 2 (1960).
142. D. C. Northrop and O. Simpson, *Proc. Roy. Soc.* (*London*), **A234**, 124 (1956); **A244**, 377 (1958).
143. P. E. Gibbons, D. C. Northrop, and O. Simpson, *Proc. Phys. Soc.*, **79**, 373 (1962).
144. D. Blanc, F. Cambou, and Y. Gervais de Lafond, *J. Phys. Radium*, **25**, 319 (1964).
145. C. A. Parker and C. G. Hatchard, *Trans. Faraday Soc.*, **57**, 1894 (1961); *Proc. Roy. Soc.* (*London*), **A269**, 574 (1962).
146. P. K. Ludwig, *Mol. Cryst.*, **4**, 147 (1968).
147. R. Schuyler and I. Isenberg, *Rev. Sci. Instr.*, **42**, 813 (1971).
148. T. A. King and R. Voltz, *Proc. Roy. Soc.* (*London*), **A289**, 424 (1966).
149. A. H. Samuel and J. L. Magee, *J. Chem. Phys.*, **21**, 1080 (1953).
150. F. Wilkinson and J. T. Dubois, *J. Chem. Phys.*, **39**, 377 (1963).
151. R. B. Cundall and P. A. Griffiths, *Trans. Faraday Soc.*, **61**, 1968 (1965).
152. *a*) R. R. Hentz and L. M. Perkey, *J. Phys. Chem.*, **74**, 3047 (1970); *b*) R. R. Hentz, D. B. Peterson, S. B. Srivastava, H. F. Barzynski, and M. Burton, *J. Phys. Chem.*, **70**, 2362 (1966).
153. J. B. Birks, *Chem. Phys. Letters*, **7**, 293 (1970).
154. P. M. Rentzepis, J. Jortner, and R. P. Jones, *Chem. Phys. Letters*, **4**, 599 (1970).
155. K. B. Eisenthal and K. H. Drexhage, *J. Chem. Phys.*, **51**, 5720 (1969).
156. K. B. Eisenthal, *Chem. Phys. Letters*, **6**, 155 (1970).
157. J. R. Novak and M. W. Windsor, *Proc. Roy. Soc.* (*London*), **A308**, 95 (1968).
158. G. Porter and M. R. Topp, *Nature*, **220**, 1228 (1968).
159. G. Porter and M. R. Topp, *Proc. Roy. Soc.* (*London*), **A315**, 163 (1970).
160. M. W. Dowley, K. B. Eisenthal, and W. L. Peticolas, *Phys. Rev. Letters*, **18**, 531 (1967).
161. R. Bonneau, J. Joussot-Dubien, and R. Bensasson, *Chem. Phys. Letters*, **3**, 353 (1969).
162. D. S. Kliger and A. C. Albrecht, *J. Chem. Phys.*, **50**, 4109 (1969).
163. J. T. Richards and J. K. Thomas, *Chem. Phys. Letters*, **5**, 527 (1970).
164. J. T. Richards and J. K. Thomas, *Trans. Faraday Soc.*, **66**, 621 (1970).
165. J. B. Birks, *Chem. Phys. Letters*, **1**, 625 (1968).
166. S. J. Formosinho, G. Porter, and M. A. West, *Chem. Phys. Letters*, **6**, 7 (1970).
167. Th. Förster, *Ann. Physik*, **2**, 55 (1948).
168. Th. Förster, *Z. Naturforsch.*, **4a**, 321 (1949).
169. G. Porter and M. W. Windsor, *Proc. Roy. Soc.* (*London*), **A245**, 238 (1958).
170. D. P. Craig and I. G. Ross, *J. Chem. Soc.*, 1589 (1954).
171. H. Kallmann and M. Furst, *Phys. Rev.*, **79**, 857 (1950).
172. S. Fisher, *Chem. Phys. Letters*, **11**, 577 (1971).
173. L. M. Theard, F. C. Peterson, and R. A. Holroyd, *J. Chem. Phys.*, **51**, 4126 (1969).
174. P. B. Cundall and W. Tippett, *Trans. Faraday Soc.*, **66**, 350 (1970).
175. N. V. Klassen, H. A. Gillis, and D. C. Walker, *J. Chem. Phys.*, **55**, 1979 (1971).
176. M. Anbar and P. Neta, *Intern. J. Appl. Radiation Isotopes*, **18**, 493 (1967).
177. J. A. Ghormley and C. J. Hochanadel, *Radiation Res.*, **3**, 227 (1955).
178. H. A. Schwarz, *J. Am. Chem. Soc.*, **77**, 4960 (1966).

179. A. Appleby, in *Chemistry of Ionization and Excitation*, G. Johnson and G. Scholes, Eds., Taylor and Francis, London, 1967, p. 264.
180. H. A. Mahlman and T. J. Sworski, *ibid.*, p. 259.
181. I. A. Taub and K. Eiben, *J. Chem. Phys.*, **49**, 2499 (1968).
182. W. P. Helman and T. A. Gregory, submitted to *Rev. Sci. Instr.*
183. C. D. Amata, Ph.D. Thesis, University of Notre Dame, Notre Dame, Ind., 1967.
184. W. H. Hamill, *J. Phys. Chem.*, **73**, 1341 (1969).
185. H. A. Schwarz, *J. Chem. Phys.*, **55**, 3647 (1971).

Some Topics in Radiation Chemical Synthesis of Organic Compounds

I. V. VERESHCHINSKII, *Karpov Institute of Physical Chemistry, Moscow, U.S.S.R.*

Contents

1. INTRODUCTION

Radiation chemical synthesis was considered promising even in the days when the principal concern of radiation chemistry was with deleterious effects of high-energy radiation. The first radiation chemical investigations in the field of organic synthesis as such were conducted in about 1950. Later, increased availability of high-energy sources made it possible to perform radiation chemical syntheses in laboratories, then on the semipilot scale and, finally, in pilot plants.

However, as the chapter by Wagner in Volume 1 of this series suggests, radiation chemical synthesis still has not become a generally practicable tool. The organic chemist does not consider this method to represent a conventional synthetic technique. Meanwhile, the radiation chemist prefers to study elementary systems in which microkinetics and, especially, mechanisms are not obscured by the complications presented by the usual reaction mixtures.

The term "radiation chemical synthesis" is here used to denote a radiation chemical process that produces certain compounds in 100-eV yields (i.e., with "G values") significantly higher than those of any other products generated. Such a limitation is rather arbitrary; consequently, this chapter does not cover a great variety of radiation chemical processes that provide no particularly directed synthesis. Values of radiation chemical yields are strongly affected by the nature of the process and its conditions.

2. RADIATION CHEMICAL INITIATION

At the small doses employed, the radiation has a significant effect only on the initiation process. The role of other radiation chemical reactions is negligibly small. Table I compares radiation initiation of chain processes with other techniques that may be used.

Radiation processes are characterized by the following features.

(1) They can be conducted at *low temperatures*. Increased solubility of the gaseous components of a reaction mixture may thus enhance the yield of a desired product. Low temperature decreases destruction and resinification of products. Hence thermally labile compounds may be obtained.

(2) Reaction mixtures are *not contaminated* by catalysts or initiators or their decomposition products.

(3) Possible noninertial control of initiation permits rapid change in dose rate and interruption of the process.

TABLE I
Potentials and Limitations of a Variety of Initiation Processes

Initiation	Temperature	Impurities introduced initially	Noninertial control	Fire and explosion
Radiation	Arbitrary	None	Possible	Impossible
Photochemical	Arbitrary	None	Possible	Possible
Chemical	Elevated	Possible	Impossible	Possible
Thermal	Elevated	Possible	Impossible	Possible

(4) There are fewer fire and explosion hazards. This advantage results from the fact that process parameters can be maintained at levels far below those leading to fire or explosion of the mixture. (Of course, in contrast to photochemical initiation, the possibility of explosion of the energy source—the ultraviolet lamp—is nonexistent.)

Radiation initiation is preferable to photochemical initiation for a variety of reasons.

(1) The process can be performed in intensively colored, opaque, turbid, or inhomogeneous systems.

(2) Radiation energy may penetrate deeply into the bulk of the reaction mixture irrespective of films or carbonaceous deposits which form on the reactor walls and which might absorb light.

(3) High-energy irradiation permits avoidance of side effects of ultraviolet illumination which may selectively decompose products having substantial molar extinction coefficients in a spectral region close to that of the initiating light.

3. HALOGENATION

Radiation halogenation of hydrocarbons has been studied in detail.

Radiation chlorination of benzene leads to the stable product, 1,2,3,4,5,6-hexachlorocyclohexane (a mixture of stereoisomers), the γ-isomer being an insecticide.[1-3] Faw, McCabe, and Isbin[4] discussed the kinetics and the probable mechanisms of the chlorination in detail. The reaction rate varies linearly with chlorine concentration and is practically constant at 15–30°C. Increase in surface does not affect the process. At higher dose rates of γ-irradiation (2.7×10^{17} eV/ml min), the reaction rate varies as dose rate raised to the power 0.2, but the power exponent approaches 0.5 at shorter contact times. The chlorine atoms that start the chemical chain process presumably also participate in the usual termination reactions and are scavenged by radiolysis products. In addition to the usual initiation

$$Cl_2 \xrightarrow{\hspace{0.8cm}} 2Cl \tag{1}$$

another possible initiation process involves participation of the first excited singlet ($^1B_{2u}$) state of benzene

$$C_6H_6{}^* + Cl_2 \rightarrow C_6H_6 + 2Cl \tag{2}$$

Radiation chlorination of heptane has been performed under static conditions. At 10^7 roentgens the products are the same as with photochemical chlorination. The G value of the product according to Krentsel[5] is about 10. Chlorination of kerosene of mean molecular weight corresponding to the C_{15} hydrocarbons has been conducted in a liquid system through which chlorine is bubbled at a γ-irradiation dose rate of 26 rads/min. The chlorination rate varies as dose rate raised to the power 0.5. The reaction rate decreases with increased conversion, while the conversion increases with temperature in the range 0–90°C. These data have been used by Zetkin and his colleagues[6] to design an industrial reactor.

Liquid-phase radiation chlorination of tetradecane has been performed by Dzhagatspanyan, Kosorotov, and Filippov[7] in the diffusion-controlled and intermediate regions at a dose rate of 0.1–36 rads/sec. Of course, properties of the liquid phase change continuously during the chlorination because chlorine absorption is accompanied by chemical reactions. At 10–30°C, the G value increases from 10^4 to 10^5 chlorine atoms absorbed per 100-eV input. The reaction rate varies as dose rate raised to the power 0.5. At 30°C increase in dose rate in the range 0.1–1.3 rads/sec accelerates the process, while further increase in dose rate does not affect the reaction course.

C_5 hydrocarbons were radiation-chlorinated by Almasy and Rona[8–9] to yield the completely chlorinated ring compound, hexachlorocyclopentadiene. Polychlorocyclopentanes were chlorinated in the liquid phase in the range 170–180°C at γ-radiation dose rates varying from 10^3 to 10^5 roentgens/hr to yield octachlorocyclopentene. The G value was indirectly estimated to be 10^4–10^5.

Lebedev and Vereshchinskii and their colleagues[11–14] report that liquid-phase radiation chlorination of 1,1,1,5-tetrachloropentane involves a set of consecutive processes

$$C_5H_8Cl_4 \rightarrow C_5H_7Cl_5 \rightarrow C_5H_6Cl_6 \rightarrow C_5H_5Cl_7 \rightarrow C_5H_4Cl_8 \rightarrow C_5H_3Cl_9 \rightarrow C_5H_2Cl_{10}$$

Penta-, hexa-, octa-, nona-, and decachloropentanes were identified in the reaction mixture. The chlorination was diffusion- or kinetically controlled depending on such conditions as temperature, pressure and dose rate. When diffusion-controlled, chlorination of polychloropentanes obeys zero-order laws, while low concentration of chlorine in the solution favors dehydrochlorination and destruction. When kinetically controlled, the process is a first-order reaction and the effects of temperature and radiation on the rate are additive. The chlorination rate varies as dose rate raised to the power 0.5. Individual highly chlorinated polychloropentanes were isolated. Decachloropentane was synthesized for the first time as a mixture of isomers having probable structural formulas $CCl_3—CCl_2—CCl_2—CCl_2—CH_2Cl$ and $CCl_3—CHCl—CCl_2—CHCl—CCl_3$.

The electron fraction of chlorine employed in the work on the polychloropentanes was rather low (e.g., 1.56 percent when tetrachloropentane was studied and 0.83 percent in the decachloropentane work). The initiation reaction

$$RCl \rightsquigarrow R + Cl \tag{3}$$

was consequently considered to play the leading role. The chain propagation reaction is described by

$$Cl + C_5H_nCl_{12-n} \rightarrow C_5H_{n-1}Cl_{12-n} + HCl \tag{4}$$

$$C_5H_{n-1}Cl_{12-n} + Cl_2 \rightarrow C_5H_{n-1}Cl_{12-n+1} + Cl \tag{5}$$

and so on

As Fig. 1 shows, radiation chemical yield of polychloropentanes decreases with increasing conversion. Each next chlorine is more difficult to introduce into the polychloropentane molecule than is the preceding one. Increase in number of chlorine atoms decreases the polarizability of C—H bonds, while the inductive effect becomes stronger; thus attack of chlorine at a C—H bond becomes the slowest step.

Laboratory-scale radiation chemical syntheses of exhaustively chlorinated polychloropentanes by the Lebedev–Vereshchinskii group self-stimulated

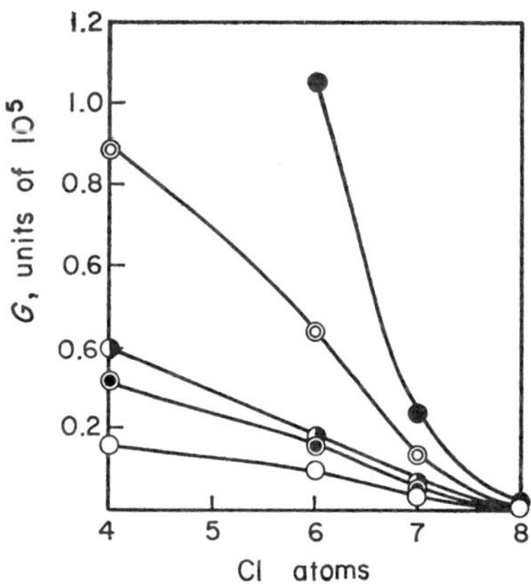

Fig. 1. The 100-eV yield of polychloropentanes versus the number of chlorine atoms in the molecule of the initial polychloropentane at various dose rates.[14] Dose in kilorads per hour is as follows: ●, 1.2; ◎, 4.4; ◑, 10; ⊙, 15.5; ○, 33.

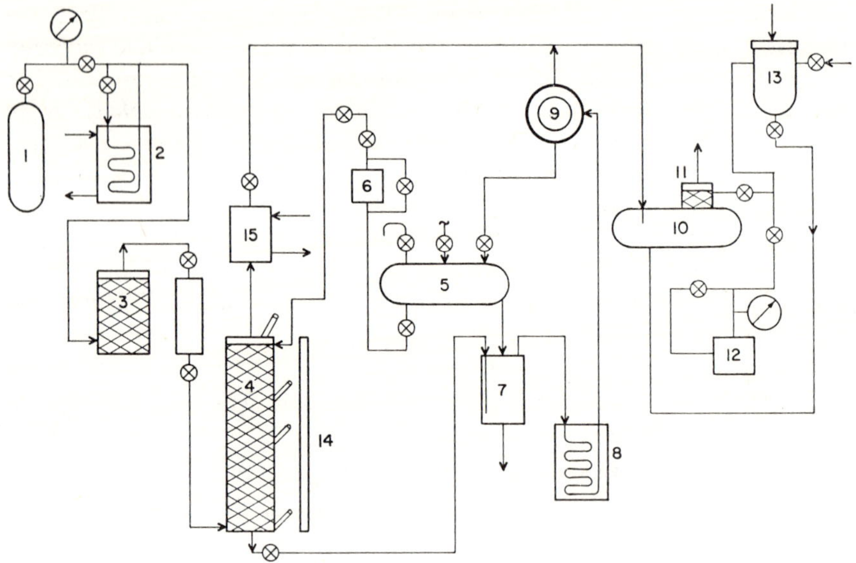

Fig. 2. Block diagram of semipilot installation for radiation synthesis of octachlorocyclopentene.[15] 1, bottle; 2, evaporator; 3, drier; 4, reactor; 5, collector; 6, pump; 7, collector; 8, cooler; 9, inspection; 10, collector; 11, scrubber; 12, pump; 13, collector; 14, cobalt-60 source; 15, cooler.

(so to speak) the study of liquid-phase radiation chemical synthesis of octachlorocyclopentene in a pilot plant.[15] The nickel reactor shown in Fig. 2 is a 14-liter column packed with Raschig rings. The mean dose rate employed is about 25 rads/sec. Tetrachloropentane is fed through the upper part of the apparatus and chlorine through the lower part. Liquid reaction products are collected in a tank and then flow to a rectifier.[15]

Radiation halogenation of the side chains of aromatic hydrocarbons proceeds in a similar manner. Synthesis of tetrabromoxylene by Plander, Formácek, and Zahálka[16] involving radiation bromination of xylene is an example of a multistage consecutive reaction. However, according to Galimba, Wilson, and Samuel,[17] toluene chlorinated by tin tetrachloride in the presence of radiation gives only isomeric chlorotoluenes. $G(C_7H_7Cl)$ does not exceed 0.2; it increases with concentration of $SnCl_4$ but does not depend on dose and temperature. The reaction products may form through an ionic mechanism.

4. SULFOCHLORINATION

When γ-irradiated, liquid cyclohexane reacts with SO_2 and Cl_2 giving high yield ($G = 4 \times 10^3$ to 2×10^6) of cyclohexylsulfonyl chloride

($C_6H_{11}SO_2Cl$) and small amounts of cyclohexanedisulfonyl dichloride [$C_6H_{10}(SO_2Cl)_2$] and chlorocyclohexane ($C_6H_{11}Cl$); see Henglein and Url,[18] as well as Schneider and Chin.[19, 20] The photochemical process is initiated by chlorine atoms, while γ-radiation induces the *initiation*

$$C_6H_{12} \leadsto C_6H_{11} + H \qquad (6)$$

followed by such *propagation* steps as

$$R + SO_2 \rightarrow RSO_2 \qquad (7)$$

$$RSO_2 + Cl_2 \rightarrow RSO_2Cl + Cl \qquad (8)$$

$$R + Cl_2 \rightarrow RCl + Cl \qquad (9)$$

$$C_6H_{12} + Cl \rightarrow C_6H_{11} + HCl \qquad (10)$$

and *termination*

$$C_6H_{11}SO_2 + Cl \rightarrow C_6H_{11}SO_2Cl \qquad (11)$$

At dose rates below 40 rads/min, the reaction rate varies as the square root of the dose rate, and the initiation is a rate-limiting step of the process. At dose rates above 400 rads/min, the reaction rate does not depend on the intensity and is governed by the flow rate of SO_2 and Cl_2. At intermediate dose rates (40–400 rads/min), reaction rate is no longer proportional to the square root of dose rate, while conversion falls.

Dzhagatspanyan found that radiation sulfochlorination of wet paraffin proceeds in similar manner.[21] On larger scales inhibition of the process in a kinetically controlled region is associated with formation of sulfuryl chloride. The chloride accumulating in the reactor interacts with atomic chlorine

$$SO_2Cl_2 + Cl \rightarrow SO_2Cl + Cl_2 \qquad (12)$$

and hydrocarbon radicals

$$SO_2Cl_2 + R \rightarrow SO_2Cl + RCl \qquad (13)$$

The RCl formed in such processes is usually an undesirable by-product.

Sulfochlorination of synthine (see Sprygayev and Mamikonyan[22]) was the first radiation chemical pilot synthesis performed in the USSR.

5. SULFONIC ACID PRODUCTION

Production of alkyl sulfonates and of aliphatic and cycloaliphatic sulfonic acids under action of ionizing radiation has been patented.[23–26] γ-Radiation initiates the reaction

$$RH \leadsto R + H$$

The generated radical propagates the chain

$$R + SO_2 \rightarrow RSO_2$$

$$RSO_2 + O_2 \rightarrow RSO_2O_2 \tag{14}$$

$$RSO_2O_2 + RH \rightarrow RSO_2O_2H + R \tag{15}$$

and persulfonic acid decomposes spontaneously giving organic sulfonic acid and new alkyl radicals

$$RSO_2O_2H \rightarrow RSO_3 + OH \tag{16}$$

$$RSO_3 + RH \rightarrow RSO_3H + R \tag{17}$$

$$OH + RH \rightarrow H_2O + R \tag{18}$$

The autoprocess observed after removal of radiation is explained by the branched chain reaction. A side interaction of persulfonic acid with water and SO_2

$$RSO_2O_2H + SO_2 + H_2O \rightarrow RSO_3H + H_2SO_4 \tag{19}$$

precedes the accumulation of persulfonic acid.

According to Black[24] and to Baxter,[26] as well as Hills and Spindler,[27] when γ-irradiated for 16 hr (cobalt-60 source, dose rate 10^5 rads/hr), dodecane reacts with SO_2 and oxygen (flow rates 75 and 40 ml/min, respectively) giving dodecylpersulfonic acid. The acid decomposes to various substances which may serve as intermediate products in manufacture of biologically decomposable detergents. To determine the kinetics of the reaction, Dalton and Bartlett measured consumption of the gases as a function of reaction time; effects of dose rate, SO_2/O_2 ratio, total pressure, and temperature were studied.[28]

In general, high-energy radiation is a better initiator for processes of this kind than is ultraviolet light because brown drops of the product absorb such light. Thermal initiation at elevated temperatures leads to rather shorter chains probably because of decrease in solubility of the gases in the hydrocarbon.

6. OXIDATION OF HYDROCARBONS

At moderate temperatures the radiation chemical oxidation of hydrocarbons is hardly interesting; the products form nonselectively and the G values are low.[29–31]

Sometimes chain oxidation may terminate with formation of hydroperoxides. For example, according to Hu,[32] hexadecane oxidized by air in the presence of γ-radiation yields mainly hexadecyl hydroperoxide and G attains a value of 490 at a dose rate of 934 rads/min. The yield is noticeably affected by temperature and reaches its maximum at 120°C. Hydroperoxide isolated from the reaction was 98 percent pure.

Radiation chemical initiation by paraffin oxidation was also attempted; in this case it is enough to irradiate the mixture at the beginning of the oxidation process.[33-35]

Oxidation of halogenated olefins, especially chloroethylenes, is much more important because it leads to valuable organic compounds employed as intermediate products in various syntheses. Laboratory-scale experiments had shown that liquid tetrachloroethylene oxidized by gaseous oxygen at 20–120°C in the presence of γ-radiation produces tetrachloroethylene epoxide and trichloroacetyl chloride.[36-38] Optimum conditions of the process were determined by Dobrov et al.[39] in a semipilot installation in which oxidation was performed in a cylindrical apparatus fitted with an axially located linear source of radiation. Average dose rate was about 100 rad/sec. Oxygen was introduced under normal pressure with the flow rate varying from 1 to 2 liters/min per liter of tetrachloroethylene, and the components were stirred vigorously. The process was carried out in batches at or above 120°C; one cycle continued from 5 to 10 hr. Trichloroacetic acid thus obtained cost one-fifth to one-fourth as much as that produced in chloral oxidation by a mixture of nitric and sulfuric acids.

Liquid trichloroethylene, when γ-irradiated, is oxidized by gaseous oxygen to trichloroethylene epoxide[40] and dichloroacetyl chloride.[41] Optimal conditions of oxidation (conducted in the semipilot plant described in the previous paragraph) were: average dose rate about 25 rads/sec, temperature 50–60°C, oxygen flow rate (normal pressure) 1–2 liters/min per liter of trichloroethylene. The process was conducted in batches; one cycle continued about 10 hr. The acyl chloride thus obtained gives methyldichloroacetate the cost of which is one-third that of the ester obtained from chloral by the usual procedure.[41]

According to Caglioti, Lenzi, and Mele,[42] liquid tetrafluoroethylene oxidation by oxygen is a typical chain process which gives significant radiation chemical yields of carbonyl fluoride (COF_2), tetrafluoroethylene epoxide [$(CF_2)_2O$], and liquid polyperoxide [$(-CF_2-CF_2-O-O)_n$.

Radiation chemical oxidation of hexachlorocyclopentadiene leads to a ketone, hexachlorocyclopentenone,[43] in the reaction

$$\tag{20}$$

An α-double bond is oxidized rather selectively.[44, 45] For example, radiation

oxidation of isoeugenol gives vanillin

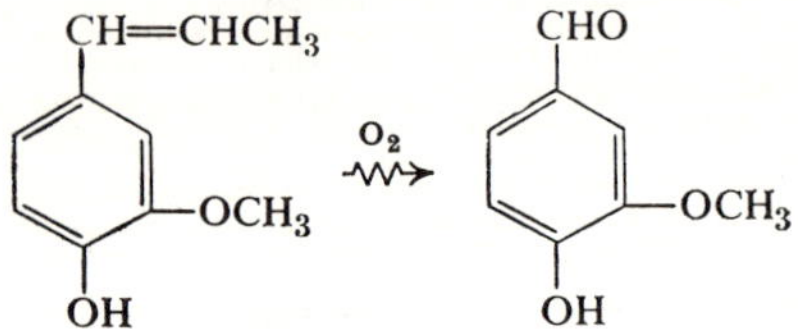

$$\text{isoeugenol} \xrightarrow{\;O_2\;} \text{vanillin} \tag{21}$$

while isosafrole gives heliotropin (piperonal)

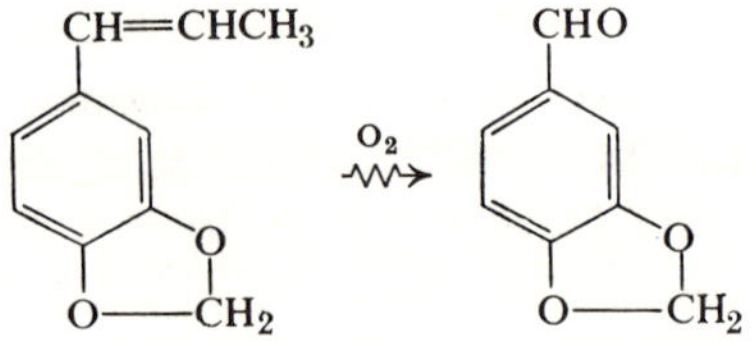

$$\text{isosafrole} \xrightarrow{\;O_2\;} \text{heliotropin} \tag{22}$$

7. HYDROHALOGENATION

Seven years of preparation resulted in the first industrial radiation chemical synthesis, namely that of ethyl bromide by Harmer and his colleagues.[46] In this process stoichiometric amounts of gaseous ethylene and hydrogen bromide are introduced via a gas separator into the bottom of a reactor filled with 150 liters of liquid ethyl bromide. An 1800-Ci cobalt-60 source is mounted in a cylindrical nickel reactor.[4] The reaction proceeds in ethyl bromide at $-2°C$ and a dose rate of 40 rads/sec; the flow of reactants is controlled to provide consumption of ethylene and HBr in the irradiation zone.

The chain reaction is initiated by bromine atoms resulting from the irradiation process

$$HBr \xrightarrow{\;\;\;\;\;} H + Br \tag{23}$$

and involves the free-radical chain

$$H + HBr \rightarrow H_2 + Br \tag{24}$$

$$Br + C_2H_4 \rightarrow C_2H_4Br \tag{25}$$

$$C_2H_4Br + HBr \rightarrow C_2H_5Br + Br \tag{26}$$

with a (*third*-order) termination

$$Br + Br \rightarrow Br_2 \tag{27}$$

The G value is as high as 3.9×10^4.

Competitiveness of the process depends on the ratio of prices of anhydrous HBr used in the radiation chemical synthesis and that of 48 percent hydrobromic acid employed in the conventional reaction with ethanol

$$C_2H_5OH + HBr \rightarrow C_2H_5Br + H_2O$$

Water formed in this reaction removes significant amounts of hydrogen bromide. The radiation chemical technique described permits production of ethyl bromide from simple compounds without any side products. Thus, it is more convenient than such other syntheses as involve, for example, peroxide initiation or use of aluminum bromide.

Rumfeldt and Armstrong[47] examined the addition of HBr to C_2H_4 at $-79°C$, with molar HBr/C_2H_4 ratio varied in the range 5:1 to 20:1 and dose rate in the range 6.4×10^{16} to 5.7×10^{18} ev/liter sec. Ethylene is consumed according to a first-order law. A similar reaction of HBr with C_3H_8 leads to isopropyl bromide but its kinetics is obscured by an induction period.[47]

According to McFarling and Kircher,[48] addition of HBr to various α-olefins (from octene to propene) in the temperature range -79 to $+25°C$ occurs, in most cases, against the Markovnikov rule and leads to primary bromoalkanes (up to 90 percent of the reaction product). The G value is about 10^3 in the gaseous phase and about 10^5 in the liquid phase.

Another example of radiation chemical hydrobromination is the synthesis of 1,1,2-trifluoro-1-bromo-2-chloroethane, an inhalation anesthetic, from halogenated ethylene.

$$\underset{F}{\overset{F}{>}}C=C\underset{Cl}{\overset{F}{<}} + HBr \rightsquigarrow F-\underset{Br}{\overset{F}{C}}-\underset{H}{\overset{F}{C}}-Cl \qquad (28)$$

The reaction can be performed at a temperature of -70 to $+20°C$, dose rate of 0.5 rad/hr to 15 Mrads/hr, and pressure up to 20 atm. The yield is purported by the inventors of the process to attain 89 percent of the theoretical value.[49]

8. SYNTHESIS OF PERCHLORINATED HYDROCARBONS

Unsaturated perchlorohydrocarbons, when radiolyzed, produce more complicated perchloro compounds.[50-54] As an example, radiolysis of tetrachloroethylene gives hexachlorobutadiene, octachlorobutene-1, and octachlorocyclohexadiene-1,3 with an overall 100-eV yield of 10. The products may result from reaction of CCl_2CCl radical, generated in the homolytic fission of C—Cl bonds with the tetrachloroethylene molecule. According to Cooper and Stafford,[50] crystalline octachlorocyclohexadiene-1,3 forms on addition of CCl_2CCl radical to $CCl_2CClCClCCl_2$ followed by cyclization and chlorine abstraction.

With perchlorohydrocarbons the G value is maximal for tetrachloroethylene and decreases in the series tetrachloroethylene > hexachloropropylene > hexachlorobutylene > octachloropentadiene. Kogan[53] found that in addition to these individual products the radiolysis yields an oil (of approximate composition C_8Cl_{10}) which is a good antiscuff additive to mineral oils. Rabovskaya and Kogan[52] report a total G value of 6 at a dose of 5×10^7 rads for conversion of hexachlorocyclopentadiene to octachlorocyclopentane along with a crystalline $C_{10}Cl_{10}$ (m.p. 110°C) which may be decachlorobiscyclopentadienyl, and a $C_{10}Cl_{10}$ oil; molecular chlorine is almost absent.

Although thermal stability decreases, the "radiolytic stability" of chlorinated hydrocarbons grows with molecular weight. Mainly, radiolysis causes abstraction of chlorine atoms followed by combination of the radicals to form products of molecular weight higher than those of the initial compounds.

9. TELOMERIZATION

High-energy radiation is also effective as a chain-reaction initiator in chlorinated the process of telomerization. Useful organic compounds of intermediate molecular weight are thus synthesized from low-molecular-weight starting materials.

9.1. Polychloroalkanes

The simplest radiation-induced telomerization is perhaps the synthesis of $\alpha,\alpha,\alpha,\omega$-tetrachloroalkanes from a CCl_4 *telogen* and an ethylene *taxogen* as reported by Beer et al.[55] Mellows and Burton[56] described a similarly simple radiation-induced preparation of trichloroalkanes [i.e., $H(CH_2CH_2)_n CCl_3$], from ethylene and chloroform, in which only products up to $n = 3$ were studied. The various steps in the work[55] with CCl_4 are presumably initiated by the overall reaction

$$CCl_4 \rightsquigarrow CCl_3 + Cl \tag{29}$$

followed by the propagation process

$$CCl_3 + CH_2{=}CH_2 \rightarrow CCl_3CH_2CH_2 \tag{30}$$

$$CCl_3CH_2CH_2 + CH_2{=}CH_2 \rightarrow CCl_3CH_2CH_2CH_2CH_2 \tag{31}$$

$$\vdots$$

$$CCl_3(CH_2CH_2)_n + CH_2{=}CH_2 \rightarrow CCl_3(CH_2CH_2)_{n+1} \tag{32}$$

or a termination process such as

$$CCl_3(CH_2CH_2)_n + CCl_4 \rightarrow CCl_3(CH_2CH_2)_nCl + CCl_3 \qquad (33)$$

followed by initiation of a new chain. The ratio of specific rates k_{33}/k_{32} for a particular value of n is called the chain transfer constant and is here designated by c_n.

Table II shows that radiation telomerization kinetics for the CCl_4–C_2H_4 system is similar to that of chemically initiated processes.

TABLE II

Chain Transfer Constants for Telomerization of CCl_4 with C_2H_4 at 70°C

Initiator	c_1	c_2	c_3
γ-Radiation	0.088	3.0	9.7
Azobisisobutyronitrile	0.105	3.0	7.1
Benzoyl peroxide and p-tolyloxymethyl-sulfone	0.097		

Under optimal conditions of 1,1,1,5-tetrachloropentane synthesis[57] (pressure $\sim$ 46 bar and temperature 75°C), two concentric-type reactors (0.2 × 0.8 × 10 m) with 7 kCi/m of cobalt-60, the source may provide a 10 ton/day efficiency of final product. According to Takehisa,[57] conversion of CCl_4 is 53 percent per cycle. For chloroform and ethylene the G values of telomerization products are 3000 and 300 at 100 and 28°C, respectively.[56]

Curiously, Hirota, Takezaki, and Hatado[58] report that telomerization of ethylene attempted with such telogens as 1,2-dichloroethane, 1,1-dichloroethane, dichloromethane, 1,2-dibromoethane, 1,1-dibromoethane, dibromomethane, and 2-bromopropane gives low yields (not more than several G units).

9.2. Higher Alcohols

Synthesis of higher alcohols from aliphatic alcohols and ethylene was studied by Hirota and co-workers.[59–61] Liquid propanol-2 irradiated in the presence of ethylene at 6–60°C gives tertiary amyl alcohol and higher alcohols. The initiation reaction

$$(CH_3)_2CHOH \xrightarrow{\text{\tiny$\sim\!\!\wedge\!\!\wedge\!\!\sim$}} (CH_3)_2COH + H \qquad (32)$$

is followed by propagation

$$(CH_3)_2COH + C_2H_4 \rightarrow (CH_3)_2C(OH)CH_2CH_2 \qquad (33)$$

$$(CH_3)_2C(OH)CH_2CH_2 + C_2H_4 \rightarrow (CH_3)_2C(OH)CH_2CH_2CH_2CH_2 \qquad (34)$$

or chain transfer

$$(CH_3)_2C(OH)CH_2CH_2 + (CH_3)_2CHOH \rightarrow (CH_3)_2C(OH)CH_2CH_3 + (CH_3)_2COH \quad (35)$$

The 100-eV yield of *tert*-amyl alcohol increases with temperature and reaches $\sim$300 at 60°C. The "effective activation energy" of 6.3 kcal/mole (for a process in which pressure, rather than concentration, of C_2H_4 is held constant with change of temperature) is close to the value 7.3 kcal/mole obtained for hydrogen abstraction from the α-carbon of propanol-2. The rotating-sector procedure[60] allows estimation of the mean concentration of radicals, $3.14 \times 10^{-3} M$, kinetic chain length, 8.3, and the specific rate of termination, 3.2×10^4 dm³ mole⁻¹ sec⁻¹.

Other products of irradiation of isopropanol (whether ethylene be present or be substituted by nitrogen) include acetone. Under irradiation by a 1.5-MeV electron beam, the 100 eV yields of both acetone and *tert*-amyl alcohol are found to be inversely proportional to the square root of the beam intensity.[60]

To obtain long-chain aliphatic alcohols, Yavorsky, Mazzocco, and Gorin[62] telomerized ethylene with methanol in the vapor phase[62] under β-radiation from strontium-90. At a dose rate of 1.69×10^6 rads/hr, the overall G value of the telomerization was 14,700, while the best radiation chemical yield of alcohols whose chains contained more than 11 carbons was 5000 to 10,000. Optimal conditions were: pressure, 320 atm; temperature, 300°C; methanol/ethylene ratio, 2:1; contact time, 5 min.

9.3. Methyl Alkyl Ketones

Hirota and Hatada and their colleagues[63-66] synthesized methyl alkyl ketones from an acetaldehyde telogen and various taxogens (ethylene, propylene, isobutylene)

$$RCH=CH_2 + CH_3CHO \rightarrow CH_3COCH_2CH_2R \quad (36)$$

through a chain process propagated by acetyl radical CH_3CO. An induction period is observed during the reaction.[64]

A mixture of acetaldehyde and cyclohexene produces methyl cyclohexyl ketone together with acetoin. The latter presumably forms via a bimolecular excited species

At higher dose rates 1-acetyl-2-cyclohexylcyclohexene[67, 68] is obtained. *n*-Hexene-1 gives methyl *n*-hexyl and methyl isohexyl ketones. The *G* value depends sharply on the ratio of components and reaches its maximum at 5 mole percent of hexene in the mixture. *G* is about 400 for methyl *n*-hexyl ketone and almost an order lower for methyl isohexyl ketone. The dose rate dependence is 0.5 and 0.75 for methyl cyclohexyl and methyl hexyl ketones, respectively.

Relative reactivities of allyl compounds in *γ*-induced telomerization with CCl_4 were thoroughly studied by Okubo.[69, 70] Telomer yields increase with temperature, especially for a telomer containing one taxogen molecule. The radiation chemical yields of the products obtained from allyl acetate telomerized at 25°C (dose rate 70×10^{11} eV/hr for 72 hr) are 190 for taxogen consumption, 14.5 for a telomer containing one taxogen molecule, and 53 for one containing two taxogen molecules. Radiation chemical telomerization of allyl esters occurs with increasing efficiency in the series: formate, acetate, benzoate. *γ*-Induced telomerization studies of the 10 different allyl esters shows that the double bond interacts with CCl_3 radical at a rate dependent on the presence of an α hydrogen in the carboxyl part, and on the nature of the substituents.

Muramatsu et al.[71] studied radiation chemical addition of tetrahydro-furan, dioxane, or ethyl ether to the chloro-fluoro-substituted ethylenes $Cl_2C{=}CFCl$, $CFCl{=}CFCl$, and $Cl_2C{=}CCl_2$. With tetrahydrofuran or dioxane, the products are formed in a high yield characterized by a 1:1 ratio of the starting components; approximately equimolar amounts of 1:1 and 1:2 adducts are formed in the case of ethyl ether.

Table III summarizes some telomerization processes involving a variety of taxogens, including boiling points of several products.

TABLE III

Telomerization Products Obtained from CCl_4 and Various Taxogens[a]

Taxogen	1:1 Telomer	2:1 Telomer
$CH_2{=}CHCH_2OH$	97°C	Decomposes during distillation
$CH_2{=}CHCH_2C_6H_5$	138–140°C	
$CH_2{=}CHCH_2COOCH_3$	91–93°C	168–171°C
$CH_2{=}CHCH_2COOH$	96–98°C	184–186°C
$CH_2{=}CHCH_2COOC_6H_5$		
$CH_2{=}CHCH_2Cl$	88–90°C	147–150°C
$CH_2{=}CHCH_2CN$		
$CH_2{=}CHCH_2OCH_3$	62–65°C	123–126°C
$CH_2{=}CHCH_2OC_6H_5$		

[a] The molar taxogen/CCl_4 ratio is 1:5.

9.4. Addition of Hydrogen Sulfide to Olefins

Dzantiyev and Shishkov[72] found that, in the gas phase, radiation chemical yield of the reaction between H_2S and C_2H_4 depends on the composition of the mixture and reaches its maximum at 30 percent ethylene. The main sulfur-containing products are ethyl mercaptan, with $G(C_2H_5SH) = 600$, and diethyl sulfide, with $G[(C_2H_5)_2S] = 250$, at a dose rate of 1.3×10^{14} eV cm^{-3} sec^{-1}. The irradiation initiates the chain process

$$H_2S \rightsquigarrow H + SH \tag{38}$$

$$C_2H_4 + SH \longrightarrow C_2H_4SH \tag{39}$$

$$C_2H_4SH + H_2S \longrightarrow C_2H_5SH + SH \tag{40}$$

Diethyl sulfide forms in a secondary process

$$C_2H_5SH \rightarrow C_2H_5S + H \tag{41}$$

$$C_2H_4 + C_2H_5S \rightarrow C_2H_4SC_2H_5 \tag{42}$$

$$H_2S + C_2H_4SC_2H_5 \rightarrow (C_2H_5)_2S + SH \tag{43}$$

Apart from such products, elementary sulfur, diethyl disulfide $[(C_2H_5)_2S_2]$, and hydrogen are also formed.

Earlier work of Sugimoto, Ando, and Oae[73] indicates that addition of H_2S to propylene proceeds in a similar manner[73] and, even earlier, Fontijn and Spinks[74] found that n-butyl mercaptan reacts with pentene-1 to give the expected dialkyl sulfide

$$RSH + CH_2{=}CHR' \rightarrow RSCH_2CH_2R' \tag{44}$$

The latter reaction was conducted in cyclohexane under X-irradiation. The yield of n-butyl amyl sulfide varied from 75 to 95 percent, and the G value reached 8×10^4.

9.5. Addition of Silanes to Ethylene

Lampe, Snyderman, and Johnston[75] found that silanes add to ethylene to give telomerization products. In the gaseous phase methylsilane or dimethylsilane forms telomers containing one or two molecules of taxogen; 100-eV yields of $CH_3CH_2CH_2SiH_3$ and $CH_3CH_2CH_2SiH_2CH_3$ are 38 and 15 at 50°C at a silane/ethylene ratio of 3:1. Near 250°C, $G(CH_3CH_2CH_2SiH_3)$ is close to several hundred. The telomerization process may involve an ionic chain mechanism, the existence of which is confirmed by a weak inhibiting effect of NO, a strong inhibiting effect of NH_3, and a low apparent activation energy ($\sim$1 kcal/mole).

According to El-Abbady,[76] radiation causes addition of triphenylsilane to the double bonds of octene-1 and cyclohexene. He irradiated a mixture of triphenylsilane (0.15 mole) and octene (0.05 mole) sealed in an evacuated ampoule at 65°C for 63 hr (at a dose rate of 810,000 roentgens/hr) and obtained 700 mg of triphenyl-1-octylsilane and 210 mg of hexaphenyldisiloxane. The reaction of triphenylsilane with cyclohexene gives triphenylcyclohexylsilane and hexaphenyldisiloxane.

10. SYNTHESIS OF ORGANOSILICON COMPOUNDS

Formation of carbon–silicon bonds via irradiation has been studied by numerous workers. Zimin and his co-workers[77-79] proposed, as an effective method to obtain substituted chlorosilanes starting from a chlorosilane containing a labile hydrogen atom, the overall process

$$(R'R'')_n SiHCl_{3-n} + RCl \rightarrow (R'R'')_n SiRCl_{3-n} + HCl \tag{45}$$

A new radical, R, is introduced into the chlorosilane molecule. This technique led to production of a variety of chlorosilanes containing different substituents in high 100-eV chemical yields; see Table IV. The synthesis is accompanied by radiolytic decomposition; e.g., in a CH_3SiHCl_2 plus C_6H_5Cl mixture, $G(- C_6H_5Cl)$ is about 7 and radiolysis of methyldichlorosilane leads to dimethyldichlorosilane.

An interesting example of radiation chemical synthesis afforded by the work of Eydokimov and his colleagues[80] is the preparation of the fluoro-arylated silane 2,3-bis(trifluoromethyl)phenyltrichlorosilane

a compound previously unknown.

Benzotrifluoride mixed with an equal volume of tetrachlorosilane was irradiated in a vacuum-sealed ampoule at an average dose rate of 87 rads/sec to producing the desired product ($G \simeq 3$). The synthesis may involve either radical or ionic reactions or both, the details of which have not been thoroughly established.

According to Dzagatspanyan, Zetkin, and Maksimov,[81] CH_3 and C_6H_5 groups may be radiationally chlorinated to phenylmethyldichlorosilane or dimethylchlorosilane, and Schoenberg[82] reports that addition of chlorosilanes to olefins is readily accomplished under the action of ultraviolet radiation.

TABLE IV
Radiation Chemical Synthesis of Substituted Chlorosilanes

Chlorosilane	Aryl halide	Components ratio	Temperature, °C	Product	G
$HSiCl_3$	C_6H_5Cl	1:1	175	$C_6H_5SiCl_3$	300
$HSiCl_3$	$p\text{-}C_6H_4Cl_2$	1:3	170	$p\text{-}C_6H_4(SiCl_3)Cl$	100
$HSiCl_3$	$p\text{-}C_6H_4Cl_2$	2:1	300	$p\text{-}C_6H_4(SiCl_3)_2$	100
$HSiCl_3$	$o\text{-}C_6F_4Cl_2$	4:1	250	$o\text{-}C_6F_4(SiCl_3)_2$	60
$HSiCl_3$	$1,2,4\text{-}C_6H_3Cl_3$	1.2:1	175	$C_6H_3Cl_2(SiCl_3)$	70
CH_3SiHCl_2	C_6H_5Cl	1:2	300	$CH_3(C_6H_5)SiCl_2$	300
$C_6H_5SiHCl_2$	C_6H_5Cl	1:1	300	$(C_6H_5)_2SiCl_2$	150
$(CH_3)_2SiHCl$	$p\text{-}C_6H_4Cl_2$	4:1	300	$p\text{-}C_6H_4[Si(CH_3)_2Cl]Cl$	80
				$p\text{-}C_6H_4[Si(CH_3)_2Cl]_2$	87
$(CH_3)_2SiHCl$	$m\text{-}C_6H_4Cl_2$	4:1	300	$m\text{-}C_6H_4[Si(CH_3)_2Cl]Cl$	80
$CH_3(C_6H_5)SiHCl$	$p\text{-}C_6H_4Cl_2$	2:1	300	$p\text{-}C_6H_4[SiCH_3(C_6H_5)Cl]_2$	20
$(C_6H_5)_2SiHCl$	$p\text{-}C_6H_4Cl_2$	2:1	300	$p\text{-}C_6H_4[(C_6H_5)_2SiCl]_2$	15

Action of ionizing radiation on ethylene mixed with trichlorosilane, dichlorosilane, methyldichlorosilane, or ethyldichlorosilane yielded a great number of compounds, some of which had been previously unprepared.[83, 84] The telomerization proceeds below 120°C.

The chain reaction is initiated by $SiCl_3$ radicals formed from trichlorosilane upon irradiation

$$HSiCl_3 \xrightarrow{\sim\!\!\!\wedge\!\!\!\wedge\!\!\!\rightarrow} H + SiCl_3 \tag{46}$$

$$RCH{=}CH_2 + SiCl_3 \longrightarrow R\dot{C}HCH_2SiCl_3 \tag{47}$$

$$HSiCl_3 + R\dot{C}HCH_2SiCl_3 \longrightarrow RCH_2CH_2SiCl_3 + SiCl_3 \tag{48}$$

Other substituted chlorosilanes react similarly. Table IV presents a summary of such processes and the products so prepared as well as of the 100-eV yields.

11. FLUORINATION

Unlike radiation chemical chlorination or bromination with free chlorine or bromine, fluorination is an indirect process.

Radiation chemical stability of C—F bonds in fluorocarbons is comparable with that of C—H bonds in hydrocarbons but, according to Feng,[85] G values are somewhat lower in fluorocarbons than in hydrocarbons. As an example, the main product of radiolysis of CF_4 is C_2F_6, formation of which may be decreased by addition of free halogen to the reaction mixture. Radiolysis of CF_3Cl leads to CF_4 and CF_2Cl_2 with G values of about 2. Free bromine does not affect the yield of CF_4 but lowers that of CF_2Cl_2 because CF_3Br and CF_2BrCl are also formed. In the case of CF_3Br radiolysis, $G(CF_4)$ is about 2 and is unchanged in the presence of free bromine. Temperature increase in the range 27–300°C is without effect on $G(CF_4)$, while $G(CF_2Br_2)$ increases from 1.6 to $\sim$3.

Fluorocarbons were used by Feng and Mamula[86] as fluorinating agents to form new carbon–fluorine bonds or to introduce a fluorine-containing group in mixtures subject to ionizing radiation. Benzene mixed with carbon tetrafluoride, when γ-irradiated in an ampule, yields nearly equal amounts of fluorobenzene and benzo trifluoride

$$C_6H_6 + CF_4 \xrightarrow{\sim\!\!\!\wedge\!\!\!\wedge\!\!\!\rightarrow} \begin{cases} C_6H_5CF_3 + HF \\ C_6H_5F + CHF_3 \end{cases} \tag{49}$$

The G values increase with CF_4 concentration and do not depend on temperature within the interval studied. The highest $G(C_6H_5CF_3)$ reported, equal to 1.3, was obtained in a mixture containing 37 percent by weight CF_4. Data for richer CF_4 mixtures were not given.

In mixtures of CF_3Br and C_6H_6, when temperature was increased from 27 to 250°C, $G(C_6H_5F)$ increased from 0.22 to 1.5 while $G(C_6H_5CF_3)$ remained about 2. Radiolysis of fluorocarbons may involve either ionic or radical mechanisms. An ionic intermediate, for example, CF_3^+, appears to act upon the C—F bond; there are also simultaneous radical processes. This conclusion is based on the activation energies found for CF_2Br_2 and $C_6H_5CF_3$ production, which are equal to 2 and 6 kcal/mole, respectively.[89]

Effects of γ-irradiation on the lower, higher, and cyclic perfluoroalkanes in *n*-hexane in the presence of N_2O were studied by Rajbenbach.[90] Lower fluorocarbons (CF_4 to C_4F_{10}) do not affect nitrogen and hydrogen yields, while perfluoropentane and higher homologs decrease the nitrogen yield significantly; hydrogen yield, however, changes only slightly. In the absence of N_2O, *lower* perfluoroalkanes do not affect the hydrogen yield; *higher* ones decrease it noticeably. Such behavior can be explained on the assumption that perfluoroalkanes compete with N_2O in the capture of electrons formed from radiolysis of *n*-hexane. Lower perfluoroalkanes differ from higher ones in that the former do not dissociate during electron capture while the latter compounds undergo dissociation of their carbon–fluorine bonds. This view is supported by the formation of hydrogen fluoride in the radiolysis of perfluoroalkanes in *n*-hexane.[91]

Several fluorinated compounds have been synthesized by radiolysis of two-component mixtures of CF_4, SiF_4, BF_3, or SF_6 with benzene, nitrobenzene, or toluene.[87] In a CF_4–$C_6H_5NO_2$ mixture, the yield of *m*-fluoronitrobenzene increases with electron fraction of CF_4 and does not depend on the irradiation temperature. With toluene, trifluoromethyl derivatives are predominant.

Matveeva, Timofeev, and Borisov[92] report that unsaturated hydrocarbons interact with fluorocarbons to give fluorinated polymer with fluorines directly adjacent to the initial double-bond sites.[92] In a mixture of acetylene with CF_4, fluorine content in the reaction products increases with CF_4 content in the starting mixture, while G of fluorine atom consumption may be as high as 4.2.

According to Zimin et al.,[93] radiation chemical fluorination of CCl_4 or $C_2H_2Cl_4$ by ionic inorganic fluorides (CaF_2, ZnF_2, AlF_3, or SbF_3) shows that in the absence of oxygen the fluorination yield is as low as $G = 4$ while in the presence of O_2 it is much higher; e.g., with SbF_3 G is then equal to 55.

12. CARBOXYLATION AND CARBONYLATION

Direct introduction of CO_2 into organic compounds has been reported in several papers. In the gas phase radiation chemical carboxylation occurs only with low yield. For instance, Cacae, Guarino, and Possagno[94] identified, among the n-pentane radiocarboxylation products, the following carboxylic acids (in decreasing order of G value): 2-ethylbutyric and 2-ethylvaleric, formic, capric, valeric, butyric, 2-methylbutyric, isobutyric, propionic, and acetic. The total G of carboxylic acids was only 7.5×10^{-3}. These acids are formed, evidently, through various free radicals appearing after the abstraction of hydrogen atoms or cleavage of the C—C bond in an excited initial molecule.

In the liquid phase carboxylation of hydrocarbons leads to carboxylic acids, so far mostly unidentified; see Table V. The G value depends on the

TABLE V

G-Values of Carboxylated Organic Compounds Obtained by Irradiation in the Liquid Phase

Initial compound	Products	Dose, Mrads	G	Reference[a]
n-Hexane	Carboxylic acids	20	0.14	Hummel . . .[100]
n-Heptane	Carboxylic acids	20	0.14	Hummel . . .[100]
n-Heptane	Carboxylic acids	130[b]	0.05	Furrow[98]
n-Heptane	Carboxylic acids	10	0.43	Kalyazin . . .[97]
n-Heptane saturated with CO_2	Carboxylic acids	0.044	0.63	Gütlbauer . . .[96]
n-Heptane with bubbled CO_2	Carboxylic acids	0.044	1.08	Gütlbauer . . .[96]
Isooctane	Carboxylic acids	20	0.19	Hummel . . .[100]
n-Nonane	Carboxylic acids	10	0.24	Kalyazin . . .[97]
Ethylene	Polymeric carboxylic acids	10	Up to 300	Furrow[98]
Isoprene	Carboxylic acids	20	0.04	Hummel . . .[100]
Cyclohexane	Cyclohexanecarboxylic acids	16	0.15	Hummel . . .[100]
		80[b]	0.07	McKusick . . .[99]
Cyclohexene	Cyclohexenecarboxylic acids	80[b]	0.11	McKusick . . .[00]
Benzene	Benzoic acid	20	0.02	Hummel . . .[100]
Toluene	Phenylacetic acid	20	0.05	Hummel . . .[100]
Toluene	Phenylacetic acid	80[b]	0.01	McKusick . . .[99]
Cumene	Carboxylic acids	10	0.07	Kalyazin . . .[97]
Methanol	Glycolic acid	0.2	0.5	Getoff . . .[95]
Ethanol	Lactic acid	80[b]	0.06	McKusick . . .[99]
Ethylamine	α-Alanine	80[b]	0.17	McKusick . . .[99]
Carbon tetrachloride	Trichloroacetic acid	10	0.05	Gütlbauer . . .[96]
Same, saturated with H_2O	Trichloroacetic acid	10	0.14	Gütlbauer . . .[96]

[a] Only the first author is cited. A series of dots indicates multiauthorship.
[b] Irradiation with accelerated electrons.

nature of initial compound. n-Heptane carboxylation is independent of the dose rate in the 0.008–2.8 Mrads/hr range and of temperature in the range -77 to $+145°C$. According to Kalyazin and Makrev,[97] at CO_2 concentration equal to 2 M the yield of carboxylic acids is maximal. Gütlbauer and Getoff[96] report that G(carboxylic acids) with CO_2 bubbling is much larger than in the case of simple saturation with CO_2.

The carboxylation mechanism in the condensed phase evidently includes the ionic reactions

$$RH^+ + CO_2^- \rightarrow [RH + CO_2]^* \begin{array}{l} \nearrow RCOOH \\ \searrow R + H + CO_2 \end{array} \tag{50}$$

Simultaneously, CO_2 can be reduced by atomic hydrogen

$$CO_2 + H \rightarrow COOH \tag{51}$$

and can thus carboxylate intermediate radicals

$$R + COOH \rightarrow RCOOH \tag{52}$$

Kalyazin and Baranova[101] report that γ-irradiation of n-alkanes in the presence of CO leads to carbonyl compounds. In methanol saturated with CO, glycolic aldehyde is formed at a dose 2.6×10^{19} eV/ml; Getoff and Seitner[102] state that $G = 0.78$. The possible mechanism includes the reactions

$$CH_3OH \rightsquigarrow CH_2OH + H \tag{53}$$

$$CO + H \longrightarrow CHO \tag{54}$$

$$CH_2OH + CHO \longrightarrow CHO \cdot CH_2(OH) \tag{55}$$

13. SYNTHESIS OF NITROGEN-CONTAINING ORGANIC COMPOUNDS

Of the great variety of the syntheses leading to nitrogen-containing compounds, those processes are of special interest that give rise to compounds with a novel C—N bond.

13.1. Synthesis of Hydrocyanic Acid

This process is interesting as a new way to fix molecular nitrogen.[103–105] In radiation synthesis of hydrocyanic acid from mixtures of nitrogen with hydrocarbons, the maximum value of G(HCN) was observed when the components were related in the ratio $RH:N_2 = $ (5–10 percent):(90–95 percent). Table VI shows the yields obtained in radiolysis of the mixtures

TABLE VI
Yield of HCN as a Function of the Nature
of the Hydrocarbon[a]

System	RH:N_2 ratio	G(HCN)
CH_4-N_2	5:95	0.83
C_2H_6-N_2	10:90	0.45
C_2H_4-N_2	20:80	0.27
C_2H_2-N_2	10:90	0.23
C_3H_8-N_2	10:90	0.06
C_3H_6-N_2	10:90	0.10

[a] Overall pressure is 2 atm throughout.[105]

at optimum ratio of the components. It is assumed now that one of the basic reactions leading to HCN is

$$CH_3 + N \rightarrow HCN + 2H \tag{56}$$

The maximum values of G(HCN) are as high as 0.83 and 4 at 2 and 100 atm, respectively.

The higher HCN yield in the N_2–CH_4 mixture is evidently a result of the fact that in this case the charge-redistribution reaction

$$N_2^+ + CH_4 \rightarrow N_2 + CH_4^+ \tag{57}$$

along with the ion-molecule reaction

$$CH_4^+ + CH_4 \rightarrow CH_3 + CH_5^+ \tag{58}$$

is an additional source of radicals. When ethylene reacts with active nitrogen, the following reaction takes place

$$N + C_2H_4 \rightarrow HCN + CH_3 \tag{59}$$

Radiolysis of nitrogen mixed with ethylene and other unsaturated hydrocarbons yields various hydrocarbon radicals which can interact with nitrogen atoms. However, in this case there is an increase in the role of a competitive reaction; i.e., radical addition to the double bond in the hydrocarbons the concentration of which far exceeds that of atomic nitrogen. No temperature dependence was observed in the N_2–C_2H_4 mixture. Variations in the magnitude of the dose do not affect G(HCN). Addition of oxygen to the N_2–CH_4 mixture inhibits the formation of HCN. In the presence of oxygen, some of the CH_3 radicals formed from methane enter the reaction

$$CH_3 + O_2 \rightarrow CH_3O_2 \tag{60}$$

Thus oxygen competes with N atoms for CH_3 radicals.

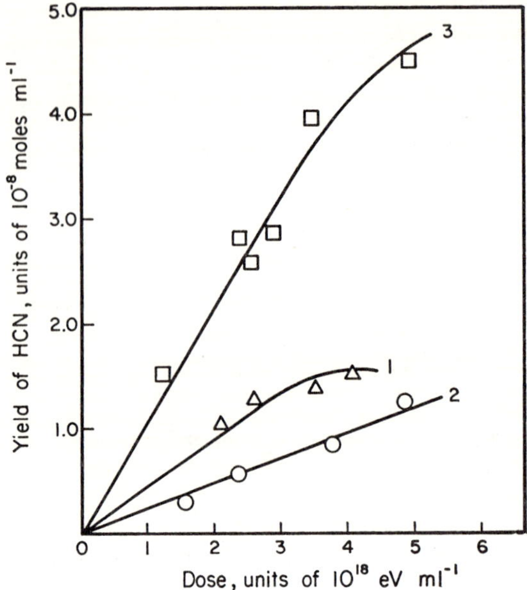

Fig. 3. Effect of additives on yield of HCN formation when nitrogen–ethylene mixtures are irradiated.[105] Overall pressure is 2 atm in all cases. Curves: 1, N_2 (80%) + C_2H_4 (20%); 2, N_2 (78%) + C_2H_4 (20%) + N_2O (2%); 3, N_2 (79.2%) + C_2H_4 (20%) + O_2 (0.8%).

Addition of oxygen to mixtures of nitrogen with unsaturated hydrocarbons (Fig. 3) increases G(HCN), probably because of formation of ketene and subsequent reaction of methylene (produced therefrom) with N

$$CH_2 + N \rightarrow HCN + H \tag{61}$$

For the optimal C_2H_4–N_2 mixture in the presence of 0.8 percent O_2 at 60 atm, G(HCN) is as high as 5.

Experimental evidence shows that in the gaseous phase the values of G(HCN) (the main product of nitrogen fixation) are as high as 4–5 under optimum conditions. Such results may favor synthesis of HCN by radiation from a nuclear reactor.

According to Dzantiev, Trubnikov, and Popov,[106] similar results were obtained in the radiative synthesis of HCN from hydrocarbons and ammonia.

13.2. Synthesis of Acetonitrile and Its Adducts with Unsaturated Compounds

Irradiation of an RH–N_2 mixture in the gas phase yields HCN as the main reaction product. Although the stationary concentration of HCN is not high, HCN molecules can react, while still in the excited state, with

TABLE VII

Radiation Chemical Yields of Acetonitrile Adducts with Unsaturated
Compounds (cobalt-60 γ-radiation, dose rate 3.84×10^5 rad hr^{-1} cm^{-3})[107]

Unsaturate	Product	Yield, %	G
Hexene-1	Heptylcyanide	35	1.2
Heptene-1	Octylcyanide	36	1.2
Octene-1	Nonylcyanide	45	1.2
Decene-1	Undecylcyanide	40	1.2
Dodecene-1	Tridecylcyanide	42	1.2
Cyclopentene	Cyclopentylacetonitrile	38	1.2
Cyclohexene	Cyclohexylacetonitrile	45	1.2
Cyclooctene	Cyclooctylacetonitrile	47	1.2
Cyclodecene	Cyclodecylacetonitrile	43	1.2
Cyclododecene	Cyclododecylacetonitrile	35	1.1
Norbornene	Norbornylacetonitrile	65	1.7
Norbornadiene	Nortricyclylacetonitrile	60	1.6
Benzene	Benzylcyanide	10–15	0.2

hydrocarbons to yield acetonitrile

$$CH_4 + HCN^* \rightarrow CH_3CN + H_2 \tag{62}$$

Podkhalyusin, Trishkin, and Vereshchinskii[104] report that, in the 5:95
CH_4–N_2 mixtures at 75 atm and in those of 18:82 composition at 50 atm,
the values of $G(CH_3CN)$ are, respectively, 0.62 and 1.07.

Rokach and Krauch found that, unlike photoaddition sensitized by
acetone or benzophenone, addition of acetonitrile to olefins in the presence
of ionizing radiation is quite successful.[107] Experiments in ampoules containing
a large excess of acetonitrile under an argon atmosphere led to a variety of
adducts, none of which was produced in industrially significant yield; see
Table VII. The addition proceeds against the Markovnikov rule in all cases

$$RCH{=}CH_2 + CH_3CN \rightarrow RCH_2CH_2CH_2CN \tag{63}$$

Nitriles are also formed in low yield (2–3 percent) as products of radical
CN addition to olefins.

13.3. Radiation Synthesis of Amines and Amino Acids and Radiation Amidation

Synthesis of aniline from benzene and ammonia[108, 109] was a pioneering
effort in the field of radiation synthesis of organic compounds in the liquid
phase.

Under γ-irradiation aliphatic and aromatic amines enter into reactions
that do not affect the C—N bonds. According to Smith and Swan,[110, 111]

abstraction of α hydrogen results in dimerization of radicals into diamines. In the case of mono-, di-, and trialkylamines, when all alkyl groups at the same nitrogen atom are identical, such overall reactions as

$$(C_2H_5)_2NCH_2CH_3 \xrightarrow{} (C_2H_5)_2NCHCH_3 + H \tag{64}$$

$$2(C_2H_5)_2NCHCH_3 \longrightarrow \begin{array}{c} H \quad H \\ CH_3C\!-\!CCH_3 \\ |\quad | \\ (C_2H_5)_2N \quad N(C_2H_5)_2 \end{array} \tag{65}$$

occur. Irradiation of amines leads to radicals of various compositions and structures, so that the sets of diamines and dimerization products become more complex.[112]

Attempts by Swan and Timmons[113] at radiation syntheses, from aryl halide, of the quaternary ammonium salts containing a phenyl group were fruitless. Trialkylamines and aryl halides form trialkyl- and dialkylammonium salts ($G < 7$), but such reactions have no preparative value.

Synthesis of amino acids from NH_3, H_2, CH_4, CO, CO_2, and H_2O in the gaseous phase in the presence of ionizing radiation is a classic experiment which has been suggested as a model for possible conditions for the genesis of life on earth.[114, 115] Getoff and Schenck[116] have discussed the synthesis of amino acids by carboxylation of different amines in aqueous solution. They give initial 100-eV yields of various products that do not exceed 1.25 in any single case and offer a variety of suggestions as to mechanism.

Complex nitrogen-containing organic compounds, in particular adenine,

have been obtained by Ponnamperuma et al.[117] by the use of 4.5-MeV electrons on mixtures of methane, ammonia, hydrogen, and water vapor.

Olefine amidation by formamides in the presence of γ-radiation or in the acetone-initiated photochemical reaction leads to the 1:1 addition products. According to Rokach, Krauch, and Flagg,[118] high chemical yield makes both techniques useful for synthetic purposes. Elad[119] found that, unlike the olefins, addition of dienes to formamide gives low yields of the 1:1 product under the action of ultraviolet light. γ-Irradiation markedly increases the yields and may find application for synthetic purposes; see Table VIII.

Carbamyl radicals $CONH_2$ result from direct irradiation of formamide, while a low concentration of diene in the reaction mixture is little affected by radiolysis.

TABLE VIII
Yield and G Values of Diene Amidation Products[a]

Initial diene	Addition product[b]	Yield, %	G
4-Vinylcyclohexene	(I)	35	3
1,5-Cyclooctadiene	(II)	50	3.5
Norbornadiene	(III)	75	5

[a] According to Krauch, Rokach, and Elad.[120]
[b]

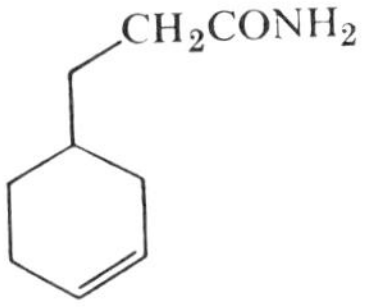

(I)

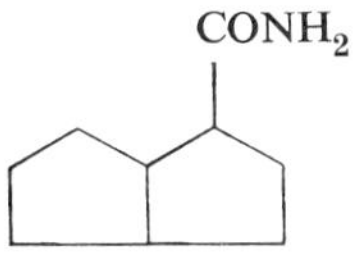

(II)

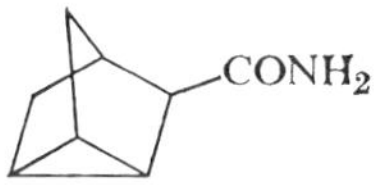

(III)

13.4. Syntheses Based on Nitrogen Oxides

Reactions with nitrogen oxides have not so far been studied in detail. Acceptance of intermediate unstable products by nitrous oxide, so extensively used in studies of radiation chemistry, has not found preparative application.

Among reactions involving nitrogen oxides, first mention should be made of the radiation chemical syntheses of nitro derivatives of halogenated methane and ethane. Henglein[121, 122] produced these by irradiation of carbon tetrachloride (or chloroform, bromoform, trichloronitromethane, trichloroethane) with electrons at a dose rate of 10^8 rad at room temperature with nitrogen oxide bubbling through at 15 liters/hr.[121, 122] The CCl_3 or CH_3CCl_2 formed in the initial radiolytic process reacts with NO

$$CCl_3 + NO \rightarrow CCl_3NO \tag{66}$$

$$CH_3CCl_2 + NO \rightarrow CH_3CCl_2NO \tag{67}$$

$G(CCl_3NO)$ is in the 4.4–6.0 range, and the chemical yield is 50 percent at 10 percent conversion. $G(CH_3CCl_2NO)$ is 0.5.

In the reaction with chloroform, dichloronitrosomethane has been synthesized along with dichloroformaldoxime

$$HCCl_3 \rightsquigarrow HCCl_2 + Cl \tag{68}$$

$$HCCl_2 + NO \rightarrow \begin{cases} CCl_2{=}NOH & (69) \\ HCCl_2NO & (70) \end{cases}$$

The latter (new) compound, stable only at low temperatures, is obtained with $G = 0.5$ and in 20 percent chemical yield.

Cyclohexane nitrosation under action of ionizing radiation differs substantially from an industrial photochemical process[82] employing nitrosyl chloride to obtain cyclohexanone oxime. Müller and Schmied[123] found that when pure cyclohexane is irradiated with accelerated electrons while a 1:3 mixture of NO and HCl is bubbled through at the rate of 10 liters/hr, cyclohexanone oxime is formed, presumably via the process

$$NO + cycloC_6H_{11} \rightarrow cyclo\text{-}C_6H_{11}NO \rightarrow C_6H_{10}NOH \tag{71}$$

In the presence of HCl, the cyclohexanone oxime (combined in the form of its hydrochloride) precipitates from the reaction zone, thus avoiding reaction with NO to form the nitrite with further decomposition to cyclohexyl radicals. At $(2.5\text{--}2.6) \times 10^7$ rads, conversion is 5 percent and yield is 46 percent. Other compounds that form in this reaction are cyclohexyl nitrate, nitrocyclohexane, chlorocyclohexane, and dicyclohexyl. In the absence of HCl, according to Burrell,[124] bisnitrosocyclohexane forms with $G = 2.6$. No photochemical process similar to the radiation chemical one is known. Cyclohexane nitrosation by nitrosyl chloride, induced by ionizing radiation, proceeds at low efficiency ($G = 0.1$); according to Hills and Johnson,[125] chlorocyclohexane and cyclohexanone are the reaction products.

Synthesis of complex compounds of the type **Ar–NOFe(CO)₃** resulting from γ-irradiation of nitro compounds and iron pentacarbonyl dissolved in benzene has been described by von Gustorf and Jun.[126] Dimeric nitrosobenzeneirontricarbonyl is formed by the reactions

$$C_6H_5NO_2 + Fe(CO)_5 \rightsquigarrow C_6H_5NO_2Fe(CO)_4 + CO \tag{72}$$

$$2C_6H_5NO_2Fe(CO)_4 \longrightarrow (C_6H_5NOFe(CO)_3)_2 + 2CO_2 \tag{73}$$

14. SYNTHESIS OF SULFUR ORGANIC COMPOUNDS

Radiation synthesis of sulfur compounds is mentioned in this chapter in the description of sulfochlorination (Section 4), sulfonic acid production (Section 5), and radiation addition at the double bond (Section 9.4). Barzynski and Hummel[127] found that sulfurization by interaction of dissolved sulfur with hydrocarbon under the action of radiation leads in the case of cyclohexane to dicyclohexyl disulfide, dicyclohexyl ether, and cyclohexyl mercaptan. According to Ando, Sugimoto, and Oae,[128] in benzene solutions of rhombic sulfur, radiolysis products are thiophenol and hydrogen sulfide. The 100-eV yield of both products (0.04) is independent of sulfur concentration and the dose absorbed, a result that suggests the occurrence of secondary reactions of sulfur with C_6H_6 radiolysis products. Vereshchinskii[129] reports that radiolysis of sulfur solutions in butyl bromide gives mercaptans, sulfide, and disulfide.

In addition to production by sulfurization, disulfides can be obtained from substituted thio alcohols. The OH-substituted thio alcohols are much more sensitive to high-energy irradiation, so that they can lead to substituted disulfides.[130] While in γ-irradiation of ethanthiol the yields of decomposition products are only 5, irradiation of β-oxiethanthiol at a dose rate of $(0.8-1) \times 10^{15}$ eV ml^{-1} sec^{-1} increases G to 400. Crystals of the disulfide

$$HOCH_2CH_2SSCH_2CH_2OH$$

(m.p. 145°C, maximum absorption at λ 255–260 nm) were isolated from the reaction mixture with yield up to 70 percent.

15. SYNTHESIS OF ORGANOPHOSPHORUS COMPOUNDS

Several directions of research can be distinguished in the field of radiation synthesis of compounds with C–P bonds. A major group of processes involves the radiation chemical reactions of aliphatic or cyclic hydrocarbons with phosphorus trichloride (PCl_3) or its alkyl (cycloalkyl) derivatives; these lead to compounds $RPCl_2$ and $RR'PCl$.

Another group of processes involves PCl_3 addition to unsaturated hydrocarbons. Use of γ-radiation is promising in the introduction of aryl groups R into tertiary phosphonium salts $RR_3'PHal$ and esters of phosphonic acids $RP(O)(OR')_2$.

15.1. Alkyl and Cycloalkyl Derivatives of Phosphorus Trichloride

Reactions of PCl_3 with several aliphatic and cyclic hydrocarbons were studied by Henglein[131] and by Babkina and Vereshchinskii.[136, 137] It is seen from Table IX that in this case alkyl- or cycloalkyl chlorophosphines (i.e., compounds in which hydrocarbon radicals substitute for chlorine in PCl_3 or its substitution products) are the main products. If such alkyl or cycloalkyl derivatives are used as initial phosphorus compounds, then the disubstituted derivatives[136] and various compounds of more complex composition[137] are synthesized. The process is conducted in a simple glass apparatus under an inert atmosphere at normal pressure and room temperature; the hydrogen chloride formed is bubbled through an absorber.

In all cases an orange, finely dispersed precipitate is also obtained in amount proportional to the dose absorbed. This product results from

TABLE IX

Yields of Organophosphorus Compounds Synthesized by γ-Irradiation[a]

Hydrocarbon	Initial phosphorus compound	Mole ratio	Dose, eV/ml	Synthesized compound	Formula	Yield, mole/liter	G (approximate)
Cyclopentane	PCl_3	1.0:1.2	3.4×10^{20}	Cyclopentyl-dichlorophosphine[b]	cyclo-$C_5H_9PCl_2$	0.5	88
Cyclohexane	PCl_3	1.0:1.2	5.0×10^{20}	Cyclohexyl-dichlorophosphine	cyclo-$C_6H_{11}PCl_2$	0.5	60
Cyclohexane	cyclo-$C_6H_{11}PCl_2$	1.3:1.0	5.0×10^{20}	Dicyclohexyl-dichlorophosphine[b]	(cyclo-$C_6H_{11}PCl)_2$	0.05	6
				Dicyclohexyl-chlorophosphine	(cyclo-$C_6H_{11})_2PCl$	0.2	24
Cyclohexane	PCl_3 and cyclo-$C_6H_{11}PCl_2$	1.0:1.3	5.0×10^{20}	Cyclohexenyl-tetrachlorodiphosphine[b]	cyclo-$C_6H_{10}(PCl_2)_2$	0.2	24
n-Hexane	PCl_3	1.0:1.2	4.0×10^{20}	Isohexyldichlorophosphine	iso-$C_6H_{13}PCl_2$	0.5	75
			5.6×10^{21}	Hexenyltetra-chlorodiphosphine[b]	$C_6H_{12}(PCl_2)_2$	0.2	2
n-Hexane	$C_6H_{13}PCl_2$	1.0:1.0	5.0×10^{21}	Dihexylchlorophosphine[b]	$(C_6H_{13})_2PCl$	0.2	2
n-Heptane	PCl_3	1.0:1.2	4.0×10^{20}	Isoheptyldichloro-phosphines	iso-$C_7H_{15}PCl_2$	0.6	90
n-Octane	PCl_3	1.0:1.2	4.0×10^{20}	Isooctyldichloro-phosphines	iso-$C_8H_{17}PCl_2$[b]	0.6	90

[a] According to Babkina and Vereshchinskii.[137]

[b] Compound had not been previously described.

104

decomposition of tetrachlorodiphosphine (P_2Cl_4); it is a polymeric compound of variable empirical composition RH_2ClP_n, where n is 20–30.

It was suggested that (in an overall sense) the reaction involves free radicals as well as excited molecules of starting components

$$RH \rightsquigarrow R + H \tag{5}$$

$$PCl_3 \rightsquigarrow PCl_2 + Cl \tag{74}$$

$$H + PCl_3 \longrightarrow HCl + PCl_2 \tag{75}$$

$$Cl + RH \rightarrow HCl + R \tag{76}$$

$$R + PCl_2 \rightarrow RPCl_2 \tag{77}$$

$$R + R \rightarrow R_2 \tag{78}$$

$$R + Cl \rightarrow RCl \tag{79}$$

$$PCl_2 + PCl_2 \rightarrow P_2Cl_4 \tag{80}$$

and also

$$RH^* + PCl_3 \rightarrow HCl + RPCl_2 \tag{81}$$

$$PCl_3^* + RH \rightarrow HCl + RPCl_2 \tag{82}$$

The reaction mixtures consist generally of $RPCl_2$, R_2PCl, and $R'(PCl_2)_2$ in ratios which are a function of the dose absorbed. With increasing dose the amounts of R_2PCl and $R'(PCl_2)_2$ likewise increase. This result suggests that they are largely the products of alkyl chlorophosphine transformations

$$RPCl_2 \rightsquigarrow \begin{cases} \cdot R'PCl_2 + H & (83) \\ RPCl + Cl & (84) \end{cases}$$

$$\cdot R'PCl_2 + PCl_2 \rightarrow Cl_2PR'PCl_2 \tag{85}$$

$$RPCl + R \rightarrow R_2PCl \tag{86}$$

Large-scale radiation syntheses of higher dichlorophosphines, various dialkyl chlorophosphines, and other related products are of interest because the customary chemical methods are complicated and inapplicable beyond the laboratory scale.

In these syntheses the chief phosphorus-containing compound used was PCl_3. Another possibility is to use (elementary) yellow phosphorus and alkyl chlorides as the reaction components. Perner and Henglein[132] report that γ-irradiation of the yellow phosphorus solutions in CCl_4 leads to several products,[132] the yields of which depend substantially on the temperature. Table X summarizes their results. Synthesis of trichloromethylphosphine dichloride is a most interesting case of compounds with a C—P bond. Formation of red phosphorus and formation of low-molecular-weight

TABLE X

Yields from γ-Irradiation of Solutions of Yellow Phosphorus in CCl_4[a]

Temperature ($^\circ$C)	G Values						
	Red P	PCl_3	CCl_3PCl_2	C_2Cl_6	Polymer $(CCl_2)_n$	Consumption CCl_4	P
25	22	1.4	4.5	3.1	1	10	28
130	0.5	9.5	41	3.1	12	100	51

[a] According to Perner and Henglein.[132]

products are competitive processes which depend differently upon temperature. At room temperature red phosphorus is the predominant product; above 100°C PCl_3 and CCl_3PCl_2 are obtained almost exclusively. The reaction mechanism involves addition of free radicals, R, generated in the radiolysis of CCl_4, (i.e., CCl_3 or Cl), to dissolved phosphorus, P_4

$$R + P_4 \rightarrow RP_4 \tag{87}$$

Further reactions of this adduct lead either to a polymeric compound of red phosphorus[132, 134]

$$RP_4 + RP_4 \rightarrow R_2P_8 \tag{88}$$

$$RP_4 + P_4 \rightarrow RP_8 \tag{89}$$

or to low-molecular-weight products

$$RP_4 + CCl_4 \rightarrow RP_4Cl + CCl_3 \tag{90}$$

The activation energy of this propagation is 8.2 kcal/mole.

Treatment of radiation-activated red phosphorus (i.e., polymer containing terminal CCl_3 groups) by chlorine, bromine, or other reactant that destroys P—P bonds might provide interesting synthetic routes. As an example, dissolution in bromine gives trichloromethylphosphine dibromide, not described earlier.

15.2. Addition Products of Phosphorus Trichloride and Unsaturated Hydrocarbons

Irradiation of liquid olefins dissolved in phosphorus trichloride causes addition

$$RCH{=}CHR' + PCl_3 \xrightarrow{} \begin{array}{c} RCH{-}CHR' \\ | \quad\;\; | \\ Cl_2P \;\;\; Cl \end{array} \tag{91}$$

by a radical-chain process. In the case of olefins not polymerizable by the free-radical mechanism (cyclic and normal olefins), 1:1 adducts, as such

α-chloroalkylphosphine dichlorides, are the main products; otherwise telomerization products form;[133, 135] see Table XI.

In order to establish optimum conditions for radiation chemical synthesis of the 1:1 adduct, Babkina and Vereshchinskii[138] studied accumulation of chlorocyclohexylphosphorus dichloride (CCPDC) by varying three parameters: temperature in the range 33–140°C, dose rate in the range 1.55–55 rads/sec, and ratio of initial components in the reaction mixture.

A linear relation between the CCPDC yield and the dose absorbed was observed only at low conversion.

At larger doses and thus larger conversions, the reaction product, CCPDC, accumulates and the ratio of components begins to differ noticeably from the optimal mixture. From the temperature dependence of the radiation chemical yield (Fig. 4), the effective activation energy was found to be 5.9 kcal/mole. The rate of CCPDC accumulation is proportional to the square root of the dose rate with each component converted according to a first-order law. The addition of PCl_3 to cyclohexene may be described by the radical-chain scheme

$$PCl_3 \xrightarrow{\hspace{0.3cm}\sim\hspace{-0.15cm}\sim\hspace{-0.15cm}\sim\hspace{0.3cm}} PCl_2 + Cl \tag{74}$$

$$C_6H_{10} \xrightarrow{\hspace{0.3cm}\sim\hspace{-0.15cm}\sim\hspace{-0.15cm}\sim\hspace{0.3cm}} C_6H_9 + H \tag{92}$$

$$C_6H_{10} + PCl_2 \longrightarrow C_6H_{10}PCl_2 \tag{93}$$

$$C_6H_{10}PCl_2 + PCl_3 \longrightarrow ClC_6H_{10}PCl_2 + PCl_2 \tag{94}$$

The chain termination reactions are

$$PCl_2 + PCl_2 \rightarrow P_2Cl_4 \tag{80}$$

$$C_6H_{10}PCl_2 + PCl_2 \rightarrow PCl_2C_6H_{10}PCl_2 \tag{95}$$

$$C_6H_{10}PCl_2 + C_6H_{10}PCl_2 \rightarrow (C_6H_{10}PCl_2)_2 \tag{96}$$

At larger absorbed dose, formation of alkenyl and alkylphosphine dichlorides is observed, probably as a result of disproportionation and dehydrochlorination reactions.

Somewhat similar kinetic results were obtained by Shostenko et al.[140, 141] in their studies of radiation chemical addition of PCl_3 to linear monoolefins in the liquid or gaseous phases. Formation of two isomers of the main product and the higher reactivity of olefins upon displacement of the double bond from a terminal to an internal carbon are explained by a bridged dichlorophosphine alkyl radical generated in the initiation step.

G values of chloroalkyl dichlorophosphine production as high as 10^4 make the γ-induced reactions of PCl_3 with unsaturated compounds promising as large-scale processes.[139, 140]

TABLE XI

Yields and 100-eV Yields of PCl_3 Adducts to Unsaturated Hydrocarbons Resulting from ^{60}Co γ-Irradiation at Room Temperature

Initial hydrocarbon	Hydrocarbon/PCl_3 volume ratio	Dose, eV/ml	Synthesized compound	Yield, mole/liter	G[a]	Reference[b]
Cyclopentene	1:2[c]	7×10^{18}	$(cyclo\text{-}C_5H_8Cl)PCl_2$	0.6		Babkina . . .[138]
	1:1				173	Renz . . .[135]
Cyclohexene	1:2[c]	8.2×10^{18}	$(cyclo\text{-}C_6H_{10}Cl)PCl_2$	0.6		Babkina . . .[138]
	1:1				240	Renz . . .[135]
	1:2[c]	1.2×10^{20}	$cyclo\text{-}C_6H_9PCl_2$	0.8		Babkina . . .[138]
		1.0×10^{20}	$(cyclo\text{-}C_6H_9ClPCl_2)PCl(cyclo\text{-}C_6H_9)$	0.07		Babkina . . .[138]
Cycloheptene	1:1		$(cyclo\text{-}C_7H_{12}Cl)PCl_2$		103	Renz . . .[135]
Isobutene	1:1		$(CH_3)_2CClCH_2PCl_2$		430	Renz . . .[135]
			$(CH_3)_2CPCl_2CH_2Cl$		270	Renz . . .[135]
2-Methylbutene-1	1:1		$C_2H_5(CH_3)CClCH_2PCl_2$		245	Renz . . .[135]
2-Methylbutene-2	1:1		$(CH_3)_2CClCH(CH_3)FCl_2$		350	Renz . . .[135]
2-Ethylbutene-2	1:1		$(C_2H_5)_2CClCH_2PCl_2$		195	Renz . . .[135]
n-Pentene-1	1:1		$CH_3(CH_2)_2CHClCH_2PCl_2$		278	Renz . . .[135]
n-Hexene-1	1:1		$CH_3(CH_2)_3CHClCH_2PCl_2$		192	Renz . . .[135]
n-Heptene-1			$CH_3(CH_2)_4CHClCH_2PCl_2$			Shostenko . . .[141]
n-Heptene-2			$CH_3(CH_2)_3CHClCHPCl_2CH_3$			Shostenko . . .[141]
n-Heptene-3			$CH_3(CH_2)_2CHClCHPCl_2CH_2CH_3$			Shostenko . . .[141]
n-Octene-2	1:1		$CH_3(CH_2)_4CHClCHPCl_2CH_3$		252	Renz . . .[135]
n-Decene-1	1:1		$CH_3(CH_2)_7CHClCH_2PCl_2$		89	Renz . . .[135]

[a] Dose rate 7×10^3 roentgens/hr.

[b] Only the first author is cited. A series of dots indicates multiauthorship.

[c] Mole ratio.

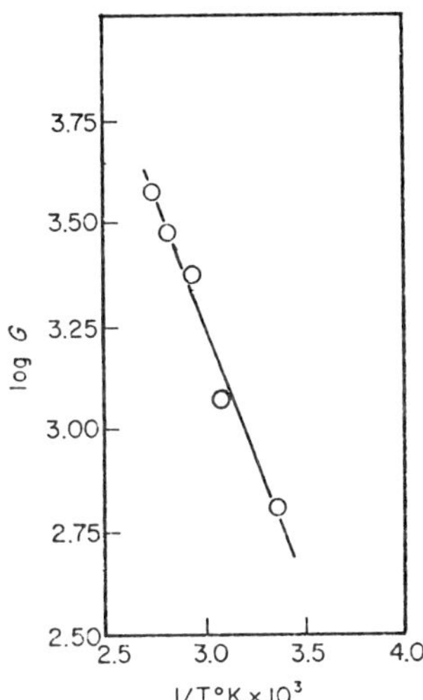

Fig. 4. Arrhenius plot for $G(ClC_6H_{10}PCl_2)$.[138] $E_a =$ 5.9 kcal/mole.

15.3. Phosphonium Salts from Aryl Halides

Although synthesis of phosphonium and other quaternary onium salts from alkyl halides is easily achieved by heating, their syntheses from aryl halides is rather complicated. γ-Irradiation of the appropriate mixtures containing halogenated *aromatic* hydrocarbons appears to be very effective for synthesis of several quaternary phosphonium salts; see Table XII. The 100-eV yields of arsonium salts are much lower, and a stibonium salt is formed in negligible quantities.

Specific features of the radiation chemical synthesis are: (1) strong dependence of the onium salt G value on the mole fraction of aryl halide in the mixture accompanied by shift of the maximum to the region of high aryl halide content and (2) substantial increase in triphenylphosphine conversion upon addition of various solvents to the reaction mixture. Curves of this conversion versus amount of the solvent added indicate a tendency toward saturation.

A mechanism suggested for dilute solutions involves participation of positive ions and electrons of the solvent **S**

$$\mathbf{S} \rightsquigarrow \mathbf{S}^+ + e^- \tag{97}$$

$$\mathbf{S}^+ + (C_6H_5)_3P \rightarrow \mathbf{S} + (C_6H_5)_3P^+ \tag{98}$$

$$e^- + C_6H_5Br \rightarrow C_6H_5 + Br^- \tag{99}$$

$$(C_6H_5)_3P^+ + C_6H_5 \rightarrow (C_6H_5)_4P^+ \tag{100}$$

TABLE XII

^{60}Co-γ-Induced Synthesis of Quaternary Onium Salts at Dose Rate of 6×10^3 to 8×10^5 roentgens/hr[a]

Reactants 1	Reactants 2	Mole % reactant 1	Onium salt	Dose, units of 10^7 rads	G	% Conversion, reactant 2	M.p., °C	Crystal color
C_6H_5Cl	$(C_6H_5)_3P$	76	$[P(C_6H_5)_4]Cl$	25	1.4	15	274–278	White
C_6H_5Br	$(C_6H_5)_3P$	73.5	$[P(C_6H_5)_4]Br$	25	3.4	43	286–294	White
C_6H_5I	$(C_6H_5)_3P$	91.7	$[P(C_6H_5)_4]I$	5.6	6.5	78	385–343	Transparent needles
C_6H_5F	$(C_6H_5)_3P$	90	$[P(C_6H_5)_4]F$	5.6	0.2	2.1	277–282	White needles
$1,3\text{-}C_6H_4Cl_2$	$(C_6H_5)_3P$	90	$[(C_6H_5)_3PC_6H_4Cl]Cl$	10	2.0	80	207–208	White
$1,4\text{-}C_6H_4Cl_2$	$(C_6H_5)_3P$	73.5	$[(C_6H_5)_3PC_6H_4Cl]Cl$	20	3.9	90	157–158	White
$1,4\text{-}C_6H_4Br_2$	$(C_6H_5)_3P$	79.7	$[(C_6H_5)_3PC_6H_4Br]Br$	2	5.1	25	134–150	White
$1,4\text{-}ClC_6H_4I$	$(C_6H_5)_3P$	74.8	$[(C_6H_6)_3PC_6H_4Cl]I$	3.5	4.4	55	219–222	Yellow
$p\text{-}ClC_6H_4CH_3$	$(C_6H_5)_3P$	92	$[(C_6H_5)_3PC_7H_7]Cl$	6	2.0	21	215–223	White
$o\text{-}ClC_6H_4CH_3$	$(C_6H_5)_3P$	85	$[(C_6H_5)_3PC_7H_7]Cl$	1.6	1.4	3	260–262	White
$o\text{-}BrC_6H_4CH_3$	$(C_6H_5)_3P$	85	$[(C_6H_5)_3PC_7H_7]Br$	1.6	2.4	7	228–237	White
$p\text{-}BrC_6H_4CH_3$	$(C_6H_5)_3P$	64	$[(C_6H_5)_3PC_7H_7]Br$	4	2.1	41	228–237	White
1-Chloro-naphthalene	$(C_6H_5)_3P$	82	$[(C_6H_5)_3PC_{10}H_7]Cl$	14	0.3	4	288–291	White
1-Bromo-naphthalene	$(C_6H_5)_3P$	82	$[(C_6H_5)_3PC_{10}H_7]Br$	14	1.0	23	243–248	White
2-Bromo-thiophene	$(C_6H_5)_3P$	80	$[(C_6H_5)_3PC_4H_3S]Br$	1.2	6.2	13	250–253	Yellow
C_6H_5Br	$(C_4H_9)_3P$	78	$[(C_6H_5)P(C_4H_9)_3]Br$	2.3	5.8	13	262–264	White
C_6H_5I	$(C_4H_9)_3P$	78	$[(C_6H_5)P(C_4H_9)_3]I$	4	2.1	13	151–152	White, Shiny
C_6H_5Cl	$(C_6H_5)_3As$	92	$[(C_6H_5)_4As]Cl$	19	0.06	2.5	256–260	White
C_6H_5Br	$(C_6H_5)_3As$	92	$[(C_6H_5)_4As]Br$	7.5	0.25	5.5	272–281	Long, colorless needles
C_6H_5I	$(C_6H_5)_3As$	88	$[(C_6H_5)_4As]I$		0.08		228–230	Dark–red
C_6H_5Br	$(C_6H_5)_3Sb$	90	$[(C_6H_5)_4Sb]Br$	25	~0.005		212–218	

a According to Drawe and Caspari [142, 143]

15.4. Phenylphosphonates

Trialkyl phosphites react with alkyl halides to yield dialkyl esters of alkylphosphonic acid (Arbuzov's rearrangement)

$$P(OR')_3 + RHal \rightarrow RP(O)(OR')_2 + R'Hal \tag{101}$$

Attempt to use a similar thermal reaction with aryl halides proves unsuccessful. On the other hand, γ irradiation gives a high yield of various phenylphosphonates; see Table XIII. The data cited show that radiation yields of phenylphosphonates are greatest when iodobenzene is one of the reaction components. In this case the reaction is also characterized by a decrease in G proportional to the square root of the dose rate as well as by a negative effect of temperature on G. The process was assumed to proceed via the chain mechanism

$$C_6H_5 + P(OR)_3 \rightarrow C_6H_5P(OR)_3 \tag{102}$$

$$C_6H_5P(OR)_3 + C_6H_5I \rightarrow [C_6H_5P(OR)_3]^+I^- + C_6H_5 \tag{103}$$

The latter transfer of iodine to the free tetravalent phosphorus radical is succeeded by decomposition of the quasi-phosphonium salt to the final product

$$[C_6H_5P(OR)_3]^+I^- \rightarrow C_6H_5P(O)(OR)_2 + RI \tag{104}$$

Second-order termination can involve interactions of phenyl radicals between themselves or with the intermediate phosphorus-containing radicals. The rate of the process is probably determined by dependence of the reaction on the strength of the C—Hal bond in the aryl halide, halogen-electron affinity, and solvation of generated ions.

16. SYNTHESIS OF ORGANOTIN COMPOUNDS

Direct radiation chemical synthesis of organometallic compounds from the free elements has not, as yet, been actually studied. Investigations are confined to the syntheses of phosphorus organic compounds from white phosphorus (Section 15.1), and sulfur-containing organic compounds from free sulfur (Section 14). However, in both cases the presence of allotropic modifications affects the course of the processes and the composition of the products.

Of the radiation syntheses of organometallic compounds reported in the literature only those of the organotin compounds have been extensively studied.[145–154] Such synthesis of compounds of the R_2SnBr_2 type is based

TABLE XIII

^{60}Co-γ-Induced Synthesis of Phenylphosphonates at Dose Rate of 2.6 × 10^5 rads/hr[a]

| Reactants | | Mole % | | | Dose, units of 10^7 | |
1	2	reactant 1	Product phosphonate	G	roentgens	B.p., °C
$(CH_3O)_3P$	C_6H_5F	50	$C_6H_5P(O)(OCH_3)_2$	0.25	9.5	247
$(CH_3O)_3P$	C_6H_5Cl	50	$C_6H_5P(O)(OCH_3)_2$	2.6	5.8	247
$(CH_3O)_3P$	C_6H_5Br	50	$C_6H_5P(O)(OCH_3)_2$	4.9	5.4	247
$(CH_3O)_3P$	C_6H_5I	50	$C_6H_5P(O)(OCH_3)_2$	250	0.1	247
$(C_2H_5O)_3P$	C_6H_5Br	50	$C_6H_5P(O)(OC_2H_5)_2$	4.0	8.0	267
$(C_2H_5O)_3P$	C_6H_5I	50	$C_6H_5P(O)(OC_2H_5)_2$	28	0.6	267
$(iso\text{-}C_3H_7O)_3P$	C_6H_5I	50	$C_6H_5P(O)(O\text{-}iso\text{-}C_3H_7)_2$	11	0.7	
$(CH_3O)_3P$	$p\text{-}ClC_6H_4I$	70	$(CH_3O)_2P(O)C_6H_4Cl$	68	0.6	264–268
$(CH_3O)_3P$	$p\text{-}BrC_6H_4CH_3$	70	$(CH_3O)_2P(O)C_7H_7$	2.5	1	262–265
$(CH_3O)_3P$	$m\text{-}C_6H_4Cl_2$	70	$(CH_3O)_2P(O)C_6H_4Cl$	1.9	1.5	265–268
$(CH_3O)_3P$	2-Bromothiophene	50	$(CH_3O)_2P(O)C_4H_3S$	9.7	1	248–255

[a] According to Caspari, Drawe, and Henglein.[144]

on the γ-induced reaction of alkyl bromide (RBr) with metallic tin. Synthesis of dibutyltin dibromide (DBTDB) has been studied in detail.

Miretskii, Vereshchinskii et al.[150] have described the synthesis of DBTDB in a semipilot installation. n-Butyl bromide of 99 percent purity and tin powder containing no less than 99.5 percent tin, with the size of particles predominantly in the 10–40 μ range, were used. To expand the overall tin surface, the powder was pickled with dilute hydrochloric acid, which removed the oxide film. In the course of the reaction, the surface of metallic tin expands proportionally to duration of irradiation and becomes >20 $m^2\,g^{-1}$. Most of the experiments were conducted at 85–90°C, at normal pressure, and with vigorous mechanical stirring. The average absorbed dose rate was about 100 rads/sec.

The interaction of n-butyl bromide with tin in the presence of γ-radiation shows clear evidence of selectivity

$$2C_4H_9Br + Sn \xrightarrow{\quad} (C_4H_9)_2SnBr_2 \tag{105}$$

The reaction results almost exclusively in DBTDB along with a small amount of TBTMB and tin dibromide as by-products. An induction period is characteristic of the DBTDB radiation chemical synthesis. However, various techniques were tried and optimal conditions were found for its practical elimination. One such technique is the treatment of tin powder as already described. Also, addition of such activators as n-butanol or water to the reaction mixture was found to be rather effective. A parameter study of the reaction, made practically free of induction period, revealed its basic macro-kinetic features.[151] Temperature affected the process significantly. From the kinetic curves of DBTDB production, in the temperature range 20–95°C, the effective activation energy was found to be 12 kcal/mole. The rate of DBTDB formation is proportional to the square root of dose rate in the interval 2.5 175 rad/sec.

Clearly, DBTDB appears to be formed by a chain mechanism. This interpretation is confirmed by (1) a G value as high as 10^3–10^4 at the high degree of conversion; (2) the characteristic behavior of the reaction system upon addition of activators; and (3) the linear relation between reaction rate and square root of the dose rate. On cessation of irradiation the reaction stops immediately but resumes again at the same rate after reintroduction of the radiation source. Another argument in support of the chain mechanism resides in the effect of inhibitors on the rate of DBTDB formation. The presence of oxygen in the reaction system or introduction of acceptors such as benzoquinone, hydroquinone, or m-dinitrobenzene slows down the reaction noticeably. The present author has suggested the following probable scheme of the chain process which involves occurrence of *all* stages of the chain propagation on the surface of the tin

Initiation

$$RBr \leadsto R + Br \tag{106}$$

$$RBr + e \longrightarrow R + Br^- \tag{107}$$

Propagation

$$R + Sn \rightarrow RSn \tag{108}$$

$$RSn + RBr \rightarrow R_2SnBr \tag{109}$$

$$R_2SnBr + RBr \rightarrow R_2SnBr_2 + R \tag{110}$$

Termination

$$R + Br \rightarrow RBr \tag{111}$$

$$R + R \rightarrow R_2 \tag{112}$$

$$R_2SnBr + Br \rightarrow R_2SnBr_2 \tag{113}$$

$$Br + Br \rightarrow Br_2 \tag{114}$$

Tributyltin monobromide and $SnBr_2$ may be formed by the reactions

$$R_2SnBr + R \rightarrow R_3SnBr \tag{115}$$

$$Br + Sn \rightarrow SnBr \tag{116}$$

$$SnBr + RBr \rightarrow SnBr_2 + R \tag{117}$$

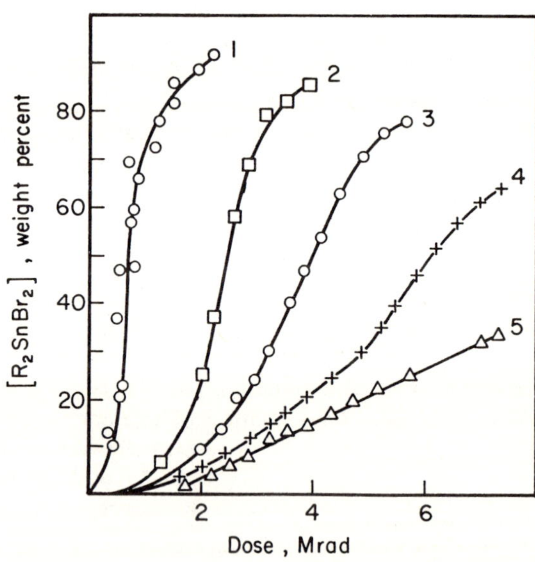

Fig. 5. Effect of the nature of *n*-alkyl radical on the rate of dialkyltin dibromide formation.[152] Curves: 1, butyl; 2, hexyl; 3, heptyl; 4, octyl; 5, nonyl. Synthesis temperature was 115°C in all cases except that of DBTDB, which was 95°C.

Study of the radiation synthesis of DBTDB homologs (from C_3 to C_9) has shown that macrokinetic dependencies of all the reactions are rather similar; see Fig. 5. Dialkyltin dibromides are formed with sufficiently high G values (from 500 to 1000) and adequate chemical yields (80–85 percent).[152]

Radiation chemical synthesis of dialkyltin dibromides displays several desirable features: one-step directed process, high yield of the desired product, fire and explosion safety, facility of control, and so on. Technical and economic estimation of the relative desirability of stabilized production of a product such as dibutyltindilaurenate at a rate of 200 tons/year employing, on the one hand, the Grignard techniques and, on the other, a combined method involving radiation chemical synthesis of the semiproduct showed that in the latter case the cost is lower by 2850 roubles per ton[152] (i.e., $3140 per ton at the U.S.S.R. official exchange).

17. ISOMERIZATION

Radiation isomerizations should be subdivided into *direct* and *sensitized* ones.

In case of sensitized isomerization, the ionizing radiation energy absorbed by a sensitizer in the solvent is transferred by some mechanism or other to a dissolved substance causing its isomerization. A detailed study was conducted of radiation-sensitized isomerization of the conjugate unsaturated steroid ketones (**IV**) and (**V**) in solution

(118)

Table XIV shows that the unsaturated ketones (**IV**) and (**V**) in the presence of sensitizers of aromatic nature undergo rearrangement, while saturated

TABLE XIV

G Values of γ-Induced Isomerization of Steroid Ketones in Benzene Solutions (Dose Rate 0.6 Mrad hr^{-1}, Temperature 30°C)[a]

Ketone, *M*	Additive	*M*	Dose, Mrad	Content, % Initial, ketone	Content, % Isomerization product	*G*
Ketone IV					Ketone V	
0.015			1.2	90	10	1.7
0.015			2.5	71	17	1.5
0.023			1.2	90	9	1.9
0.023			1.8	87	12	1.8
0.023			2.4	83	15	1.6
0.093			2.4	87	5	2.3
0.093			3.6	89	7	2.1
0.088			7.2	55	10	2.1
0.39			3.6	99	3	2.7
0.39			11.2	90	6	2.5
0.39			18.4	87	11	2.7
0.015	*p*-Terphenyl	0.017	1.2		5.5	0.9
0.015	*p*-Terphenyl	0.017	2.4	87	12	1.1
0.088	*p*-Terphenyl	0.02	7.2	84	13	1.1
0.088	*cis*-Butene	0.33	7.2	79	10	1.4
Ketone V					Ketone VI	
0.088			7.2	90	6	1.5
					Ketone VII	
0.022			7.2	61	3	1.3
					Ketone VI	
0.088	*p*-Terphenyl	0.02	7.2	92	5	1.0
Ketone VIII						
0.088				87		0

[a] According to Hoigne, Schaffner, and Wenger.[155]

ketone (VIII)

is not affected by γ-radiation. The 100-eV yield of the rearrangement, ∼2, changes little with variation in concentration. Decomposition is negligible.

This example illustrates the possibility of application of γ-irradiation to preparative sensitized isomerization of complex organic compounds.

Radiation-induced cis-trans isomerization, repeatedly studied[156–158] in order to determine mechanism of radiative transformations, is similar to the ultraviolet-induced isomerization and may be hardly of interest in preparative radiation chemistry.

Another line of research is the direct radiation isomerization effected by γ irradiation. On such irradiation terpene, α-pinene, isomerizes into dipentene ($G = 0.9$) and a mixture of hydrocarbons including ocimene ($G = 2.8$) and alloocimene.[159, 160] Similarly, β-pinenes isomerize into a mixture of dipentene ($G = 1.2$) and miricene ($G = 2.0$).[161] Both isomerizations lead to a trimer ($G = 5.2$) in the first case, and a polymer, with $G(-$ monomer$) \sim 850$, in the second case.

Compounds containing phenolic and carbonyl groups are formed by rearrangement of aromatic esters.[162]

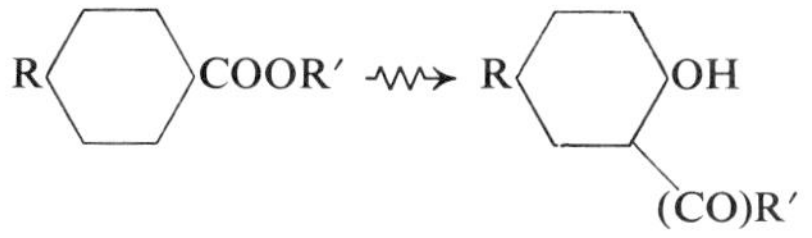

$$R\text{—C}_6\text{H}_4\text{—COOR}' \rightsquigarrow R\text{—C}_6\text{H}_3(\text{OH})(\text{CO})R' \tag{119}$$

Solid o-nitrobenzaldehyde γ-irradiated *in vacuo* isomerizes according to[163]

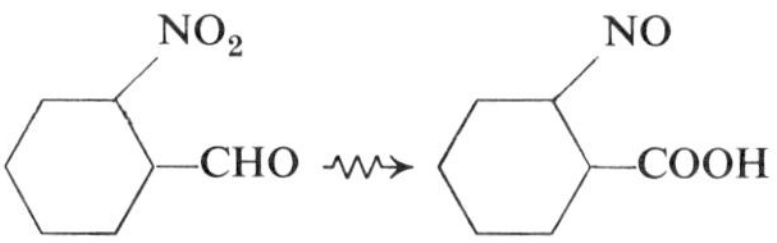

$$o\text{-}O_2N\text{-}C_6H_4\text{-CHO} \rightsquigarrow o\text{-}ON\text{-}C_6H_4\text{-COOH} \tag{120}$$

o-Nitrobenzoic acid is formed with $G = 5.8$, while G of gaseous radiolysis products is 0.2. Isomerization in the presence of γ-radiation is similar to the photochemical one; both cases generate excited o-nitrobenzaldehyde.

Rearrangement of hirsutic acid p-bromophenacyl ester illustrates exceptionally effective solid-state radiative isomerization ($G \geq 57$). The intramolecular rearrangement results in rupture of the epoxide ring and carbonyl group formation. Irradiated crystals contain molecules of both kinds (**IX** and **X**).

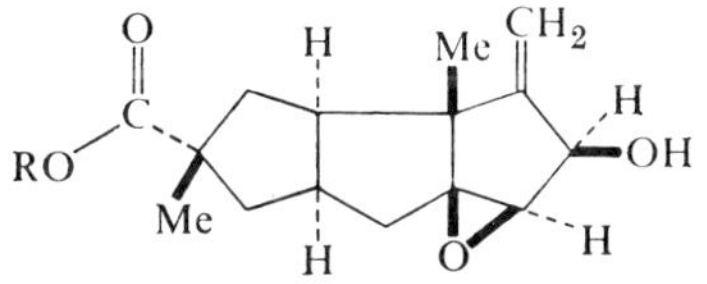

(IX) 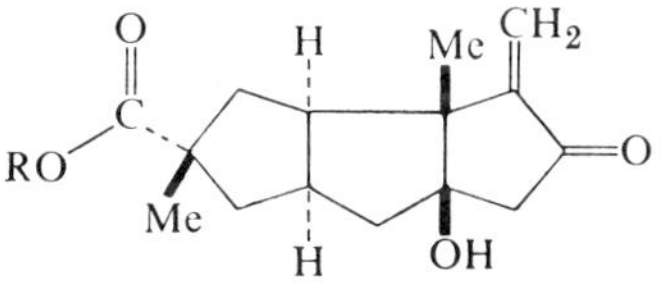   (X)

The high value of G suggests either a chain mechanism or highly effective energy transfer.[164]

Discussion of radiation isomerization processes occurring in the gas phase at elevated temperature or in the presence of catalysts is beyond the scope of this chapter.

18. CONCLUSIONS

Further research in radiation chemical synthesis will probably develop in several directions:

(1) Development of industrial radiation chemical syntheses. Certain advantages, discussed in Section 2, make it possible to consider such processes in the light of their competitiveness.

(2) Syntheses of various organic compounds (including intermediate products) that are hardly available or entirely unavailable via conventional techniques.

(3) Exploratory studies. Extension of the number of systems under study and application of the radiation chemical synthetic methods to new classes of organic compounds are no less a challenge than is radiation synthesis at low and, probably, ultralow temperatures. In addition, synthesis in matrices and clathrates may prove of interest.

Many aspects of radiation synthesis are described in the review articles of the last decade[165–175] in the table of references.

References

1. V. P. Zimakov, E. V. Volkova, and L. A. Krasnousov, in *Trudy Vsesoyuznogo Soveshchaniya po Radiatsionnoy Khimii*, Izdatel'stvo AN SSSR, Moscow,1958, p. 224.
2. P. R. Hills, R. Roberts, and M. W. Spindler, in *Large Radiation Sources in Industry*, Vol. 2, I.A.E.A., Vienna, 1959, p. 61.
3. L. A. Krasnousov, V. P. Zimakov, and E. V. Volkova, in *Trudy II Vsesoyuznogo Soveshchaniya po Radiatsionnoy Khimii*, Izdatel'stvo AN SSSR, Moscow,1962, p. 426.
4. P. E. Faw, J. M. McCabe, and H. S. Isbin, *Nucl. Appl.*, **1**, 548 (1965).
5. V. A. Krentsel, *Khlorirovaniye Parafinov*, Izdatel'stvo Nauka, Moscow, 1964.
6. V. I. Zetkin, G. M. Panchenkov, E. T. Zakharov, I. M. Kolesnikov, and R. V. Dzhagatspanyan, *Khim. Prom.*, **41**, 733 (1965).
7. R. V. Dzhagatspanyan, V. I. Kosorotov, and M. T. Filippov, *Khim. Vysok. Energii*, **1**, 453 (1967); R. V. Dzhagatspanyan and V. I. Kosorotov, *Khim. Vysok. Energii*, **2**, 418 (1968); R. V. Dzhagatspanyan, V. I. Kosorotov, and L. I. Kheifets, *Khim. Vysok. Energii*, **3**, 173 (1969).
8. A. Almasy and V. Rona, in *Radiation Chemistry, Proceedings of the 1962 Tihany Symposium*, Akademiai Kiadó, Budapest, 1964, p. 17.
9. A. Almasy and V. Rona, in *Industrial Uses of Large Radiation Sources*, Vol. 1, I.A.E.A., Vienna, 1963, p. 101.
10. A. Almasy, V. Rona, G. Koranyi, M. Kovacs, and B. Raskai, Hungarian Patent 150,904.

11. D. D. Lebedev and I. V. Vereshchinskii, in *Proceedings of the Second Tihany Symposium on Radiation Chemistry*, Akademiai Kiadó, Budapest, 1967, p. 365.
12. D. D. Lebedev and I. V. Vereschchinskii, *Zh. Org. Khim.*, **3**, 2247 (1967).
13. D. D. Lebedev and I. V. Vereshchinskii, *Khim. Vysok. Energii*, **3**, 267 (1969).
14. I. V. Vereshchinsky, D. D. Lebedev, V. Y. Miretsky, and A. T. Podkhalyusin, *Advan. Chem. Ser.*, **82**, 475 (1968).
15. F. Z. Raychuk, D. D. Lebedev, I. V. Vereshchinskii, and V. S. Blinova, *Vsesoyuznaya Nauchno-Tekhnicheskaya Konferentsiya XX Let Proizvodstva i Primeneniya Izotopov i Istochnikov Yadernykh Izluchenii v Narodnom Khozyaistve SSSR, Sektsiya "Radiatsionnaya Khimya"* (Tezisy Dodladov), Atomizdat Publishers, Moscow, 1968, p. 34.
16. E. Plander, V. Formácek, and J. Zahálka, Czechoslovakian Patent 119,713 (1965).
17. F. B. Galimba, W. E. Wilson, and A. H. Samuel, *J. Phys. Chem.*, **68**, 647 (1961).
18. A. Henglein and H. Url, *Z. Physik. Chem. (Frankfurt)*, **9**, 285 (1956).
19. A. Schneider and Chu Ju Chin, *I. & E.C. Process Design Develop.* **3**, 164 (1964).
20. A. Schneider and Chu Ju Chin, *Am. Inst. Chem. Eng. J.*, **10**, 930 (1964).
21. R. V. Dzhagatspanyan. See p. 30 in Ref. 15.
22. I. F. Sprygayev and S. V. Mamikonyan, *Atomn. Energiya*, **23**, 469 (1967); English translation: *Soviet Atomic Energy*, **23**, 1227 (1967).
23. J. F. Black and E. F. Baxter, *Proc. 2nd U.N. Intern. Conf.*, Vol. 29, Geneva, 1958, p. 162.
24. British Patent 810,574; French Patent 1,215,577; German Patent 1,139,116; J. F. Black, Australian Patent 225,893.
25. French Patent 1,320,018.
26. E. F. Baxter, Australian Patent, 239,634.
27. P. R. Hills and M. W. Spindler, *U.K. At. Energy Auth., Res. Group, Atomic Energy Res. Estab. Rept.*, No. AERE-R 5434, 18 pp. (1967); through *Chem. Abstr.*, **67**, 69400 (1967).
28. F. L. Dalton and J. A. Bartlett, in *Abstracts 3rd International Congress Radiation Research, Cortina Italy, June 1966*, p. 246.
29. M. A. Proscurnin, Yu. L. Khmel'nitskiy, E. V. Barelko, A. T. Slepneva, and I. I. Melekhonova, *Dokl. Akad. Nauk SSSR*, **112**, 886 (1957).
30. V. A. Antonovskii, I. A. Kuznetsov, and Yu. Ya. Mekhryushev, *Neftekhimiya* **4**, 863 (1964).
31. P. N. Komarov, *Zh. Fiz. Khim.*, **40**, 243 (1966).
32. P. C. C. Hu, *Hua Hsueh Hsueh Pao*, **32**, 237 (1966); through *Chem. Abstr.*, **66**, 15462 (1967).
33. I. Drimus, G. Ioanid, A. Dragut, P. Vasilesco, and V. Dumitresco, in *Proc. 2nd U.N. Intern. Conf.*, Vol. 29, Geneva, 1958, p. 152.
34. G. Ioanid, A. Dragut, I. Drimus, V. Dumitresco, and I. Stoian, in *Large Radiation Sources in Industry*, Vol. 2, I.A.E.A., Vienna, 1959, p. 139.
35. Yu. L. Khmelnitskiy, N. I. Melekhonova, and V. V. Nesterovskii, *Neftekhimiya*, **2**, 368 (1962).
36. J. W. Sutherland and J. W. T. Spinks, *Can. J. Chem.*, **37**, 79 (1959).
37. V. A. Poluektov, I. V. Dobrov, N. G. Ageev, S. A. Lyapina, and M. E. Yershov, USSR Patent, 176,286, Byulleten' Izobretenii No. 22 (1965).
38. V. A. Poluektov, I. V. Dobrov, and S. A. Lyapina, *Khim. Vysok. Energii*, **2**, 536 (1968).
39. I. V. Dobrov, V. A. Poluetktov, S. A. Lyapina, B. I. Konobeev, and M. A. Fandeev. See p. 32 in Ref. 15.
40. V. A. Poluektov, I. V. Dobrov, Yu. Ya. Mekhryushev, and S. A. Lyapina, USSR Patent, 196,772, Byulleten' Izobretenii No. 12 (1967).

41. V. A. Poluektov, I. V. Dobrov, Yu. Ya. Mekhryushev, S. A. Lyapina, G. N. Lisov, D. M. Torgovitskii, S. G. Stoenko, A. V. Chertorizhskii, F. Z. Raichuk, A. Kh. Breger, and V. A. Zorin. See p. 33 in Ref. 15.

42. V. Caglioti, M. Lenzi, and A. Mele, *Nature*, **201**, 610 (1964).

43. A. T. Podkhalyuzin and A. Ye. Trishkin, USSR Patent 197,560, Byulleten' Izobretenii No. 13 (1967).

44. S. Kinoshita and T. Sunada, Japanese Patent 22,463 (1965); through *Referativnii Zh. Khim.*, **12**, R470 P (1967).

45. S. Kinoshita and T. Sunada, *9th Symposium on Perfume, Terpene and Essential Oil of the Chemical Society of Japan*, 1965.

46. D. E. Harmer, J. S. Beale, C. T. Pumpelly, and B. W. Wilkinson, in *Industrial Uses of Large Radiation Sources*, Vol. 2, I.A.E.A., Vienna, 1963, p. 205.

47. R. C. Rumfeldt and D. A. Armstrong, *Can. J. Chem.*, **44**, 2219 (1966).

48. J. L. McFarling and J. F. Kircher, *Intern. J. Appl. Radiation Isotopes*, **18**, 345 (1967).

49. J. Zahálka and M. Hudlický, Czechoslovakian Patent 116341 (1965).

50. W. Cooper and W. H. Stafford, in *Proc. 2nd UN Intern. Confer.*, Vol. 29, 1958, p. 118.

51. L. M. Kogan, N. S. Rabovskaya, and S. I. Volfkovich, *Dokl. Akad. Nauk SSSR*, **157**, 131 (1964).

52. N. S. Rabovskaya and L. M. Kogan, *Dokl. Akad. Nauk SSSR*, **165**, 112 (1965).

53. L. M. Kogan, *Glubokoe Khlorirovanie Uglevodordov C_4 i C_5 i Nekotorye Svoystva Produktov Reaktsii*. Avtoreferat Dissertatsii na Soiskanie Uchenoi Stepeni Doktora Khimcheskikh Nauk, Moscow 1965.

54. N. S. Rabovskaya, L. M. Kogan, and A. N. Nikolaeva, *Vestn. Mosk. Gos. Universiteta, Seriya Khim*, No. 4, 57 (1965).

55. A. A. Beer, P. A. Zagorets, V. F. Inozemtsev, G. S. Povkh, and A. I. Popov, *Neftekhimiya*, **2**, 617 (1962).

56. F. W. Mellows and M. Burton, *J. Phys. Chem.*, **66**, 2164 (1962).

57. M. Takehisa, *Kogyo Kagaku Zasshi*, **69**, 419 (1966).

58. K. Hirota, J. Takezaki, and M. Hatada, *Ann. Rept. JARRP*, **4**, 235 (1962).

59. K. Hirota and M. Hatada, *Bull. Chem. Soc. Japan*, **34**, 1644 (1961).

60. K. Hirota and M. Hatada, *Ann. Rept. JARRP*, **4**, 215 (1962).

61. K. Hirota, M. Hatada, and N. Ohtsuka, *Ann. Rept. JARRP*, **4**, 219 (1962).

62. P. M. Yavorsky, N. J. Mazzocco, and E. Gorin, *U.S. Atomic Energy Comm.*, NYO-2978-44 (1965); through *Chem. Abstr.*, **67**, 77841 (1967).

63. K. Hirota and M. Hatada, *Bull. Chem. Soc. Japan*, **33**, 1682 (1960).

64. K. Hirota, H. Ochi, and M. Hatada, *Isotopes Radiation (Tokyo)*, **3**, No. 6 (1966).

65. K. Hirota and M. Hatada, *Bull. Chem. Soc. Japan*, **36**, 817 (1963).

66. M. Hatada, J. Takezaki, and K. Hirota, *Bull. Chem. Soc. Japan*, **37**, 166 (1964).

67. K. Hirota, J. Takezaki, and M. Hatada, *Ann. Rept. JARRP*, **4**, 223 (1962).

68. K. Hirota, A. Shingyouchi, and M. Hatada, *Ann. Rept. JARRP*, **4**, 227 (1962).

69. M. Okubo, *Nippon Kagaku Zasshi*, **87**, 1191 (1966).

70. M. Okubo, *Nippon Kagaku Zasshi*, **87**, 1196 (1966).

71. H. Muramatsu et al., *Nagoya Koguo Gijutsu Shkensho Hokoku*, **14**, 213 (1965); through *Chem. Abstr.*, **66**, 90120 (1967).

72. B. G. Dzantiyev and A. V. Shishkov, *Khim. Vysok. Energii*, **1**, 111 (1967).

73. B. Sugimoto, W. Ando, and S. Oae, *Bull. Chem. Soc. Japan*, **37**, 365 (1964).

74. A. Fontijn and J. W. T. Spinks, *Can. J. Chem.*, **35**, 1384 (1957).

75. F. W. Lampe, J. S. Snyderman, and W. H. Johnston, *J. Phys. Chem.*, **70**, 3934 (1966).

76. A. M. El-Abbady, *J. Chem. UAR*, **9**, 281 (1966).

77. A. V. Zimin, L. P. Sidorova, and M. N. Khasidovich, USSR Patent 134,690, Byulleten' Izobretenii No. 1 (1961).

78. A. V. Zimin, A. D. Verina, and A. V. Gubanova, USSR Patent 173,229, Byulleten' Izobretenii No. 15 (1965).
79. A. V. Zimin, L. P. Sidorova, A. D. Verina, A. V. Gubanova, and V. I. Savushkina, USSR Patent 138,360, Byulleten' Izobretenii No. 10 (1961). German (BRD) Patent 1,232,579 (1967); French Patent 1,415,110 (1965).
80. V. F. Evdokimov, I. Ya. Poddubnyi, M. A. Sokolova, I. A. Kuzin, I. B. Zaitsev, and I. Yu. Tsereteli, *Khim. Vysok. Energii*, **1**, 46 (1967).
81. R. V. Dzhagatspanyan, V. I. Zetkin, and M. P. Maksimov, USSR Patent, 118,560, Byulleten' Izobretenii No. 6 (1959).
82. A. Shoenberg, *Preparativnaya Organicheskaya Fotokhimiya*, Izdatinlit Publishers, Moscow, 1963.
83. A. V. Zimin, A. D. Verina, L. P. Sidorova, and A. V. Gubanova, *Dokl. Akad. Nauk SSSR*, **144**, 576 (1962).
84. A. M. El Abbady and L. C. Anderson, *J. Am. Chem. Soc.*, **80**, 1737 (1958).
85. P. Feng, in *Proc. 2nd U.N. Intern. Conf.*, Vol. 29, Geneva, 1958, p. 166.
86. P. Feng and L. Mamula, *J. Chem. Phys.*, **28**, 507 (1958).
87. P. Feng, in *Proc. 4th Conf. Radioisotopes, Tokyo, October 1961*, p. 445.
88. P. Feng, U.S. Patent 3,081,243 (1963).
89. P. Feng and W. Bahmet, in *Abstracts 3rd International Congress Radiation Research, Cortina, Italy, June 1966*, p. 324.
90. A. Rajbenbach, in *Abstracts 3rd International Congress Radiation Research, Cortina, Italy, June 1966*, p. 742.
91. W. G. Askew. Effects of radiation of mixtures containing fluorocarbons and the identification of the perfluoroheptane isomers. Univ. Microfilms, Order No. 66-11, 103, 122 pp.; through *Chem. Abstr.*, **66**, 70821 (1967).
92. L. P. Matveeva, V. D. Timofeev, and E. A. Borisov, *Khim. Vysok. Energii*, **3**, 130 (1969).
93. A. V. Zimin, S. V. Churmanteev, and A. D. Verina, *Trudy i Vsesoyuznogo Soveshchaniya po Radiatsionnoi Khimii*, Akad. Nauk SSSR Publishers, Moscow, 1958, p. 221.
94. F. Cacae, A. Guarino, and E. Possagno, *Cass. Chim. Ital.*, **89**, 1837 (1959).
95. N. Getoff, F. Gütlbauer, and G. O. Schenck, *Intern. J. Appl. Radiation Isotopes*, **17**, 341 (1966).
96. E. Gütlbauer and N. Getoff, *Österr. Chem.-Ztg.*, **67**, 373 (1966).
97. E. P. Kalyazin and V. I. Makrov, *Neftekhimiya*, **3**, 95 (1964).
98. C. D. Furrow, U.S. Patent 3,168,456 (1965).
99. B. C. McKusick, W. E. Mochel, and F. W. Stacey, *J. Am. Chem. Soc.*, **82**, 723 (1960).
100. D. Hummel and H. Barzynski, *Z. Elektrochemie*, **64**, 1015 (1960).
101. Ye. P. Kalyazin and N. M. Baranova, *Neftekhimiya*, **4**, 275 (1964).
102. N. Getoff and D. Seitner, *Z. Physik. Chem.*, (*Frankfurt*), **51**, 27 (1966).
103. F. D. Osterholtz, B. H. Hamling, T. J. Myron, F. J. Morse, and P. A. King, *151st Meeting ACS, March 1966*, J42.
104. A. T. Podkhalyuzin, A. Ye. Trishkin, and I. V. Vereshchinskii, *Khim. Vysok. Energii*, **3**, 178 (1969).
105. A. T. Podkhalyuzin, A. Ye. Trishkin, and I. V. Vereshchinskii, *Khim. Vysok. Energii*, **3**, 401 (1969).
106. B. G. Dzantiev, V. P. Trubnikov, and V. N. Popov, *Izv. Akad. Nauk Belorussk. SSR*, No. 3, 5 (1966).
107. J. Rokach and C. H. Krauch, *Naturwissenschaften*, **54**, 166 (1967).
108. A. V. Zimin, S. V. Churinanteev, and A. D. Verina, *Sbornik Rabot po Radiatsionnoi Khimii*, Akad. Nauk SSSR Publishers, Moscow, 1955, p. 249.

109. E. Schwarz, in *Atomstrahlung in Medizin und Technik*, Verlag Karl Thiemig, K.G., 1964, p. 284.
110. G. Smith and G. A. Swan, *J. Chem. Soc.*, 886 (1962).
111. G. Smith and G. A. Swan, *Chem. Ind.*, 1221 (1961).
112. G. A. Swan, P. S. Timmons, and D. Wright, in *Proc. Sec. U.N. Intern. Conf.*, Vol. 29, Geneva, 1958, p. 115.
113. G. A. Swan and P. S. Timmons, *J. Chem. Soc.*, 9 (1959).
114. S. L. Miller, *J. Am. Chem. Soc.*, **77**, 2351 (1955).
115. S. L. Miller, *Biochim. Biophys. Acta*, **23**, 480 (1957).
116. N. Getoff and G. O. Schenck, *Radiation Res.*, **31**, 486 (1967).
117. C. Ponnamperuma, R. M. Lemmon, R. Mariner, and M. Calvin, *Proc. Natl. Acad. Sci., U.S.*, **49**, 737 (1963).
118. J. Rokach, C. H. Krauch, and D. Elad, *Tetrahedron Letters*, 3253 (1966).
119. D. Elad, *Israel J. Chem.*, **2**, 233 (1964).
120. C. H. Krauch, J. Rokach, and D. Elad, *Tetrahedron Letters*, 5099 (1967).
121. A. Henglein, in *Large Radiation Sources in Industry*, Vol. 2, I.A.E.A., Vienna, 1959, p. 139.
122. A. Henglein, *Intern. J. Appl. Radiation Isotopes*, **8**, 149 (1960).
123. E. Müller and G. Schmied, *Chem. Ber.*, **94**, 1364 (1961).
124. E. J. Burrell, *J. Phys. Chem.*, **66**, 401 (1962).
125. P. B. Hills and R. A. Johnson, *Intern. J. Appl. Radiation Isotopes*, **12**, 80 (1961).
126. E. Koerner, V. Gustorf and M. -J. Jun, *Z. Naturforsch.*, **20b**, 521 (1965).
127. H. Barzynski and D. Hummel, *Z. Physik. Chem.*, **33**, 352 (1962).
128. W. Ando, K. Sugimoto, and S. Oae, *Bull. Chem. Soc. Japan*, **36**, 893 (1963).
129. I. V. Vereschinskii, *20th International Congress of Pure and Applied Chemistry, USSR, Moscow.* Abbreviated English Translations of Scientific Papers Presented in Russian, B45 (1965).
130. E. M. Nanobashvili, G. G. Chirakadze, and G. A. Mosashvili, *Khim. Vysok. Energii*, **2**, 403 (1968).
131. A. Henglein, *Intern. J. Appl. Radiation Isotopes*, **8**, 156 (1960).
132. D. Perner and A. Henglein, *Z. Naturforsch.*, **17b**, 703 (1962).
133. K. Wunder, H. Drawe, and A. Henglein, *Z. Naturforsch.*, **19b**, 999 (1964).
134. K. D. Asmus, A. Henglein, G. Meissner, and D. Perner, *Z. Naturforsch.*, **19b**, 549 (1964).
135. M. Renz, K. Wunder, and H. Drawe, *Z. Naturforsch.*, **22b**, 486 (1967).
136. E. I. Babkina and I. V. Vereshchinskii, *Zh. Obshch. Khim.*, **37**, 513 (1967).
137. E. I. Babkina and I. V. Vereshchinskii, *Zh. Obshch. Khim.*, **38**, 1772 (1968).
138. E. I. Babkina and I. V. Vereshchinskii, *Khim. Vysok. Energii*, **3**, 450 (1969).
139. E. I. Babkina and I. V. Vereshchinskii. See p. 25 in Ref. 15.
140. A. G. Shostenko, P. A. Zagorets, and A. M. Dodonov. See p. 27 in Ref. 15.
141. A. G. Shostenko, P. A. Zagorets, A. M. Dodonov, and A. A. Greiish, *Khim. Vysok. Energii*, **4**, 357 (1970).
142. H. Drawe and G. Caspari, *Angew. Chem.*, **78**, 331 (1966).
143. G. Caspari and H. Drawe, *Z. Naturforsch.*, **22b**, 574 (1967).
144. G. Caspari, H. Drawe, and A. Henglein, *Radiochemica Acta*, **8**, 102 (1967).
145. L. V. Abramova, N. I. Sheverdina, and K. A. Kocheshkov, *Dokl. Akad Nauk SSSR*, **123**, 681 (1958).
146. L. V. Abramova, N. I. Sheverdina, and K. A. Kocheshkov, USSR Patent 136,313, Byulleten' Iozbretenii No. 5 (1961).
147. L. V. Abramova, N. I. Sheverdina, and K. A. Kocheshkov, *Trudy II Vsesoyuznogo Soveshchaniya po Radiatsionnoy Khimii*, Akad. Nauk SSSR Publishers, Moscow, 1962, p. 394.

148. L. V. Abramova, I. V. Vereshchinskii, K. A. Kocheshkov, V. Yu. Miretskii, V. V. Pozdeev, Yu. S. Ryabukhin, and N. I. Sheverdina, in *Industrial Uses of Large Radiation Sources*, Vol. 1, I.A.E.A., Vienna, 1963, p. 83.

149. I. V. Vereshchinskii, V. V. Pozdeev, V. Yu. Miretskii, and A. Yu. Ivanov, USSR Patent 181,104; Byulleten' Izobretenii No. 9 (1966).

150. V. Yu. Miretskii, I. V. Vereshchinskii, A. Yu. Ivanov, K. A. Kocheshkov, V. V. Pozdeev, and N. I. Sheverdina, *Khim. Prom.*, No. 4, 258 (1968).

151. V. Yu. Miretskii, I. V. Vereshchinskii, and A. Yu. Ivanov, *Khim. Vysok. Energii*, **3**, 231 (1969).

152. V. Yu. Miretskii, I. V. Vereshchinskii, N. Ye. Nikolayev, and E. A. Ryabov. See p. 26 in Ref. 15.

153. A. F. Fentiman, R. E. Wyant, D. A. Jeffrey, and J. F. Kircher, *J. Organometal. Chem.*, **4**, 302 (1965).

154. A. F. Fentiman, R. E. Wyant, J. L. McFarling, and J. F. Kircher, *J. Organometal. Chem.*, **6**, 645 (1966).

155. J. Hoigné, K. Schaffner, and R. Wenger, *Helv. Chem. Acta*, **48**, 527 (1965).

156. M. A. Golub and C. L. Stephens, *J. Phys. Chem.*, **70**, 3576 (1966).

157. J. Fajer and D. R. MacKenzie, *J. Phys. Chem.*, **71**, 784 (1967).

158. W. G. Burns, R. B. Cundall, P. A. Griffiths, and W. R. Marsh, *Trans. Faraday Soc.*, **64**, 129 (1968).

159. W. G. Burns, in *Radiation Chemistry; Proceedings of the 1962 Tihany Symposium*, Akademiai Kiadó, Budapest, 1964, p. 31.

160. T. H. Bates, J. V. F. Best, and T. F. Williams, *J. Chem. Soc.*, 1521 (1962).

161. T. H. Bates, J. V. F. Best, and T. F. Williams, *J. Chem. Soc.*, 1531 (1962).

162. H. Kobza, *J. Org. Chem.*, **27**, 2293 (1962).

163. A. Sugimori, K. Nakata, and H. Hirai, *Bull. Chem. Soc. Japan*, **39**, 2613 (1966).

164. D. C. Walker, *Chem. Commun.*, 1050 (1967); F. W. Comer and J. Trotter, *J. Chem. Soc. (B)*, 13 (1966).

165. A. Henglein, *Angew. Chem.* **72**, 603 (1960).

166. H. Krauch, *Atomwirtschaft*, **5**, 265, 411 (1960).

167. H. Mohler, *Chem.-Ztg.*, **84**, 175 (1960).

168. A. J. Swallow, *Brit. Chem. Eng.*, January (1963).

169. A. Henglein, in *Application of Large Radiation Sources in Industry*, Hague, Netherlands, 1964, p. 23.

170. J. Hoigné, *Chimia*, **19**, 18 (1965).

171. S. Rosinger, *Chem.-Ing.-Tech.*, **37**, 128 (1965).

172. J. Kuchera and B. Chutný, *Radiation Synthesis of Organic Compounds*, UJV 925/63, Informačni středisko pro jadernou energii ÚJV ČSAV, Řež-Praha, 1965.

173. G. R. Hall and P. G. Clay, *Chem. Engr.*, **204**, CE 336-43 (1966); (*Trans. Inst. Chem. Engr.*, **44**, N 10).

174. I. V. Vereshchinskii, *Khim. Vysok. Energii*, **4**, 483 (1970).

175. I. V. Vereshchinskii, *Uspekhi Khim.*, **39**, 880 (1970).

176. A. F. Trotman-Dickenson and E. W. R. Steacie, *J. Chem. Phys.* **19**, 329 (1951).

Radiation Chemical Mechanisms in Radiation Biology

G. E. ADAMS, *Cancer Research Campaign, Gray Laboratory, Mount Vernon Hospital, Northwood, Middlesex, England*

Contents

1. INTRODUCTION

1.1. General

Inasmuch as all forms of reversible and irreversible damage in biological material exposed to high-energy radiation stem from the molecular events of excitation or ionization or both, the concepts and mechanisms of fundamental radiation chemistry are relevant to radiobiology. The processes, however, by which the initial molecular damage in a biological system is sustained, repaired, or enhanced, are many and varied. They depend upon the interplay of a wide variety of internal and external parameters, physical, chemical, and biological. Thus there are many problems both in the design of appropriate and relevant radiation chemical models for radiobiological systems and in the construction of radiation chemical hypotheses adequate to account for given radiobiological responses.

In the past the two fields of radiation chemistry and radiobiology have diverged—quite naturally so. As radiation chemical research progressed, greater emphasis was laid on the nature of the primary chemical species produced in condensed media, particularly water. The mechanisms of the energy absorption process and the translation of the initial physical perturbation into thermally stable, transient chemical species occupied the attention of most radiation chemists. In the pursuit of these objectives, the chemist was generally restricted, therefore, to the use of one- or two-component systems in order to explore and understand the complexities even of irradiated water. However, the radiobiologist faced with an existing complex phenomenon was obliged to take account of many other parameters. As these increased with the development of radiobiology, the apparent irrelevance of most simple one- or two-component radiation chemical systems tended to underline the divergence of the two fields. Thus the application of radiation chemistry to radiobiology, which appeared to be so promising in the early period, became somewhat empirical.

However, the enormous advances in both fields ultimately, and quite naturally, produced a reversal in the trend. With respect to the post-thermalization stage, radiation chemistry has solved, in the main, its quantitative problems of yields and reactivities—certainly in aqueous solutions.

The absolute reactivities of the fundamental water radicals, the hydroxyl radical, OH, the hydrated electron, e_{aq}^-, and the hydrogen atom, H, are known accurately for a large number of solutes of widely differing properties, and, for those that are not so well known, reliable methods are now available for their determination. Further, the development of fast-time resolution techniques, including pulse radiolysis, has done much to remove the restrictions imposed by the requirement to use simple one- or two-solute systems. Consequently, interest in the application of radiation chemical techniques and concepts to radiobiology is increasing rapidly.

In this chapter an attempt is made to illustrate some of the many advances in the field of molecular radiobiology. While it would be impossible in this presentation to review comprehensively a field that is large and rather ill-defined, the literature selected does embrace both radiation effects in aqueous solutions of amino acids, peptides, enzymes, proteins, nucleotides, and other macromolecules of interest, and, also, the radiation responses of intact cellular systems.

1.2. Nature of the Problem

Water constitutes about 70–80 percent of the mass of bacterial and mammalian cells; water radiolysis is therefore a major consequence of the overall energy absorption process. However, the internal structure of the cell is extremely heterogeneous, particularly at the micromolecular level and therefore, strictly speaking, there are no grounds *a priori* for supposing that any radiobiological damage arises directly from the products of water radiolysis. Nevertheless, water *is* the major energy absorber, and the rôle of the indirect effect in radiobiological damage has been and is still the object of much study. In the bacterial spore it has been established beyond doubt that the overall biological response, for example, survival, is extremely dependent upon water content.[1] Again, as is discussed in Section 5.4, certain dose modification phenomena in cellular systems are very sensitive to temperature in the region of the freezing point of cellular water. One fundamental problem of mechanistic radiobiology is, therefore, the relative importance of the direct and indirect effects with respect to a given biological phenomenon. The application of matrix isolation techniques to the study of model systems should help in this respect. The ionization process in most cases imparts to the medium an increased redox capacity. The ejected electron, provided it escapes from the electrostatic field of its geminate ion, acquires thermal stability with respect to the medium. Depending upon the nature of the medium, this entity may react as a powerful reducing agent. The positive ion is generally the precursor of any subsequent oxidative capacity, although the fate of this ion is also very sensitive to the nature of the medium. In water the positive ion breaks down before dielectric relaxation can provide the solvation energy necessary for stabilization, and thus the powerfully oxidizing radical hydroxyl, OH, is formed.

Although the dual consequences of the ionization event in water and some other media, that is, the induction of separate oxidation and reduction pathways, is clearly evident and recognized as such in the design of radiation chemical experiments, it is often ignored in the more complex phenomena of radiobiological systems. While it may often be impossible or impracticable to formulate actual radiation chemical reactions that occur in such systems, a classification of a given phenomenon into an oxidative or a reductive response is useful, particularly in attempts to explain dose modification phenomena (see Section 5).

Pulse radiolysis studies in liquid systems are being conducted with time resolution down to and beyond the nanosecond region. Both singlet and triplet excited molecules can be observed directly in a variety of solvents under conditions that permit the direct study of their formation by charge neutralization and by energy transfer processes.[2] Solid-state pulse radiolysis techniques have been applied to the study of intramolecular energy transfer phenomena in normal and substituted DNA. Such work, of great promises with respect to the question of energy-migration distances and effective "target size," is described in Section 5.3.5.

1.3. Appropriate Techniques of Radiation Chemistry

"Classic," or stationary-state, methods for the evaluation of a radiation chemical mechanism depend on the accurate determination of yields and distribution of the stable end products of the various reactions. The radiation time is usually very large compared with the lifetimes of the unstable intermediates of these reactions (i.e., neutral and charged free radicals and excited molecules), and therefore their stationary-state concentrations during the irradiation are far too low for direct observation. Thus the rôle of these species in the overall mechanism must be extrapolated from the analysis of chemically stable end products.

However, in pulse radiolysis the radiation time is very short compared with the lifetimes of the transient species, and thus the concentrations of these intermediates suffice for their direct observation by spectroscopic or other techniques.

In some respects the techniques are complementary. In the former the overall products can be measured, but the difficulties in interpretation of the intermediate processes ultimately prohibit the unequivocal interpretation of the radiolysis mechanism for most multicomponent systems. It is this restriction that has tended, in the past, to limit the relevance—or usefulness— to radiobiology of many radiation chemical model systems.

Pulse radiolysis has its own limitations. Techniques employing absorption spectroscopy require transparent and, preferably, homogeneous test systems. The unequivocal identification of a transient absorption spectrum is sometimes difficult to achieve and, further, the dose rates required for sufficiently sensitive analysis are usually several orders of magnitude larger than those conventionally employed in most radiobiological experiments. The great advantage is, of course, its exploitation of reaction time scales. With conventional techniques the time resolution has been improved down to and beyond the time scale of diffusion-limited reactions between thermally stable fragmentary molecules occurring under conditions that are both relevant and convenient for their accurate measurement. Because of this facility for separating in time the several processes that may be involved in the overall phenomenon, systems that may be studied profitably by pulse radiolysis methods can be, and often are, quite complex. When combined with final product analysis, or bioassay data from biochemical or biological systems, pulse radiolysis becomes even more versatile. Frequent reference is made in this chapter to the various applications of this technique.

Although it, too, has its limitations, electron spin resonance (ESR) spectroscopy (see Fessenden and Schuler in Volume 2 of this series) has been applied for several years to the study of molecular damage in biological systems generally. It is undoubtedly a powerful method. It detects and measures molecular damage with a high degree of resolution. Often the molecular structures of a free radical or other unstable species can be defined with a precision unavailable with other techniques. Unlike the requirements of pulse radiolysis, transparency of medium is not necessary. Although in conventional use, in order to extend the lifetimes of the species under investigation, the method is restricted to solid materials or frozen solutions, fast-flow techniques do, to some extent, permit the investigation of liquid systems at room temperatures. Fast-response ESR techniques have been combined with irradiation for direct investigation of short-lived transients produced in liquid media at room temperature. Such a technique is of particular interest because of the short-time resolution achieved.

2. AMINO ACIDS AND PEPTIDES

2.1. Introduction

An extensive literature exists on the radiation-induced inactivation of enzymes in dilute aqueous solution, in which the damaging process arises from attack by products of water radiolysis. Much of the early work has been reviewed,[3, 4] although some of it is difficult to assess in the light of more recent progress in aqueous radiation chemistry. For a given enzyme, reported inactivation yields often vary by as much as a factor of 10; the discrepancies are almost certainly the result of lack of uniformity of such experimental conditions as concentration, pH, gaseous environment, and the nature and concentration of the substrate used for the assay. Nevertheless, several theories of radiation-induced inactivation have emerged based either upon the radiation chemical properties of the constituent amino acids and peptides or upon the bulk properties associated with a macromolecule possessing a well-defined three-dimensional structure.

Much of the basic information obtained from studies of the radiation chemistry of amino acids in aqueous solution is probably of little relevance to mechanisms of the inactivation of enzymes in solution. However, there is considerable evidence that certain types of reactions of the more reactive amino acids, particularly those containing sulfur or ring systems, are very important in this respect.

Before discussion, therefore, of some of the mechanisms of the radiation-induced inactivation of enzymes, the general features of the radiolysis of amino acids and simple peptides are summarized. More detailed information can be found in a review by Garrison.[5]

2.2. Simple Aliphatic Amino Acids

2.2.1. *Glycine and Alanine*

It is now well established that deamination is a major process in the radiolysis of dilute aqueous solutions of simple α-amino acids, for example, glycine, $(NH_2)CH_2COOH$, and alanine, $CH_3CH(NH_2)COOH$, and, in the absence of oxygen, can occur by both oxidative and reductive mechanisms.[6] In these reactions the corresponding keto and fatty acids are produced together with small amounts of hydrogen, carbon dioxide, aldehydes, and substances of higher molecular weight. In neutral solution the products derive from reactions of both the hydrated electron, e_{aq}^-, and the hydroxyl radical, OH.

Provided the solute concentration is sufficient to ensure complete scavenging of e_{aq}^- and OH, the distribution of the final products can be accounted for[6-8] by the reactions

$$e_{aq}^- + NH_3^+CH(R)COO^- \rightarrow NH_3 + \dot{C}H(R)COO^- \tag{1}$$

$$\rightarrow NH_2 + CH_2(R)COO^- \tag{2}$$

$$\rightarrow H + NH_2CH(R)COO^- \tag{3}$$

$$OH + NH_3^+CH(R)COO^- \rightarrow H_2O + NH_3^+\dot{C}(R)COO^- \tag{4}$$

$$H + NH_3^+CH(R)COO^- \rightarrow H_2 + NH_3^+\dot{C}(R)COO^- \tag{5}$$

For alanine solutions the addition of the OH scavenger formate ion, $HCOO^-$, reduces the yield of ammonia by almost 50 percent and eliminates completely the yield of the keto acid product (pyruvic acid). Thus the major pathways for removal of the α-carbon radical formed in (4) are reactions (6) and (7), both of which are presumed to involve the formation of an intermediate imino cation

$$2NH_3^+\dot{C}(R)COO^- \rightarrow NH_4^+CH(R)COO^- + NH_2^+{=}C(R)COO^- \tag{6}$$

$$NH_3^+\dot{C}(R)COO^- + \dot{C}H(R)COO^- \rightarrow CH_2(R)COO^- + NH_2^+{=}C(R)COO^- \tag{7}$$

followed by

$$NH_2^+{=}C(R)COO^- + H_2O \rightarrow NH_4^+ + RCOCOO^- \tag{8}$$

As predicted by the mechanism, when the pH of the solution is sufficiently low, protonation of the hydrated electron

$$e_{aq}^- + H_3O^+ \rightarrow H + H_2O \tag{9}$$

suppresses reactions (1), (2), and (3) in favor of (5), thus eliminating the fatty acid yield from reaction (7) and increasing the yield of the keto acid. In the presence of oxygen, e_{aq}^- is captured to form the hydroperoxide radical anion O_2^- (known also as the superoxide ion), which is unreactive with the amino acid, and the carbon radicals are oxidized to the peroxy radicals

$$NH_3^+\dot{C}(R)COO^- + O_2 \rightarrow NH_3^+C(\dot{O}_2){\cdot}RCOO^- \tag{10}$$

The overall approximate stoichiometry for glycine and alanine is

$$G(OH) \approx G(NH_3)$$

2.2.2. Substituent Effects in Irradiated Aliphatic Acids

Data on the effect of various substituent groups on the radiation yields from simple aliphatic amino acids are collected in a review by Garrison.[5]

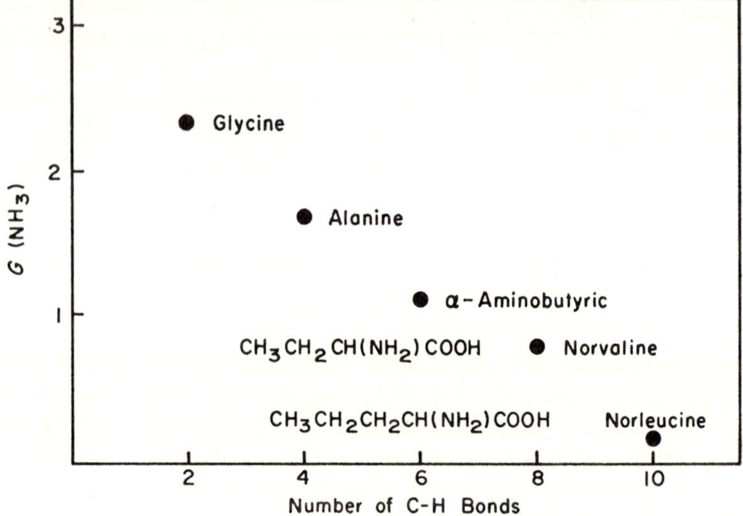

Fig. 1. Effect of carbon-chain length on the yield of ammonia in γ-irradiated solutions of aliphatic amino acids. (Holian and Garrison.[9])

As might be expected, the site of radical attack, particularly by OH, depends upon the nature of the substituent. With increasing length of the side chain, OH attack becomes multifocal and the effect is shown by a decrease in ammonia yield. Figure 1 reproduces some data of Holian and Garrison,[9] who determined the ammonia yields from a range of simple amino acids. The solutions were oxygenated and the concentrations of the amino acids were such as to ensure quantitative scavenging of e_{aq}^{-} by oxygen and of OH by the amino acids. $G(NH_3)$ falls from 2.3 for glycine to < 0.2 for norleucine (i.e., α-amino-n-pentanoic acid). This drop in $G(NH_3)$ reflects the decrease in the yield of the α-amino radical formed by abstraction of hydrogen from this position; that is, reaction (4). These results are consistent with the finding that in the radiolysis of various aliphatic amino acids products of higher molecular weight result from dimerization of radicals produced by hydrogen abstraction from the β, γ, and other positions.[10]

Multifocal attack of OH is also observed in aliphatic alcohols. Pulse radiolysis studies[11] on rate of oxidation by ferricyanide ion of alcohol radicals formed by reaction of OH showed that the percentage attacks at the α-carbon atom in ethanol, n-propanol, and n-butanol are, respectively, 97, 61, and 34

$$R_1-\underset{\underset{H}{|}}{\overset{\overset{R_2}{|}}{C}}-OH \xrightarrow{OH} R_1-\underset{\cdot}{\overset{\overset{R_2}{|}}{C}}OH \tag{11}$$

In substituted alcohols in which the side chain is branched (e.g., isopropanol and isobutanol), the relatively larger inductive effect leads to an increase in the electron density at the α-carbon atom. This increase is reflected in an increase in the probability of attack at the α position, (89 and 75 percent, respectively).

Some simplification of the radiation chemistry of simple amino acids in aqueous solution can be effected by the use of inorganic oxidizing agents, for example, Cu^{II} and Fe^{III}.[12] Copper II chelates with glycine and the complex reacts with e_{aq}^- to form copper I without decomposition of the glycine.[12] The radical $NH_3^+\dot{C}HCOO^-$ formed in reaction (4) is oxidized by Cu^{II} according to

$$Cu^{II} + NH_3^+\dot{C}HCOO^- + H_2O \rightarrow Cu^I + NH_4^+ + CHOCOO^- + H^+ \qquad (12)$$

Reductive deamination occurs also in irradiated solutions of the simple peptide glycylglycine, $NH_2CH_2CONHCH_2COOH$, in which over 80 percent of the ammonia yield is attributable to reactions of e_{aq}^-. A direct demonstration of the stoichiometry between $G(e_{aq}^-)$ and $G(NH_3)$ was accomplished[13] by use of the electron scavenger chloracetate ion. This solute reacts with e_{aq}^- by dissociative electron attachments to yield free chloride ion

$$e_{aq}^- + ClCH_2COO^- \rightarrow Cl^- + \cdot CH_2COO^- \qquad (13)$$

In oxygen-free solutions of 0.2 M glycylglycine, irradiated at pH 6.5, $G(NH_3)$ decreases with increasing concentration of added chloracetate ion. Over the same concentration range, $G(Cl^-)$ increases, consistent with the competition between the two solutes for e_{aq}^-.

Oxidative attack at the terminal carbon atom in glycylglycine and, probably, other simple aliphatic peptides to give radicals of the type $\diagdown C-NH_3^+ \diagup$ is relatively unimportant. The main reaction appears to take place at the carbon atom alpha to the peptide nitrogen.

$$NH_3^+-CH_2(CONH)CH_2COO^- \xrightarrow{OH} NH_3^+CH_2(CONH)\dot{C}HCOO^- \qquad (14)$$

2.3. Aromatic and Heterocyclic Amino Acids

2.3.1. *General*

It is clear from the low G values for ammonia production (< 0.5) in irradiated aqueous solutions of either phenylalanine, $C_6H_5CH_2CH(NH_2)COOH$, tyrosine, $p\text{-}(HO)C_6H_4CH_2CH(NH_2)COOH$, or histidine (see Table I for formula) that both reductive and oxidative deamination are not predominant processes in these systems. The rate constants for reaction with e_{aq}^- are fairly large, particularly so for histidine, although the reactions for all three solutes are pH-dependent.

Hydroxyl radicals react at rates that approach that for a diffusion-limited reaction, and these amino acids are particularly vulnerable to oxidative attack in many enzyme structures. The marked sensitivity to OH-induced inactivation of several enzymes is discussed in detail in Section 3.

Pulse radiolysis techniques have proved of considerable value in the elucidation of the mechanisms of OH attack in ring amino acids, with respect to both the measurement of the absolute rate constants for the reactions and also to the locus of attack in the molecules.

It is now well established that OH radicals add directly to the aromatic ring of a large number of benzene and heterocyclic derivatives. The rate constants are usually in the range 10^9–10^{10} dm^3 mole^{-1} sec^{-1} and are influenced by the electron-donating or electron-withdrawing properties of any substituent group in the molecule. The first quantitative study of this type of reaction was the radiation-induced oxidation of benzene, in which case formation of the spectrum of the hydroxycyclohexadienyl radical was directly observed[14]

$$OH + \langle \bigcirc \rangle \rightarrow \text{(adduct)} \tag{15}$$

The rate constant of 7.9×10^9 dm^3 mole^{-1} sec^{-1} is surprisingly high in view of the considerable loss of resonance energy involved in formation of the adduct. Nevertheless, subsequent work has amply verified the large rate constants for reactions of this type. Substituted hydroxycyclohexadienyl radicals possess strong transient absorption spectra with maxima usually in the 300–400 nm region. These have been used to study the types of reactions undergone by such species. In general, two types of reaction occur, (1) bimolecular association or disproportionation, and (2) unimolecular elimination of some fragment of the molecule, often water. The relative extent of either reaction depends upon the total radical concentration; therefore, in pulse radiolysis experiments, unusually low dose rates are often required for investigation of the latter type of reaction.

Water elimination occurs in the OH adducts of various hydroxylated phenols[15, 16] according to reactions of the type

$$OH + \text{(phenol)} \rightarrow \text{(adduct)} \rightarrow \text{(product)} + H_2O \tag{16}$$

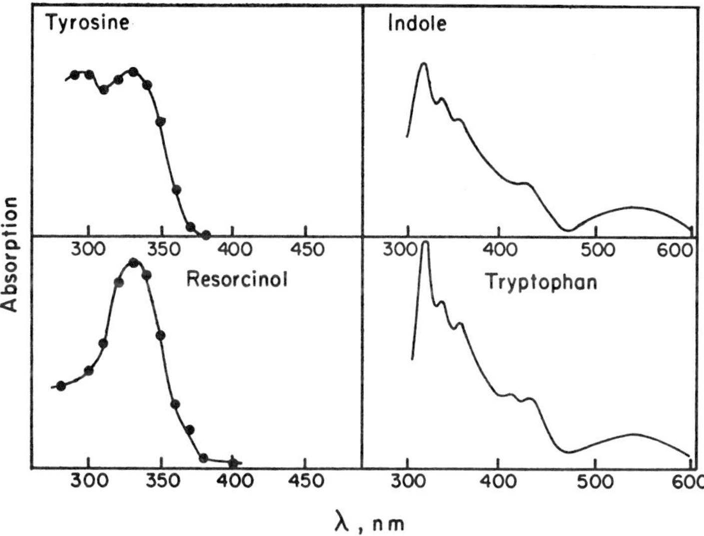

Fig. 2. Transient spectra of the radicals formed by addition of hydroxyl radicals to some aromatic compounds.

In Fig. 2, the spectra of the OH adducts of the amino acids, tyrosine and tryptophan, are compared with those of their respective analogs, resorcinol and indole. Unimolecular elimination of H_2O from the tyrosine radical produces a substituted phenoxyl. Although at pH 6.2 the time scale of the reaction is several hundred microseconds, the reaction is much faster in both acid and alkaline media.

2.3.2. *Pulse Radiolysis of Tryptophan*

As seen in Section 3, there is strong evidence for the involvement of tryptophan damage in the radiation inactivation of various enzymes, in particular, lysozyme.

The radiation chemistry of the free amino acid has been studied by various groups using stationary-state methods, although some aspects of the mechanism remain obscure. Such obscurity is partly the result of lack of data from oxygen-free systems. Early stationary-state studies established that a variety of products is produced during radiolysis of oxygenated solutions of tryptophan. These include NH_3 $(G \sim 0.6)$, anthranilic (i.e., 2-aminobenzoic) acids, kynurenine, formyl kynurenine, tryptamine, and aliphatic amino acids. The detection of the small yields of kynurenine, and also the formyl derivatives, led Jayson, Scholes, and Weiss[17] to propose a mechanism involving the addition of OH radical to the 2-position of the heterocyclic ring

$$
\text{Tryptophan} \xrightarrow{\text{OH}} \qquad (17)
$$

However, the variety of products suggested that OH attack occurs at several sites on the molecule. Indeed, studies by Eich and Rochelmeyer[18] on tryptophan degradation by OH radicals produced in the Fenton reaction indicate that the 5- and 7-positions on the aromatic nucleus are also susceptible to attack. However, a comprehensive and elegant study by Armstrong and Swallow[19] of the pulse- and γ radiolysis of tryptophan and related compounds has done much to clarify the mechanism, particularly with respect to the site of OH attack.

Different transient spectra were obtained from deaerated solutions of tryptophan irradiated under a variety of conditions. Good spectral resolution was achieved by use of an intense light source and narrow bandwidths; rigorous spectral analysis showed that the composite spectrum has three components, attributable to the OH and H adducts and also to the electron adduct, possibly in its protonated form. Further, it was found that the first two spectra change considerably with pH, again an effect of protonation.

The isolated spectrum of the OH adduct reproduced in Fig. 2 shows, in the fine structure, three maxima located at wavelengths 310, 325, and 345 nm. A major accomplishment in this work was the recognition that this OH-adduct spectrum itself has three components and that the three peaks in the 300–350 nm band represent OH attack at these different positions in the indole moiety. Figure 3 illustrates the transient spectra obtained on pulse radiolysis of neutral, N_2O-saturated solutions of indole, L-tryptophan, *N*-methylindole, 2-methylindole, 3-methylindole, 5-methyltryptophan, 7-methylindole, and naphthalene. For the seven indoles, the spectra are qualitatively similar, although the differences in the fine structures, particularly in the region around 300–350 nm, are clearly evident. The rationale behind this experiment was as follows.

Substitution of a methyl group on various positions of the indole ring should restrict electrophilic attack by OH at that position. Complications arising from changes in the electron density distribution in the molecule because of the inductive effect of the methyl group appear to be negligible, for the molecular spectrum of indole is not very different from the molecular spectra of the substituted compounds. The relative changes in the three ultraviolet maxima in the various radical spectra reflect the suppression of

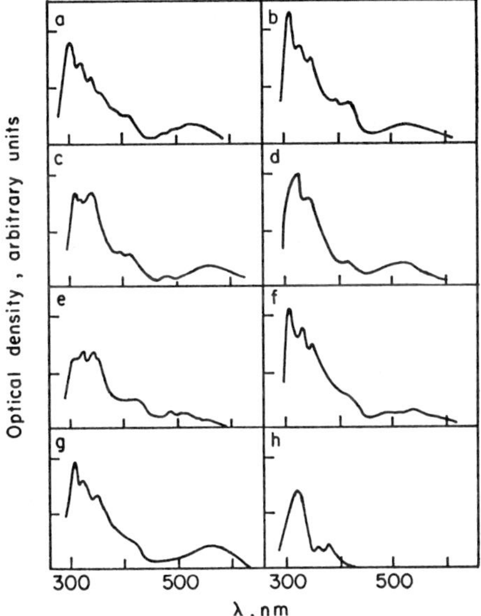

Fig. 3. Transient spectra of the hydroxyl radical adducts of some derivatives of indole. (*a*) Indole; (*b*) L-tryptophan; (*c*) *N*-methylindole; (*d*) 2-methylindole; (*e*) 3-methylindole; (*f*) 5-methyltryptophan; (*g*) 7-methylindole; (*h*) naphthalene. (Armstrong and Swallow.[19])

OH attack at the substituted position. It was concluded that the peaks at 345 and 325 are attributable, respectively, to OH addition at C-2 and C-3 on the heterocyclic ring, and that the peak at 310 nm is the result of reaction at some site on the aromatic nucleus.

345 nm   325 nm   310 nm

Kinetic data showed that the OH reactivities at the three sites are of the same order of magnitude ($\sim 10^{10}$ dm^3 mole^{-1} sec^{-1}) from which it was concluded that in tryptophan the three OH adducts are formed in approximately equivalent yield.

The success of this technique in resolving multiple reaction sites in a single molecule has obvious potential applications in the study of molecular damage in both monomeric and polymeric structures. Further reference to these results is made in the discussion of radiation damage in lysozyme in Section 3.

Parallel stationary-state studies, also by Armstrong and Swallow,[19] confirmed that radical attack at the alanine moiety in tryptophan is relatively unimportant and also showed that reconstitution reactions that occur in deaerated solution are responsible for the low net yield of tryptophan destruction ($G = 0.7$).

2.4. Sulfur-Containing Amino Acids

The radiation chemistry of the simple thiol amino acid cysteine[20-27] and also its oxidation product, cystine,[28-30] $[-S-CH_2CH(NH_2)COOH]_2$, is very well understood, largely because of the considerable attention devoted to the subject during the last decade. Early inconsistencies in reported yields were the result of sensitivity of cysteine radiolysis to traces of oxygen and metal impurities.

Cysteine reacts rapidly with the hydrated electron ($k = 8.7 \times 10^9$ dm^3 $mole^{-1}$ sec^{-1}) in the dissociative electron capture reaction

$$HS-CH_2CH(NH_2)COO^- + e_{aq}^- \rightarrow \cdot CH_2CH(NH_2)COO^- + SH^- \qquad (18)$$

$$SH^- + H^+ \rightarrow H_2S \qquad (19)$$

The amino acid radical then undergoes a repair reaction in which α-amino propionic acid is formed

$$\cdot CH_2CH(NH_2)COO^- + RSH \rightarrow CH_3CH_2(NH_2)COO^- + RS\cdot \qquad (20)$$

Repair reactions of this type and their relevance to mechanisms of radiobiological protection are discussed later in Section 5.4.3. In acid solution the yield of H_2S falls, and the hydrogen yield shows a complementary increase. Such behavior is the result of suppression of reaction (18) by protonation of the hydrated electron [reaction (9)]. However, the H_2S yield does not decrease to zero. Although H atoms react with cysteine, partly by hydrogen abstraction to form the RS· radical

$$RSH + H \rightarrow H_2 + RS\cdot \qquad (21)$$

some H_2S is produced by the reaction sequence

$$RSH + H \rightarrow RH + SH \qquad (22)$$

$$SH + RSH \rightarrow H_2S + RS\cdot \qquad (23)$$

Product analysis by Al-Thannon, Peterson, and Trumbore[26] led to the estimate $k_{22}/k_{21} = 0.33$. Hydroxyl radicals oxidize cysteine to cystine by abstraction of the sulfhydryl hydrogen atom [$k = 1.3 \times 10^{10}$ dm^3 $mole^{-1}$ sec^{-1} (pH. 1)]

$$RSH + OH \rightarrow RS\cdot + H_2O \qquad (24)$$

Under nonacid conditions the intermediate RS radical complexes with a solute molecule to form the ion radical RSSR$^-$ according to the equilibria

$$RSH \rightleftharpoons RS^- + H^+ \tag{25}$$

$$RS\cdot + RS^- \rightleftharpoons RSSR^- \tag{26}$$

Cystine is probably formed by dimerization of unassociated RS radicals. Support for this mechanism is provided by pulse radiolysis studies of Adams, McNaughton, and Michael[31] on the related thiol, cysteamine,

$$NH_2CH_2CH_2SH$$

which have shown that the rate of decay of the corresponding RSSR$^-$ ion radical is extremely pH-dependent in the pH region corresponding to the pK of the SH group.

The very high yields for consumption of RSH in the presence of oxygen indicate occurrence of a chain reaction. It was suggested by Packer and Winchester[27] that RS radicals formed in reaction 24 react with O_2 to form the RSO_2 radical that propagates the chain by the H-atom transfer reaction

$$RSO_2 + RSH \rightarrow RSO_2H + RS \tag{27}$$

Ultimately, cystine is produced.

Reduction of disulfides to thiols proceeds initially by formation of the ion radical RSSR$^-$, which decays unimolecularly to form the simple RS radical [equilibrium (26)]. For the corresponding radical from the analogous disulfide cystamine, $(NH_2CH_2CH_2S{-})_2$, the decay constant is 8×10^5 sec^{-1} and is probably not very different for cystine.

According to Purdie,[30, 32] an unusual insertion reaction occurs in the OH-induced oxidation of cystine and other disulfides. The trisulfide RSSSR is formed in fairly high yield, particularly in deaerated solution and originally was thought to result from the reaction

$$RS\cdot + RSSR \rightarrow RSSSR + R\cdot \tag{28}$$

Later work indicated, however, that the reaction probably involves the intermediate, RSSOH

$$OH + RSSR \rightarrow RSOH + RS \tag{29}$$

$$OH + RSSR \rightarrow RSSOH + R \tag{30}$$

$$RSSOH + RSH \rightarrow RSSSR + H_2O \tag{31}$$

2.5. Free-Radical Reactivity of Amino Acids and Peptides

2.5.1. *The Hydrated Electron*

Tables I and II list bimolecular rate constants for the reaction of e_{aq}^- with various amino acids and peptides.

TABLE I

Bimolecular Rate Constants for Reactions of e_{aq}^- with Simple Amino Acids[a]

Amino acid	Formula[b]	pH	k, units of 10^7 dm^3 mole^{-1} sec^{-1}
Alanine		6.8	$\leq$0.5
L-Arginine	$H_2NC(:NH)NH(CH_2)_2CH(NH_2)COOH$	6.1	15
		11.5	6.3
L-Asparagine	$H_2NCOCH_2CH(NH_2)COOH\cdot H_2O$	7.3	15
		11.7	2.4
DL-Aspartic	$HOOCCH_2CH(NH_2)COOH$	7.3	<1.0
		10.5	<0.5
L-Glutamic	$HOOC(CH_2)_2CH(NH_2)COOH$	5.7	~2.0
L-Hydroxyproline	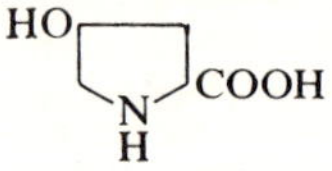	7.0	4.5
		10.8	1.1
L-Leucine	$(CH_3)_2CHCH_2CH(NH_2)COOH$	6.5	<1.0
DL-Lysine	$H_2N(CH_2)_4CH(NH_2)COOH$	7.8	~2.0
L-Proline	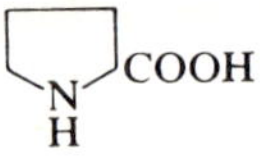	6.7	2.0
DL-Serine	$HOCH_2CH(NH_2)COOH$	6.1	$\leq$3.0
DL-Threonine	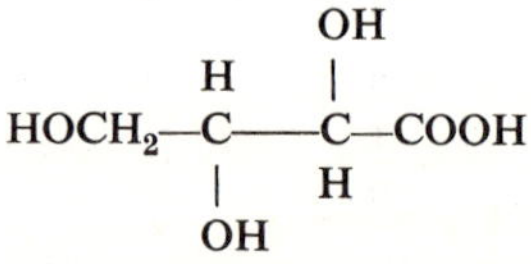	6.2	$\leq$1.0
DL-Valine	$(CH_3)_2CHCH(NH_2)COOH$	6.4	<0.5
DL-Methionine	$CH_3S(CH_2)_2CH(NH_2)COOH$	6.0	3.5
L-Cysteine·HCl		6.3	870
		11.6	7.5
L-Cystine		6.1	1300
		10.7	250
DL-Phenylalanine	$C_6H_5CH_2CH(NH_2)COOH$	6.3	11.0
		11.2	1.7
DL-Tryptophan		6.8	41
L-Tyrosine	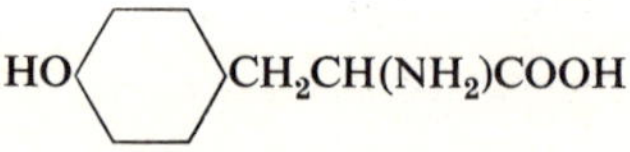	5.8	16
DL-Histidine·2HCl	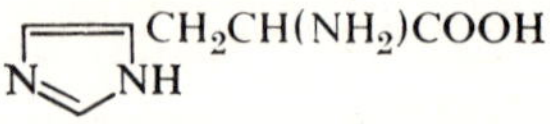	6.0	387
		6.7	141
		8.6	4.5

[a] Data from Braams.[33]

[b] See also text.

140

TABLE II
Bimolecular Rate Constants for Reactions of e_{aq}^- with Simple Peptides[a]

Peptide[b]	pH	k, units of 10^8 dm^3 mole^{-1} sec^{-1}
Alanylalanine	6.2	1.3
Glycylalanine	6.2	2.9
Leucylalanine	6.1	1.6
Glycylasparagine	5.3	5.4
Alanylglycine	6.2	2.1
Glycylglycine	6.4	2.5
Glycylglycylglycine	6.0	9.0
Leucylglycine	6.1	1.1
Histidylhistidine	5.5	79
	6.8	24
	7.3	13
	8.4	2.9
	11	0.51
Alanylleucine	6.4	1.3
Glycylleucine	6.5	2.8
Leucylleucine	6.0	0.9
Glycylphenylalanine	6.7	1.6
Glycylproline	6.7	11
Glycyltryptophan	6.4	4.5
Glycyltyrosine	6.1	4.1
Glycylvaline	6.0	2.6
Glutathione (red)	6.4	0.032
Glutathione (ox)	8.2	46

[a] Data from Braams[34] and Scholes et al.[35]

[b] Glycine is H_2NCH_2COOH; glutathione is $HOOCH(NH_2)$-$(CH_2)_2CONHCH(CH_2SH)CONHCH_2COOH$. For other basic formulas see text and Table I.

The data obtained by direct measurement using the pulse radiolysis technique have been classified by Braams[33] into three categories comprising amino acids of low, moderate, and high reactivity. The first group contains most of the common aliphatic amino acids; the second, asparagine, arginine, tryptophan, tyrosine, and phenylalanine, and the third, cysteine, cystine, and histidine. The rate constants for all compounds are pH-dependent, an effect of the various ionic equilibria in the structure. At or around neutral pH, all amino acids exist mainly in the zwitterionic form (**I**), which is the more reactive structure

$$NH_3^+\!-\!CH\!-\!COO^- \rightleftharpoons NH_2\!-\!CH\!-\!COOH \qquad (32)$$
$$\qquad\quad |\qquad\qquad\qquad\qquad\quad |$$
$$\qquad\quad R\qquad\qquad\qquad\qquad\quad R$$
$$\textbf{(I)}$$

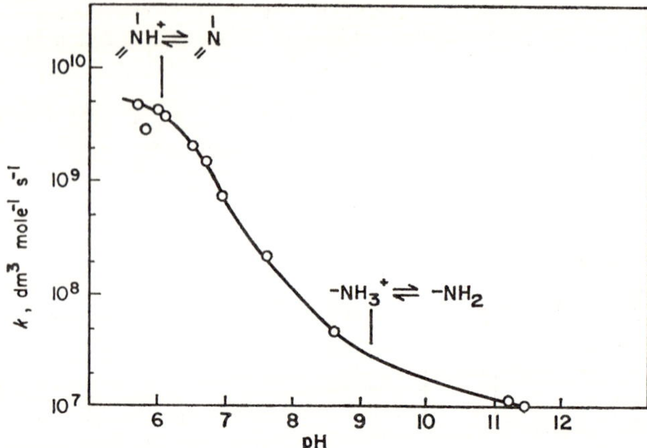

Fig. 4. Effect of pH on the reactivity of histidine with the hydrated electron. (Braams.[33])

As the pH increases, the reactivity falls—markedly so in the region of the pK_a of the cationic group. For most of the aliphatic compounds, the major locus of reductive attack is the amino group; changes of electron density in the molecules are responsible for the effect on the rate constant for deamination [reaction (1)].

In the intermediate group the reactivities of the two aliphatic amino acids, asparagine and arginine, are about an order of magnitude higher than those of the first group. Braams[33] attributes the increased reactivity of asparagine relative to aspartic acid, to the influence of the NH_2 group in the amide structure. In arginine the guanidine residue probably represents a major point of attack because the reactivity of free guanidine, which is positively charged at neutral pH, is 2.5×10^8 dm^3 mole^{-1} sec^{-1}.

In the highly reactive group, the disulfide cystine reacts with e_{aq}^- in neutral solution at a rate which is probably diffusion controlled. As mentioned previously, the immediate product of the reaction is the radical ion $RSSR^-$ in which the net negative change is associated with the disulfide group. The rate constant decreases with increasing pH, but the effect is not marked. The large pH dependence for the reactivity of the SH compound, cysteine, is consistent with observations from the numerous stationary-state studies that the major reaction is electron capture followed by loss of the SH^- ion [reaction (18)]. Because pK_a for the SH group in cysteine is 8.4, the structure at higher pH is unfavorable for the reaction and the rate constant would be expected to fall appreciably in alkaline solution

$$RSH \rightarrow RS^- + H^+ \tag{33}$$

For histidine also, as Fig. 4 shows, there is a very large pH effect on the electron reactivity. Two ionic equilibria involve the nitrogen atoms in the imidazole ring and in the amino group of the side chain. The respective values of pK_a are 6.00 and 9.17

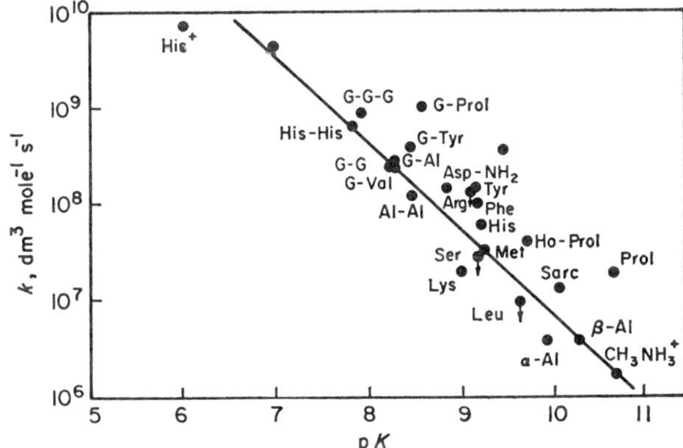

The net charge on the three structures is $+1$, 0, and -1. It is not surprising therefore that the electron reactivity decreases over the pH range where the two equilibria are important. The curve in Fig. 4, when analyzed in terms of the contribution of the reactivities of the separate structures to the overall rate constant, yields the following data[33]

$$k \text{ (positive ion)} = 7 \times 10^9 \text{ dm}^3 \text{ mole}^{-1} \text{ sec}^{-1}$$
$$k \text{ (zwitterion)} = 6 \times 10^7 \text{ dm}^3 \text{ mole}^{-1} \text{ sec}^{-1}$$
$$k \text{ (negative ion)} \sim 10^7 \text{ dm}^3 \text{ mole}^{-1} \text{ sec}^{-1}$$

Figure 5 shows the effect of protonation on the electron reactivity of amino acids generally. There is no doubt that as the pK_a of the amino group decreases the electron reactivity increases. Whether the relationship is a simple linear one, as suggested by Braams,[34] is, however, somewhat speculative.

Fig. 5. Relationship between the electron reactivity of some amino acids and the dissociation constants for the protonated amino groups. G is the symbol for glycine. The other symbols are clear from Table III. (Braams.[33])

2.5.2. *The Hydroxyl Radical*

Tables III and IV list the rate constants for reaction of OH with the amino acids and several simple peptides.

As is the case for the electron rate constants, the OH data for the amino acids vary considerably. For all solutes the rates vary with pH because of the influence of the ionic equilibrium. The simple aliphatic compounds, glycine and alanine, are the least reactive. However, when OH groups or branched chains are present, inductive effects tend to increase their reactivity. As suggested by Section 2.3.1, the ring-structured amino acids are the most reactive because of the ease with which OH addition to the π-electron system of the ring can occur.

Similar trends are evident in the peptides, although the peptide link appears to increase the overall reactivity relative to that of the free acids. The

TABLE III

Bimolecular Rate Constants for Reaction of OH
with Amino Acids[a]

Amino acid	pH	k, units of 10^8 $dm^3\,mole^{-1}\,sec^{-1}$
Alanine	6	0.77
Arginine	7	35
Aspartic	7	0.75
Cysteine	1	132
Cystine	2	53
Glutamic	2	1.32
Glycine	6	0.17
Histidine	7	50
Hydroxyproline	2	3.5
Isoleucine	2	17.5
Leucine	6	16.3
Lysine	2	6.2
Methionine	7	87
Phenylalanine	6	58
Proline	2	3.0
Serine	6	3.2
Threonine	2	3.8
Tryptophan	6	142
Tyrosine	2	97
Valine	2	7.0

[a] The data of Scholes et al.[35] are normalized according to Willson et al.[74]

pH dependence is illustrated by the data for the glycyl derivatives shown in Table IV. As expected, in peptides in which the component acids have different reactivities in the free state (e.g., glycyl tyrosine) the rate constants are comparable to those of the more reactive component. Simic, Neta, and Hayon[36] have reported some results of a pulse radiolysis study of some simple peptides in which transient absorption spectra, resultant from OH attack at the carbon atom alpha to the peptide bond, were measured (see Section 2.2.2). The radical species produced absorb strongly in the ultraviolet with maxima in the region around 260 nm and the spectra are pH-dependent because of the acid-base properties of the radicals.

TABLE IV

Bimolecular Rate Constants for Reaction of OH with Simple Peptides[a]

Peptide[b]	pH	k, units of $10^9 \ dm^3 \ mole^{-1} \ sec^{-1}$
Alanylglycine	2	0.16
Glutathione (ox)	1	14.7
G-alanine	6	0.35
G-glycine	7	0.27
G-G-glycine	6	0.33
	5.4	0.73[c]
	10.6	5.2[c]
G-G-G-glycine	2	0.23
	6	0.45
	7.8	1.2
	9.6	3.0
G-isoleucine	2	2.3
G-leucine	2	2.5
G-phenylalanine	2	8.5
G-proline	2	1.4
G-serine	2	5.7
G-tyrosine	2	9.3
G-valine	2	1.17
Histidylhistidine	6	9.0
Acetylglycine	8.7	0.42[c]
Acetylalanine	9.2	0.46[c]
Acetylglycylglycine	8.6	0.78[c]

[a] According to Scholes et al.[35] except as noted.

[b] The symbol "G" is used for "glycyl."

[c] According to Simic, Neta, and Hayon.[36]

3. ENZYMES AND PROTEINS

3.1. General Aspects of Free Radical Inactivation

3.1.1. *Amino Acid Damage*

Although many different amino acids constitute the primary structure of an enzyme, relatively few are involved in the "active center." These may be instrumental either in binding the substrate to the enzyme or in the actual process of catalytic reaction. Other amino acids serve to maintain the secondary and tertiary structures of the molecule which are essential to its activity. Such an effect can operate either through interchain molecular bonds, for example, the disulfur bridges, or through more general hydrogen bonding between the various amino acids.

Suggested mechanisms of free-radical inactivation therefore often involve direct attack at a vital amino acid present in the active center. This attack prevents one or both of the functions of substrate binding and subsequent catalysis. Other mechanisms postulate damage to a subunit which, although distant from the active center, results in cleavage of intramolecular bonds. Thus, by affecting the tertiary structure, a gross change occurs in the topography of the active center.

Following the work of Barron and his co-workers,[37–39] much emphasis has been placed on inactivation studies of the role of the sulfur-containing amino acid cysteine. Generally, it appears that SH-containing enzymes are more susceptible to radiation inactivation than are the sulfur-free enzymes, although there are discrepancies. It was shown, for example, that the SH-binding agent, *p*-chloromercuribenzoate protects against radiation by reversible complexing of the SH residue. Furthermore, in that work the X-ray inactivation of glyceraldehyde-3-phosphate dehydrogenase (GAPD) and adenosine triphosphatase could be reversed by the postirradiation addition of glutathione. This effect was confirmed by Tanaka et al.,[40] who used urease and glutamic dehydrogenase. However, Lange, Pihl, and Eldjarn[41] were not able to reproduce the reactivation of GAPD nor observe reactivation of yeast alcohol dehydrogenase.

Direct measurement of the loss of titratable SH groups, following enzyme irradiation, by Lange and Pihl[42, 43] was correlated by them with loss of enzymic activity; in the case of lactate dehydrogenase (LDH), Adelstein and Drakes[44] observed a one-to-one correspondence between the ability of the enzyme to bind to its coenzyme and the removal of SH groups. Results from other enzyme systems have indicated, however, that only a fraction of the total SH groups are destroyed even after complete inactivation.[45–48]

Data for papain, which contains a single SH group, are conflicting. Pihl and Sanner related a high inactivation yield ($G = 2.6$) directly to SH attack at the active site.[49] Some postirradiation reactivation was observed in the presence of cysteine; further, compounds that reversibly complex with SH groups were found to afford a high degree of protection. However, a much lower yield ($G = 0.4$) has been reported by Myers and Abernethy.[50]

Radiation produces changes in the ultraviolet optical absorption spectra of enzymes and proteins. Barron, Ambrose, and Johnson[51] attributed these effects to specific attack on the aromatic and heterocyclic ring amino acids, although conformational changes in the irradiated molecule may also cause changes in the overall absorption spectra.

Analysis of constituent amino acids modified or destroyed during irradiation of solutions of enzymes has revealed some interesting trends. In a variety of enzymes and proteins, it has been found that the sulfur-containing amino acids and the ring-structured compounds are the most radio-labile residues. This observation is consistent with the high reactivity of these amino acids with the OH radical. Unfortunately, as the result of their destruction during the analytical procedure, data for tryptophan and cysteine are often lacking.

3.1.2. *Structural Changes*

Apart from specific damage to amino acids in the active site, the topography of this region can be modified by gross changes in the secondary and tertiary structures. These structures are stabilized generally by interchain disulfide bonds and by hydrogen or hydrophobic bonds. Measurements of such properties as viscosity, optical rotation, sedimentation coefficients, light scattering, and gelation provide some evidence that changes in the secondary and tertiary structures can occur following irradiation.

Aggregation takes place more readily after irradiation in the absence of oxygen, an effect typical of the inhibition of free-radical dimerization by peroxy radical formation. It should be indicated, however, that cross-linking may occur without necessarily producing critical alteration of the enzyme conformation.

Jayko, Weeks, and Garrison[52] have suggested that scission of a primary peptide bond can lead to inactivation. However, although dialyzable peptide fragments have been obtained by Brown and Akounoglou[53] from myoglobin irradiated with large doses, studies by Liebster and Kopoldova[54] indicate that peptide cleavage is not a major process in the radiation-induced inactivation of enzymes, except possibly in oxygenated solution. A possible explanation is the low free-radical reactivity of the peptide bond relative to that of the ring-structured amino acids.

There is evidence[55] that peptide scission need not necessarily lead to loss of activity. Indeed, the formation of active chymotrypsin from the inactive zymogen involves the breakage of four peptide bonds.

It has often been suggested that cleavage of sulfur bridges can be a mechanism of radiation inactivation.[56, 57] The evidence on this point is conflicting. In irradiated dilute solutions of lysozyme, little evidence could be found by Dose[58] for the contribution of S—S bond rupture to the inactivation. However, Mee, Navon, and Stein[59] found the rate of inactivation of trypsin by H atoms to occur at the same rate as S—S bond cleavage, although this result was true also for tryptophan destruction. Recent pulse radiolysis findings relevant to this topic are discussed in the next section.

In a reaction between an enzyme and its substrate, the reaction velocity, for example, is given by the Michaelis–Menten equation

$$r = r_{\max}\bigg/\left(1 + \frac{K_m}{[S]}\right) \tag{3.1}$$

where r is the velocity when the substrate concentration is [S] and $r_{\max}$ is the maximum velocity when the substrate concentration is sufficient to saturate the enzyme. The Michaelis constant, K_m, in the simple Michaelis theory of enzyme reactions,[211] is the equilibrium constant for the formation of the enzyme–substrate complex.

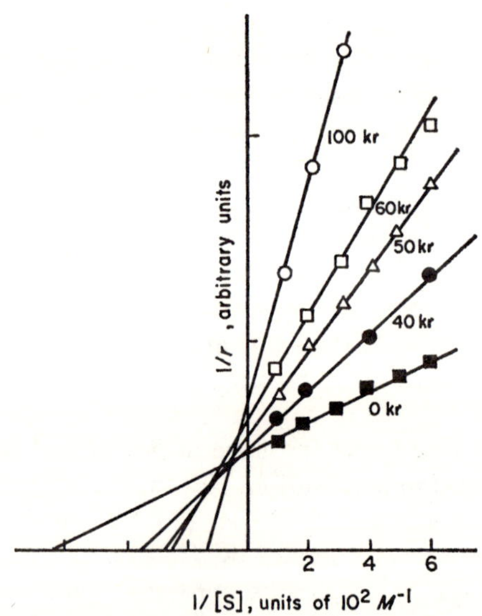

Fig. 6. Effect of radiation dose on the kinetic parameters and K_m for chymotrypsin irradiated in dilute aqueous solution. (Mee.[60])

In some enzymes, particularly chymotrypsin,[60, 61] radiation affects the substrate-binding efficiency (k_m) in addition to reducing the velocity of the substrate reaction ($r_{\max}$). Denaturation of this enzyme, either by heat or by chemical reaction with urea, occurs without destruction of constituent amino acids and without any change in K_m. Although similar behavior has been observed after H-atom attack, OH radicals produced in the Fenton reaction affect both kinetic parameters.[58, 62]

Lineweaver and Burke[212] expressed the rate law in the form

$$\frac{1}{r} = \frac{1}{r_{\max}}\left(\frac{K_m}{[S]} + 1\right) \tag{3.2}$$

for which a plot of $1/r$ against $1/[S]$ yields a straight line. Figure 6 shows kinetic data for irradiated chymotrypsin plotted according to Eq. 3.2. For each plot the intercepts on the abscissa and ordinate are numerically equal to $1/K_m$ and $1/r_{\max}$, respectively. Clearly, irradiation affects *both* parameters in this system.

3.2. Pulse Radiolysis of Enzyme Solutions

3.2.1. *General Features*

The first example of the application of the pulse radiolysis technique to the study of enzyme inactivation was an attempt by Braams[34] to explain the overall electron reactivity of enzymes in terms of the reactivities of the individual component amino acids. It was considered that the rate constant k_{34} for the protein reaction

$$e_{aq}^- + P \xrightarrow{\ k_{34}\ } \text{products} \tag{34}$$

might be equal to the algebraic sum of the reactivities of the individual amino acids, that is,

$$k_{34} = n_1 k_1 + n_2 k_2 + \cdots + n_i k_i \tag{3.3}$$

where n_1 is related to the frequency with which a given amino acid of reactivity k_1 occurs in the structure.

Three proteins were studied by the methods of pulse radiolysis: ribonuclease, lysozyme, and the linear molecule gelatin. Comparison of observed with calculated rate constants for gelatin and lysozyme is shown in Table V. The agreement between theory and experiment is good for gelatin, but poor for lysozyme. The pronounced trend with pH shown by the latter does not appear at all in the calculated rate constants. Poor agreement also found (but not shown) for ribonuclease was ascribed to the influence of protein charge on the encounter frequency and to the effect of chain folding on the effective collision diameter for the reaction. Refinement of the calculation

TABLE V
Calculated Reactivities of e_{aq}^- with Gelatin and Lysozyme

Protein	pH	Calculated k, units of 10^{10} dm^3 mole^{-1} sec^{-1}	Observed k, units of 10^{10} dm^3 mole^{-1} sec^{-1}
Gelatin	5.85	5.5	6.1
	6.2	4.5	5.0
	10.5	2.1	2.8
Lysozyme	6.2	6.1	7.5
	10.1	5.7	2.7
	10.7	5.7	1.8
	11.1	5.7	1.3
	11.8	5.7	0.8

to take account of the charge factor revealed the pH trend in both lysozyme and ribonuclease. However, throughout the pH range studied, calculated values were still high by about an order of magnitude.

The effect of chain folding on reactivity of enzymes with e_{aq}^-, was demonstrated directly by Braams and Ebert[63] in a pulse radiolysis experiment involving ribonuclease. It was known that this enzyme can undergo reversible changes with increasing temperature, a process that involves uncoiling of the ordered structure and transition into a random configuration. Figure 7 shows that the reactivity of this enzyme with e_{aq}^- increases about threefold

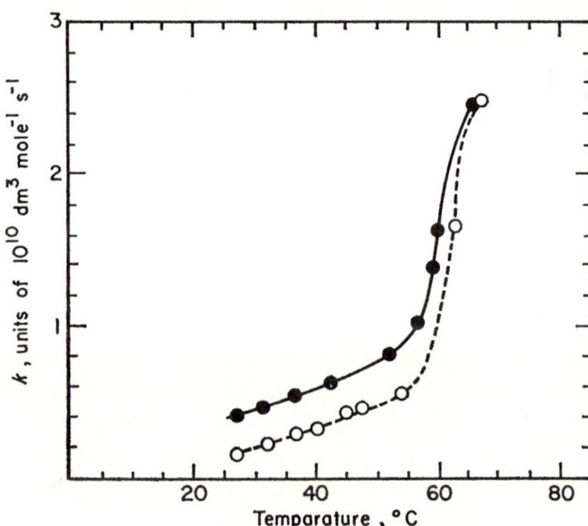

Fig. 7. Effect of temperature on the rate constant for reaction of e_{aq}^- with ribonuclease. Solid line, increasing temperature; broken line, decreasing temperature. (Braams and Ebert.[63])

near the transition temperature for unfolding; by contrast, the reactivity with OH shows little change. The effect is mainly reversible, although some hysteresis is observed. Braams and Ebert suggested that the disulfide bridges, which are by far the most reactive centers for electron attack, become more exposed during the temperature transition. However, the observed slight increase in OH reactivity merely reflects the increase in encounter frequency between OH and the multiple reaction sites on the enzyme where fast reaction can occur.

3.2.2. *Reduction Attack*

Electron localization at disulfur bridges in irradiated enzymes has been observed directly.[64–66] Figure 8 shows transient absorption spectra observed microseconds after pulse radiolysis of aqueous solutions of some enzymes. The solutions each contained *tert*-butanol, which reacts with OH radicals to form a radical which in turn does not react rapidly with the enzyme

$$OH + (CH_3)_3COH \rightarrow H_2O + (CH_3)_2\underset{|}{\overset{}{C}}OH \tag{35}$$
$$\cdot CH_2$$

The spectra are those of the one-electron adducts at the sulfur bridges in the enzyme (PSSP), formed either by direct attack

$$e^-_{aq} + PSSP \rightarrow PS^-SP \tag{36}$$

or by *intra*molecular electron transfer following initial attack elsewhere in the molecule.

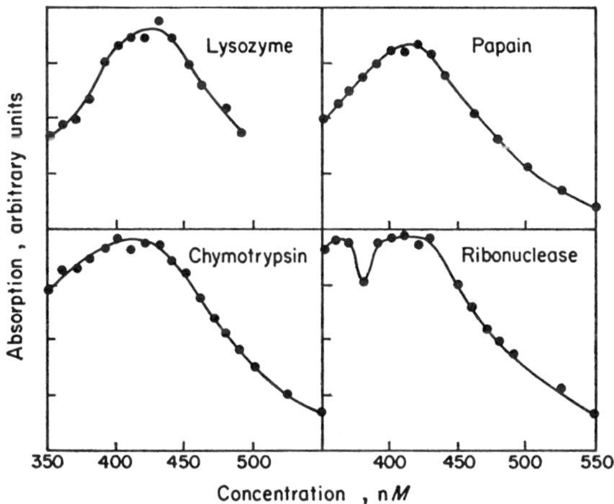

Fig. 8. Transient absorption spectra assigned to the electron adducts at the sulfur bridges in some enzymes. (Adams, Cundall, and Wilson.[66])

Similar spectra can also be formed by *inter*molecular electron transfer, for example, from the electron-donor radical $\cdot COO^-$.[66] This radical is formed by reaction of OH in N_2O-saturated solutions of formate ion

$$OH + HCOO^- \rightarrow \cdot COO^- + H_2O \tag{37}$$

In the presence of enzymes, at concentrations too low to allow direct reaction between the enzyme and either e_{aq}^- or OH, the spectrum of PS⁻SP is still observed on pulse radiolysis. It is formed in the reaction

$$\cdot COO^- + PSSP \rightarrow CO_2 + PS^-SP \tag{38}$$

Gabor and colleagues[67] report that, in the pulse radiolysis of lysozyme solutions containing excess N_2O and formate ion, the final absorption intensity is both increased and more rapidly attained if the solution is preirradiated with several pulses.

Initially, the rate constant for formation of the absorption attributed to PS⁻SP, by electron transfer from CO_2^-, is about 10^9 dm³ mole⁻¹ sec⁻¹. This value is smaller than that measured for direct reaction of e_{aq}^- with lysozyme. Further, the rate constant of the latter reaction does not vary significantly if the solution is preirradiated. It was suggested[67] that the increases both in the total transient absorption and also in the rate of formation represent the effect of preirradiation on the conformation of the enzyme. In this system both e_{aq}^- and OH radicals are removed by the solutes, and therefore H atoms are probably responsible for this component of the damage.

These results suggest therefore that reduction of the sulfur bridges by electron transfer from a donor radical can occur only if this donor can approach within reaction radius of the disulfide links (Fig. 9) and suggest a possible method of studying radiation effects on enzyme conformation. The

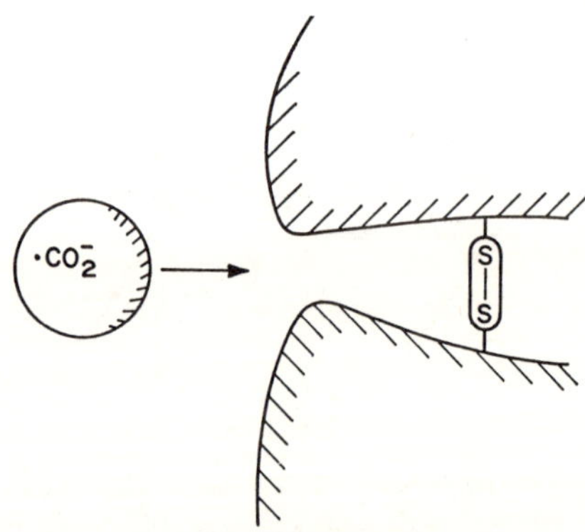

Fig. 9. Structural inhibition of electron transfer from the formate radical ion to the S—S bridges in enzymes (see text).

larger rate constant for reaction of e_{aq}^- and also the lack of any effect of pre-irradiation might suggest that the hydrated electron can penetrate into the interior of the enzyme (Fig. 8). However, it is clear that there are many alternative sites in lysozyme at which e_{aq}^- can react. A more plausible explanation of these results[67] is that, initially, attack by e_{aq}^- does indeed occur at multiple sites on the enzyme, but that such attack is immediately followed by fast *intra*molecular electron transfer along conduction bands in the molecule to the electron-affinic sulfur bridges. There is some direct evidence from flash photolysis[68] indicating that at least some of the total number of electrons liberated by photoionization in tryptophan residues do not solvate but migrate instead to the sulfur bridges by a direct intramolecular conduction process.

3.2.3 *Selective Free-Radical Probes*

Stationary-state inactivation data, particularly when combined with information from pulse radiolysis studies, show decisively that in aqueous solution some enzymes, particularly lysozyme and ribonuclease, are inactivated by OH radical attack. There is no doubt that OH radicals do react with the ring amino acids and there is the evidence, reviewed in Section 2.3, that oxidative inactivation probably involves, at least in part, attack at some of these residues. However, the OH radical is not specific in its reactivity, and thus it is likely that this species can attack a large number of sites on the enzyme and lead to reactions only some of which affect the activity of the enzyme. The problem is to determine which reactions are relevant; it constitutes a major problem in the elucidation of the mechanism of inactivation.

Clearly, more specific techniques are required in order to identify which residue, or residues, if damaged by free-radical attack, would lead to inactivation of the enzyme.

Classical methods used routinely in enzyme chemistry for the identification of amino acid residues, particularly those crucial to activity, involve the use of reagents that attack specific functional groups. Recently, it has been shown that some inorganic free radicals themselves possess similar properties.

One such species, found accidently, is the ion radical $(CNS)_2^-$

$$CNS^- + OH \rightarrow CNS\cdot + OH^- \tag{39}$$

$$CNS\cdot + CNS^- \rightleftharpoons (CNS)_2^- \tag{40}$$

It was confirmed[65, 69] that, during a combined inactivation and pulse radiolysis study of lysozyme, OH radicals are mainly responsible for the radiation-induced inactivation in aqueous solution. For example, several OH radical scavengers protect the enzyme against inactivation. Thiocyanate ion is, however, an exception.

When an aqueous, N_2O-saturated solution of lysozyme containing a high concentration of thiocyanate ion is irradiated, the inactivation efficiency is unaffected. Under these conditions the hydrated electron is removed and all the hydroxyl radicals are scavenged by the thiocyanate. It was concluded therefore that the $(CNS)_2^-$ radical can inactivate the enzyme with an efficiency at least equivalent to that of the OH radical it replaces

$$(CNS)_2^- + \text{lysozyme} \rightarrow \text{inactive product} \qquad (41)$$

Because the $(CNS)_2^-$ radical absorbs strongly in the visible region of the spectrum ($\varepsilon^{480} = 7 \times 10^3$ cm^2 mole^{-1} where, as is customary, ε is defined on the decadic scale), this system is conveniently studied by pulse radiolysis. Such study (Fig. 10) showed directly that the radical reacts rapidly with lysozyme[66, 69] with $k = 6.6 \times 10^8$ dm^3 mole^{-1} sec^{-1}.

Lysozyme contains 20 different species of amino acids. The next step therefore was to measure directly the rate constants for reaction of $(CNS)_2^-$ with each of these solutes. It was found that of the 20 amino acids represented in lysozyme 19 either did not react with $(CNS)_2^-$, or reacted very slowly. The exception was tryptophan, for which the rate constant was not very different from that for the native enzyme and at least 60 times greater than any of the other values (Fig. 10). Since there are eight tryptophan residues in lysozyme, the conclusion was obvious. Free radical-induced damage to one or more of these residues leads directly to inactivation. Subsequent study[69] of lysozyme in which tryptophan-108 had been selectively oxidized by iodine[70] showed

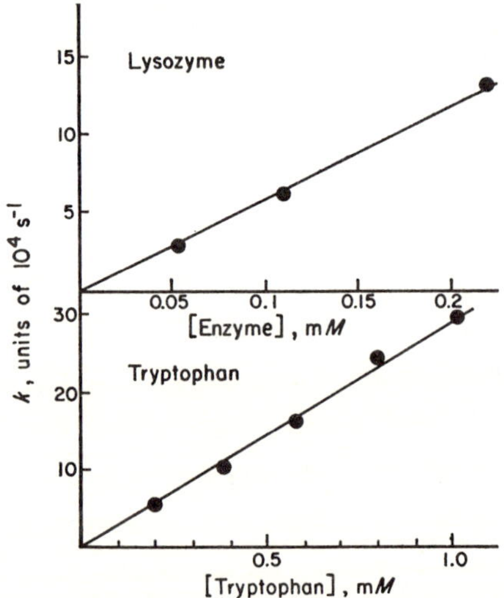

Fig. 10. Rate constants for reaction of the thiocyanate radical $(CNS)_2^-$ at pH 7 showing first-order dependence on solute concentration. (Aldrich et al.[69])

that this treatment lowers the reactivity of the modified enzyme with $(CNS)_2^-$ by a factor of two. This result confirms previous crystallographic evidence implicating tryptophan-108 integrity as crucial to the enzymic activity.

Although confirmatory in nature, the work with lysozyme was interesting in that it suggested that other free-radical reagents may exist that also exhibit selective reactivity with other amino acids. If the inference is correct, such species should prove useful as specific "probes" in experiments designed to identify amino acids present in the active centers and crucial to the activity.

In a search for such selective reagents,[71] some success has been achieved in investigation of radicals derived from halide ions and the carbonate ion. These solutes (with the exception of Cl in nonacidic medium) each react with OH quantitatively and rapidly, for example

$$OH + Br^- \rightarrow Br\cdot + OH^- \tag{42}$$

$$Br\cdot + Br^- \rightleftharpoons Br_2^- \tag{43}$$

$$OH + CO_3^= \rightarrow CO_3^-\cdot + OH^- \tag{44}$$

Similarly to the thiocyanate radical, all the transient species, Cl_2^-, Br_2^-, I_2^-, and CO_3^- possess intense visible absorption spectra which are sufficiently long-lived to permit the reactions between the radicals and some amino acids to be measured directly. Investigation of the reactivity of these species with all the common amino acids showed only six to be reactive above the limit of measurement (about 10^6 dm^3 $mole^{-1}$ sec^{-1}). Table VI shows bimolecular rate constants for these solutes. The rate constants for Br_2^-, $(CNS)_2^-$, and

TABLE VI

Bimolecular Rate Constants for Reactions of Free-Radical Anions with Some Amino Acids [units of $(1 \pm 0.1) \times 10^7$ dm^3 $mole^{-1}$ sec^{-1}] at $22°C^{a,b}$

Radical anion	Tryptophan	Tyrosine	Histidine	ϕ-Alanine	Cysteine	Methionine
Br_2^-	77	2.0 (pH 7.5)	1.5 (pH 7.6)	<0.1	18 (pH 6.6)	1.1 (pH 7.3)
$(CNS)_2^-$	27	0.5	<0.1	<0.1	5 (pH 6.6)	0.2
I_2^-	<0.1	<0.1	<0.1	<0.1	11 (pH 6.8)	<0.1
CO_3^- (pH 11.2 ± 0.3)	44	29	0.7	<0.1	27	12
Cl_2^- (pH 2)	260	27	1.4	0.6	85	0.7

[a] Adams et al.[71]

[b] pH 7.0 ± 0.2 except where otherwise stated; inorganic salt concentration, 10^{-1} M, except iodide (5×10^{-1} M).

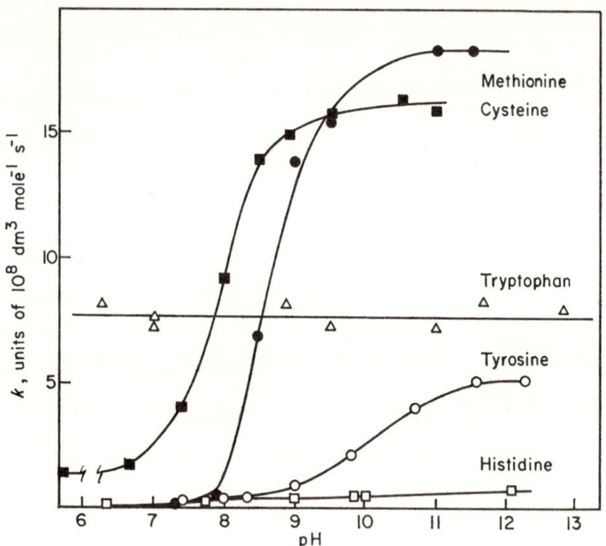

Fig. 11. Effect of pH on bimolecular rate constants for reaction of Br_2^- with some amino acids at $[KBr] = 0.1\ M$. (Adams et al.[71])

I_2^- refer to neutral pH; those for CO_3^- are limited to pH 11. Data for Cl_2^- pertain to pH 2 because at higher pH the rate of formation of Cl_2^- is too slow to follow its subsequent reactions. If the pH differences are ignored, the order of reactivity with both tryptophan and tyrosine decreases in the order

$$Cl_2^- > Br_2^- > I_2^-$$

This result is consistent with the trend in their oxidative properties and, if meaningful, would imply that the pseudohalogen radical, $(CNS)_2^-$, occupies a position intermediate between Br_2^- and I_2^-. For all ring-structured amino acids, the reactivities with $(CNS)_2^-$, Br_2^-, and CO_3^- decrease in the order

$$\text{tryptophan} > \text{tyrosine} > \text{histidine} > \text{phenylalanine}$$

In the case of I_2^- the reactivities are too small, however, to detect this trend.

3.2.3.1. pH Effects. The reactivities show a considerable variation with pH. Figure 11 depicts an example of the data for Br_2^- in which, with the exception of the reaction with tryptophan, all reactivities increase with pH. Data for $(CNS)_2^-$ are quantitatively similar, although in this case tryptophan reactivity does increase gradually with pH. It is interesting that the increase for tyrosine and methionine occurs over the range where deprotonation of the amino group occurs

$$R{-}NH_3^+ \rightleftharpoons R{-}NH_2 + H^+ \tag{45}$$

3.2.3.2. Spectral Characteristics. For a given *amino acid* the transient spectrum produced by the reaction of the radical is *independent* of the nature of the inorganic species; Fig. 12 reproduces some data for tryptophan. However, for a given *radical*, the spectra are characteristic of the amino acid.

The main points that emerge from these studies are that there are three characteristics that indicate the potential usefulness of the method for studying enzyme damage:

(1) The reactivities of these radicals are highly selective.
(2) Most of the rate constants are pH-dependent.
(3) The spectra of the transient products of the reactions are characteristic of the amino acids.

Two main possibilities exist with regard to the mechanisms of these reactions. They are either simple electron transfer processes of the type

$$RH + X_2^- \rightarrow R\cdot + H^+ + 2X^- \tag{46}$$

or, where relevant, direct addition to the ring structure

$$RH + X_2^- \rightarrow (RH\cdot X_2)^- \tag{47}$$

This technique has been applied in further study of lysozyme[71] and also of ribonuclease[72] by Adams et al.

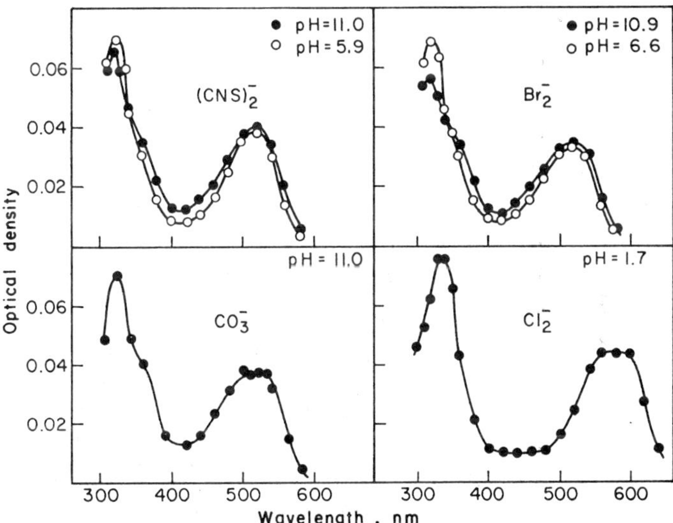

Fig. 12. Transient spectra from pulse radiolysis of N_2O-saturated solutions of tryptophan (10^{-3} M) and thiocyanate, bromide, carbonate, and chloride ions. The small displacement of the low-energy peak for the chloride solution may represent the effect of high acidity. (Adams et al.[71])

Because tryptophan damage is implicated in the just-described study of lysozyme–thiocyanate mixtures, it may be expected that the transient absorption spectra of the reaction product of this enzyme and $(CNS)_2^-$ would be the same as that formed by reaction with tryptophan. Figure 13 compares the two spectra measured under similar conditions of pH and radiation dose. They are clearly identical.

In ribonuclease, in which tryptophan is not represented, the picture is a little more complicated.[72] In this enzyme Br_2^- has proved to be a useful reagent because it inactivates in neutral solution with an efficiency somewhat greater than that of the OH radical it replaces

$$RNAase + Br_2^- \rightarrow inactive\ product \tag{48}$$

Br_2^- reacts with methionine, tyrosine, and histidine, all of which are present in the structure. However, the effect of pH on both the reactivity of the enzyme and the efficiency of inactivation strongly indicates that the reaction with histidine is responsible for the inactivation.

It has been known for some time that various iodine compounds, including iodide ion, influence the efficiency of radiation-induced inactivation of several enzymes.[73] In some cases the enzymes are sensitized to damage,

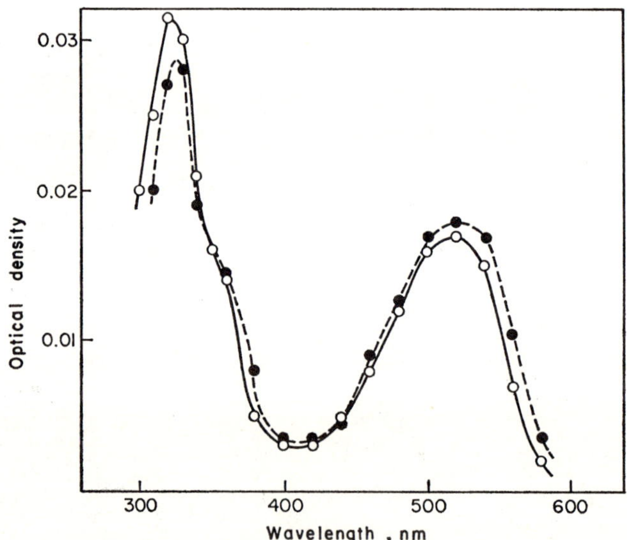

Fig. 13. Comparison of spectra of the products of reaction of $(CNS)_2^-$ with tryptophan and with lysozyme. Solid line, 10^{-3} M tryptophan, 10^{-1} M KCNS, pH 5.9, N_2O-saturated; broken line, lysozyme 2 mg/ml, 10^{-1} M KCNS, pH 5.4, N_2O-saturated; spectra measured 50 μsec after pulse. Dose in each case was 1050 rads. (Adams et al.[71])

whereas in others protection occurs. Davies, Ebert, and Quintiliani[73] suggested that the iodine atom, as I_2^-, can inactivate the enzymes. Clearly,the efficiency of inactivation relative to that produced by the OH radicals that are replaced should influence whether protection or sensitization would be observed. It appears therefore that the study of the reactivities of these enzymes within the available range of free-radical reagents would be profitable.

4. RADIATION CHEMISTRY OF AQUEOUS SOLUTIONS OF NUCLEIC ACID DERIVATIVES

4.1. Introduction

The relationship between intracellular damage to DNA and the response of the cell (lethal or otherwise) to radiation is one of the fundamental problems of molecular radiobiology. Absolute sensitivities of cells to radiation killing vary widely but, in general, mammalian cells are particularly sensitive. In oxic suspensions (i.e., suspensions under oxygenated conditions), doses of the order of a few hundred rads suffice to prevent division of the large majority of the cells. From purely radiation chemical considerations, the amount of molecular damage that can arise from the absorption of such small amounts of radiation is extremely small. Most of the large macromolecules such as proteins and enzymes, that are essential to the functioning of the cell, are represented many times in its structure. It would not be expected that damage to a few of these molecules would bring about such a drastic response to radiation. However, there are not many nucleic acids in that structure. The DNA molecule and its sequences of base is fairly unique. It is not surprising therefore that damage to DNA was suspected as being the main cause of radiation lethality in cells.

There is now considerable evidence, albeit much of it circumstantial, that damage to DNA is indeed one of the major factors in cell inactivation. Data from the many studies of the sensitizing effect of bromodeoxyuridine incorporation (see Section 5) strongly support this inference. In consequence, much attention has been directed to the study of the general radiation chemistry of nucleic acids and the component free bases, nucleotides, and nucleosides, from which it is constituted. Much of this work, up to and including 1967, has been reviewed in detail by Garrison[5] and by Scholes.[75] However, the application of pulse radiolysis techniques to this field is providing much diagnostic and kinetic evidence which is not only substantiating the results of earlier stationary-state studies of simple molecules but also, in some instances, is clarifying the mechanism of indirect damage in DNA itself.

The question is often asked, however, "How relevant is the radiation chemistry of dilute solutions of nucleic acid bases and other fragments to the mechanism of DNA damage *intracellularly?*" If the question relates to whether or not such indirect processes are involved in cell-inactivation, then the answer is, "We do not know." If, however, the question is rephrased, "Is the information gained from the study of simple aqueous solutions concerning the nature and properties of free radicals in DNA constituents helpful in the interpretation of *intra*cellular DNA damage?" the answer is "Yes." An example follows.

It is now established that the 5-6 double bond in the pyrimidine bases is particularly sensitive to free-radical attack. The lifetimes and the course of subsequent reaction of the radical products depend both on the intrinsic properties of the radicals themselves and also on the particular environment. In some instances these properties can be related to the radical site on the heterocyclic base, for example, the C-5 or C-6 position. Specific effects of damage at these positions have been observed in cellular systems. Kieft[76] reports that in the fruit fly *Drosophila* an increase in mutation frequency follows incorporation of tritium-labeled uridine. It was found that when the uridine was labeled at the 5-position the mutagenic effect was about double that found for uridine labeled at C-6.

An even greater differential effect of tritium labeling at C-5 relative to C-6 in uracil has been reported by Person[77] for bacteria in which the uracil is probably incorporated as cytosine. The mutagenic efficiency for C-5 is about sixfold greater than that for C-6 tritiation. It was concluded for the bacterial system, that, while *lethality* is the result of local irradiation, the mutagenic effect is caused by nuclear transmutation. Tritium decays by β-particle emission to form a helium ion. Because the carbon–tritium bond is lost in the process, a free radical results. The subtle differences in the subsequent chemistry of the C-5 radical compared with that of the C-6 species are responsible for the differences in mutagenic efficiency in these two widely different biological species. Study of the radiation chemistry of dilute aqueous solutions of nucleic acids needs no further justification on the score of relevance.

4.2. Early Studies

4.2.1. *Pyrimidines*

Much of the early work on radiation damage in nucleic acids and related substances was conducted before the three-dimensional structures of double-stranded DNA were elucidated. Nevertheless, many of the conclusions from these early studies have since been confirmed.

It was shown that thymine, cytosine, uracil, and dimethyl uracil in oxygenated dilute aqueous solution are bleached following irradiation and that the increase in light transmission in the irradiated solutions is proportional to absorbed dose.[78, 79] The chromophore in these pyrimidines is the 5-6 double bond. The G value for loss of this chromophore was found to be about 2.7 (i.e., corresponding to 5.4 radical centers); it was therefore concluded that saturation of this bond was the major product of indirect radiation action. This conclusion was confirmed by measurement of residual double bonds by bromination with free bromine after irradiation. Scholes and co-workers[80, 81] found that hydroxyhydroperoxides are formed in oxygenated thymine solutions. After synthesis of both the cis and trans isomers of 5-hydroxy-6-hydroperoxy thymine and 6-hydroxy-5-hydroperoxy thymine,[82] it was shown chromatographically by Latarjet et al.[83] that the radiation-induced peroxy compound is a mixture of the cis and trans isomers of the 6-hydroxy derivative. The reaction sequence is probably

$$
\begin{array}{c}
\overset{\displaystyle CH_3}{\underset{\displaystyle H}{\overset{C}{\underset{C}{\|}}}}
\quad \xrightarrow{\ OH\ } \quad
\overset{\displaystyle CH_3}{C\cdot}\ \ \overset{OH}{\underset{H}{C}}
\quad \xrightarrow{\ O_2\ } \quad
\overset{\displaystyle CH_3}{\underset{O_2\cdot}{C}}\ \ \overset{OH}{\underset{H}{C}}
\quad \xrightarrow{\ H\ } \quad
\overset{\displaystyle CH_3}{\underset{O_2H}{C}}\ \ \overset{OH}{\underset{H}{C}}
\end{array}
$$

Hydroxyhydroperoxides are also formed in solutions of uracil and 1-3-dimethyluracil, although a study by Scholes and Weiss[84] of the reactivity of the products with iodine indicated the presence of two components. It is possible therefore that both the 6-hydroxy-5-hydroperoxy and the isomeric 5-hydroxy-6-hydroperoxy derivatives are formed in these instances. In cytosine solutions, because of their instability at neutral pH, only small amounts of the peroxy compounds are found. This thermal instability leads to difficulty in interpreting radiation chemical data from oxygenated systems irradiated under stationary-state conditions. It has been shown that the thermal decomposition of the hydroperoxide leads mainly to *vic*-glycols with trace amounts of other products.[85, 86]

An interesting application of the aerated thymine system employed by Scholes and Willson[87] is its use as a reference standard for the measurement of hydroxyl radical rate constants. The method is based on kinetic analysis of the competition

$$OH + thymine \rightarrow loss\ of\ chromophore \tag{49}$$

$$OH + solute \rightarrow products \tag{50}$$

in which the extent of bleaching of the thymine absorption is used to measure the relative rate constant ratio k_{49}/k_{50}. Subsequent work has shown that OH rate constants obtained by this method are, in the main, fairly reliable.[74]

At high pH (>9), according to Myers et al.,[88] the mechanism of aerated thymine radiolysis changes. As the pH is raised, a gradual increase occurs in the yield of 5-hydroxymethyl uracil. It was concluded that at high pH the major site of OH reaction shifts from addition at the 5-6 double bond to substitution at the 5-methyl group

$$
\begin{array}{ccc}
\text{CH}_3 & \xrightarrow[\text{alkali}]{\text{OH}} & \text{CH}_2\text{OH}
\end{array}
\tag{51}
$$

Some simplification of the radiolysis mechanism of aerated pyrimidines was achieved by Holian and Garrison[89] by the use of cupric ion as a radical oxidant. In deaerated solution the OH adduct of the pyrimidine base, for example, uracil, is oxidized directly to a *vic*-glycol

$$
\xrightarrow[\text{H}_2\text{O}]{\text{Cu}^{\text{II}}}
\tag{52}
$$

A small yield ($G \leqslant 0.5$) of the 5-hydroxypyrimidine, isobarbituric acid, is attributed to the side reaction

$$
+ \text{Cu}^{\text{II}} \longrightarrow + \text{CuI} + \text{H}^+
\tag{53}
$$

4.2.2. *Purines*

Compared with work on pyrimidines, the radiation chemistry of the purine derivatives has received relatively little attention, doubtless because of the greater complexity of these compounds.

In general, purines are more resistant to radiation-induced degradation. Early experiments by Scholes and Weiss[90, 91] showed that irradiation of aqueous solutions leads to gross changes, such as deamination, ring opening and, at high doses, the formation of fragment molecules, for example, urea and oxalic acid. It was proposed that one of the main initiating reactions is OH addition to the 4-5 double bond

$$\text{(54)}$$

However, according to Conlay,[92] some attack elsewhere in the molecule is indicated by the production of low yields of 8-hydroxyadenine and 4,5,6-triaminopyrimidine in this system. Confirmation of the 4-5 double bond as the main site of OH attack derives from the quantitative studies of Holian and Garrison[93, 94] using several purine bases. It was argued, on the basis of conventional purine chemistry, that oxidation at this position should give rise to products which on hydrolysis should yield 4,5-diketo derivatives. In the case of xanthine, this product would be alloxan formed according to the stoichiometry

$$+ 4H_2O + O_2 \longrightarrow$$

$$+ 2NH_3 + HCOOH + H_2O_2 \qquad \text{(55)}$$

Analysis of irradiated oxygenated xanthine solutions showed the formation of alloxan and ammonia in relative yields required by this stoichiometry. For the purines, for example, hypoxanthine and uric acid, the degradation products vary, although quantitative addition of OH at the 4-5 bond is indicated in each case.

4.2.3. *DNA*

Early studies established that in dilute aqueous salt solution DNA is denatured by irradiation. The optical absorption of undenatured DNA in solution is about 40 percent less than that calculated from the absorption coefficients of the individual bases. This "hyperchromic effect" is the result of dipole-dipole interaction between the base pairs stacked in the helical structure. On irradiation the optical absorption of the solution first increases and then falls at higher dose levels.[95] If the DNA is denatured before irradiation, a continual decrease is alone observed. It was proposed that the initial increase represents denaturation caused by breakage of hydrogen bonds, a process that occurs simultaneously with free radical-induced degradation of the base chromophores. At higher doses the latter process predominates with a resultant fall in optical absorption.

According to Scholes and Weiss,[90] elimination of free bases also occurs in irradiated DNA solutions, as the consequence of free-radical attack on the sugar "backbone." The free-base yield, measured after hydrolysis of the irradiated DNA, is about 20 percent of the total base damage. An interesting observation by Collins et al.[95] was the formation of singly bound phosphate groups which were assayed by use of the enzyme phosphomonoesterase. This method led to an estimate of the number of breaks in the sugar-phosphate backbone. It had been shown previously by Peacocke and Preston,[96] by a light-scattering technique, that the mean molecular weight of irradiated DNA in solution is inversely proportional to the square of the radiation dose. This was one of the early indications that single-strand breaks in DNA do not lead to complete rupture.

Studies by Emmerson et al.[97] with nucleoprotein solutions showed that the nucleic acid component appears to be greatly protected from free-radical damage. This result contrasts with the marked degradation of DNA that occurs when DNA and protein mixtures are irradiated in solution. The protection of DNA in nucleoprotein is consistent with the proposed structure for the molecule in which the protein chains envelop the double helix of the DNA.

These early studies, particularly the contributions of Weiss, Scholes, and their co-workers, did much to lay the foundation upon which the present-day knowledge of the radiation chemistry of the nucleic acids was built. In recent years more emphasis has been placed on the detailed study of the free-radical reactivity of these molecules and on the molecular structure of the intermediates. Contributions of the pulse radiolysis technique to this field are discussed in Section 4.3.

4.3. Pulse Radiolysis Studies

4.3.1. *Free-Radical Reactivity*

The earliest survey of the reactivity of free basis, nucleosides, and nucleotides with hydrated electrons was conducted by Scholes and Simic[98] who employed a stationary-state competition method in which nitrous oxide was the reference solute. However, the rapid development of the pulse radiolysis technique

TABLE VII

Bimolecular Rate Constants for Reaction of e_{aq}^- with Nucleic Acid Derivatives[a]

Solute	pH	k, units of $10^9\,dm^3\,mole^{-1}\,sec^{-1}$
Thymine	5.5	17
Thymidylic acid	6.7	1.5
Uracil	5.5	7.7
	7	15[b]
Dihydrouracil	7	4.5[b]
Cytosine	6.0	8
5-Methylcytosine	7.7	10
Cytidine	12	11
Orotic acid	6.5	15
Adenine	6	30[b]
	5.5	10
Adenosine	12	10
Adenosine 5′-phosphate	7	3.8
Guanidine	7	0.25[c]
Hypoxanthine	6.6	17

[a] Data from Scholes[75] except as noted.
[b] According to Greenstock, Hunt, and Ng.[99]
[c] According to Braams.[33]

provided a simpler and more direct method for the measurement of electron reactivity. Table VII summarizes rate constants obtained by this method.

In general, the reactivities are high and there appears to be little difference between the free pyrimidine and the free purine bases. However, for the nucleotides, the rate constants are significantly lower, an effect that has been ascribed by Scholes[75] to the phosphate negative charge.

The two carbonyl groups in the pyrimidine ring are potential sites for reaction of e_{aq}^-. Resonance in the molecule leads, however, to some electron delocalization in the adduct. Studies of reactivity as a function of pH have been informative. According to Greenstock, Hunt, and Ng[99] the rate constants for uracil and the uridylic acids decrease in alkaline solution; these investigators concluded that the effect is the result both of negative-charge repulsion and also of reduction of carbonyl character caused by tautomeric changes of the type

$$\rightleftharpoons \qquad + \; H^+ \qquad\qquad (56)$$

The transient spectra of the one-electron reduction products of several pyrimidines and purine derivatives have been reported by several groups.[35, 75, 99–103] Although the spectra are rather weak and not well resolved, agreement among them is generally fair. The electron adduct spectrum of orotic acid reported by Greenstock[101] is an exception. It exhibits a strong absorption maximum at 330 nm ($\varepsilon^{max} = 8500$ cm^2 mole^{-1}) and has proved to be useful as a marker for electron transfer studies in these systems (see Section 4.3.3).

Myers and Theard[104] have resolved a problem of radical identity. The product of hydrogen atom addition of thymine is *not* the same as the product of protonation of the electron adduct. The spectrum of the species formed by hydrogen-atom addition in acid solution is identical with that of the radical obtained by hydrogen-atom abstraction from dihydrothymine, presumed to be the 5-yl radical ($\lambda_{max} = 400$ nm)

$$\tag{57}$$

It differs greatly from the spectrum of the protonated electron adduct ($\lambda_{max} = 340$ nm).

Table VIII lists rate constants for reactions of the hydroxyl radical with various pyrimidine and purine derivatives. Structure has little effect on the specific rates, all of which appear to be near-diffusion-controlled. This high reactivity is similar to that of OH with benzene derivatives. Greenstock, Hunt, and Ng[100] report that in uracil the rate of formation of the radical product is identical with the rate of bleaching of the 5-6 double bond chromophore, confirming that this bond is the major site for OH attack. Unlike the electron reactivities, there appears to be little effect of tautomerism on the OH rate constants.

In general, the transient spectra[35, 75, 100–109] of several OH adducts, similar to H adducts, have broad absorption maxima near 400 nm.

Other radicals also form adducts with pyrimidines. The ethanol radical CH$_3\dot{C}$HOH adds to the 5-6 double bond of thymine;[110, 111] analysis establishes that this moiety locates at the 6-position in the stable end product[112]

$$\tag{58}$$

TABLE VIII

Bimolecular Rate Constants for Reactions of OH with Free Bases, Nucleosides, and Nucleotides[a]

Solute	pH	k, units of 10^9 dm^3 mole^{-1} sec^{-1}	Reference[b]
Thymine	7.2	5.3	Scholes[35]
	7	5.1[c]	Greenstock[99]
	7	4.6	Willson[74]
	7	4.6[c]	Willson[74]
Uracil	7 4	5.3	Scholes[35]
		6.5[c]	Greenstock[100]
Cytosine	7.5	5.0	Scholes[35]
Adenine	7.4	5.1	Scholes[35]
Thymidine	7.5	4.7	Scholes[35]
	7	4.7[c]	Greenstock[99]
Cytidine	7.3	4.7	Scholes[35]
Adenosine	7.7	4.3	Scholes[35]
Thymidylic acid	7	5.4	Scholes[35]
Deoxycytidylic acid	6.9	5.1	Scholes[35]
Deoxyadenylic acid	6.5	3.6	Scholes[35]
Deoxyguanylic acid	6.8	7.0	Scholes[35]
Adenosine 5′-phosphate	5.4	3.1	Scholes[35]
Cytosine 5′-phosphate	7.5	4.5	Scholes[35]

[a] The thiocyanate data from Scholes et al.[35] have been corrected on the basis $k(\text{OH} + \text{CNS}^-) = 1.1 \times 10^7$ dm^3 mole^{-1} sec^{-1} (Willson et al.[74]).

[b] All the references are multiauthored.

[c] The data so marked are the result of absolute measurement by pulse radiolysis; the other data are obtained from competition studies using both pulse radiolysis and stationary-state methods.

According to Fielden and Stevens,[113] this reaction does not occur at the high dose rates used in pulse radiolysis experiments. Evidently, the rate constant for the addition reaction is too low and radical-radical reactions predominate under these conditions.

Ward and Kuo[114] investigated the effect of chloride ion on the radiolysis of nucleosides, nulceotides, and DNA. In acid solution Cl_2^- ions are produced in the reactions

$$\text{OH} + \text{Cl}^- \rightarrow \text{OH}^- + \text{Cl}\cdot \tag{59}$$

$$\text{Cl}\cdot + \text{Cl}^- \rightleftharpoons \text{Cl}_2^- \tag{60}$$

The Cl_2^- ions react with the single base derivatives to form radicals which are different from those produced by OH radical attack. In neutral solution, however, chloride ions have no effect because, at this pH, reaction (59) is slow and OH radicals are not scavenged. With uridine, cytidine, and the corresponding nucleotides, but not with derivatives of adenine and thymine, it appears that Cl_2^- attack results in cleavage of the glycosidic bond to yield the pyrimidine base. With *deoxy*nucleotides, however, this reaction does not occur. Interestingly, DNA in neutral solution is protected by chloride ions even though Cl_2^- is not produced under these conditions. It is suggested that the effect is attributable to modification of the decay of DNA radicals because of the influence of chloride ions on the macromolecular structure.

4.3.2. *Addition Reactions of Radicals with* DNA *Derivatives*

4.3.2.1. Oxygen. A reaction of some interest with respect to the mechanism or mechanisms of the "oxygen effect" in radiobiology is the addition of oxygen to radicals derived from free bases, nucleotides, and nucleosides

$$B\cdot + O_2 \rightarrow BO_2\cdot \tag{61}$$

Considerable evidence that reactions of this type are invariably fast is confirmed by pulse radiolysis experiments of Willson.[115] Rate constants for reactions of 17 derivatives, including DNA, were measured directly by observation of the effect of oxygen on the rate of decay of the transient absorption of the respective OH adducts. The rate constants for the nucleotides and nucleosides are all reported to be about 1.0×10^9 dm^3 mole^{-1} sec^{-1} with slightly higher values for the free bases (2.0×10^9 dm^3 mole^{-1} sec^{-1}).

4.3.2.2. TAN. In view of the cellular radiosensitizing properties of the stable nitroxyl free radical, triacetoneamine-N-oxyl (TAN) described by Emmerson,[116] there is much interest in its reactions with free-radical derivatives of DNA. According to Emmerson and his colleagues,[117–119] it reacts rapidly with the OH adducts of a variety of deoxyribonucleotides, including DNA

$$\tag{62}$$

The reactions are rapid, with rate constants in the range $(1–2) \times 10^8$ dm^3 mole^{-1} sec^{-1}, that is, about an order of magnitude lower than those for the corresponding reactions with oxygen.

Figure 14 shows pulse radiolysis oscillograms illustrating the reaction of TAN with the OH adduct of calf thymus DNA. The trace on the left,

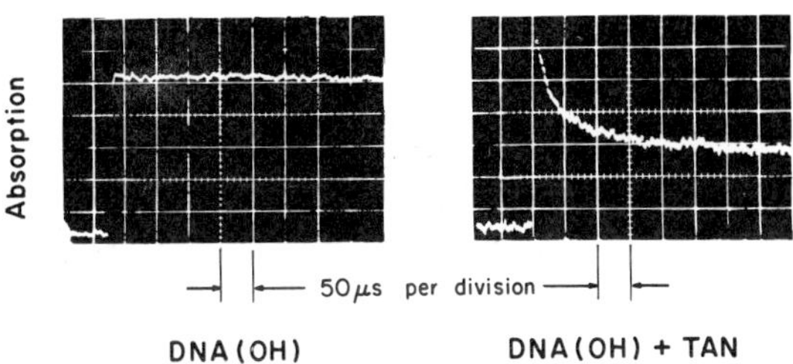

Fig. 14. Oscillograms showing reaction of the sensitizer TAN with the OH adduct of DNA. Left: transient absorption of the DNA–OH adduct formed in pulse radiolysis of DNA (0.06 percent) in N_2O-saturated solution. Right: transient decay in the presence of 150 μM TAN. (Willson and Emmerson.[118])

measured at 310 nm, shows the transient absorption attributed to this radical. In the absence of TAN, there is little decay over the time scale recorded (almost 500 μsec). However, with 1.5×10^{-4} M TAN present (trace on the right) the absorption decays rapidly and appears to reach a lower limiting value which is about 40–50 percent of the initial absorption. The first decay is the result of reaction (62) and the residual absorption is probably that of the TAN–DNA adduct. However, kinetic analysis by Fielden and Roberts[120] has indicated that the addition reaction may be more complicated than that originally proposed. Similarly, reaction of radicals of DNA derivatives with other cellular radiosensitizers, particularly N-ethylmaleimide (NEM) and p-nitroacetophenone (PNAP), are being investigated by pulse radiolysis. These reactions involve both addition and electron transfer mechanisms. This subject is discussed in more detail in the section on radiosensitization (Section 5).

4.3.3. *Electron Transfer Reactions*

In DNA, intramolecular energy transfer is facilitated by the electron interaction between the pyrimidine and the purine bases, a property that is responsible for the hyperchromicity of DNA and also its electrical conductivity. Intramolecular electron transfer has been implicated in the mechanism, or mechanisms, of radiosensitization (see Section 5.3.4). There is therefore some interest in the study of model electron transfer reactions in aqueous solutions of DNA and its constitutent bases, processes which in some cases can be studied by pulse radiolysis techniques. An example early discussed is the reaction of the electron adduct of thymine with the electron acceptor benzophenone.[121] Pulse radiolysis of an aqueous solution containing thymine and a low concentration of benzophenone revealed formation of the spectrum of the benzophenone ketyl radical ion. Under these conditions thymine, the major solute, reacts with e_{aq}^- to form the electron adduct, which then undergoes fast electron transfer to benzophenone to form the ketyl radical ion of benzophenone

$$[\text{thymine}]^- + \text{(benzophenone)} \rightarrow \text{thymine} + \text{(benzophenone ketyl radical ion)} \tag{63}$$

Subsequently, other reactions of this type were observed[122, 123] for a range of pyrimidine and purine bases, nucleotides, and nucleosides with benzophenone, nitroacetophenone, and 2-methylnaphthaquinone used as acceptors. With few exceptions the rate constants for electron transfer were found to be of the order $(2–5) \times 10^9$ dm³ mole⁻¹ sec⁻¹, values that are probably diffusion-limited. Systematic pulse radiolysis studies by Dorfman, Shank, and Arai[124] of one-electron transfer reactions involving aromatic compounds in non-aqueous solvents indicate that reactions of this type are probably generally diffusion-limited.

Although there are differences between the electron affinities of the various pyrimidines and purine derivatives, it is difficult to observe directly one-electron transfer processes between these compounds even by pulse radiolysis because the absorption spectra of the electron adducts are rather weak and not sufficiently well resolved. Orotic acid, 6-carboxyuracil (OA), is, however, an exception since, as stated in Section 4.2.1, its electron adduct shows a strong maximum[101] at 330 nm ($\varepsilon^{\max} = 8500$ cm² mole⁻¹). It has a higher electron affinity than most other purine and pyrimidine derivatives and acts as a ready electron acceptor in bimolecular reactions with free radicals derived from these compounds.[123] In dilute solutions containing *tert*-butanol, orotic acid, and one of the other bases or nucleotides, pulse radiolysis reveals the formation of the electron adduct of orotic acid even when its concentration is much lower than that of the other base. Under these conditions hydroxyl radicals are scavenged by the alcohol, and the hydrated electrons react exclusively or predominantly with the second base, for example, thymine

$$\text{(thymine)} + e_{aq}^- \rightarrow [\text{thymine}]^- \tag{64}$$

The electron adduct of orotic acid is then formed by electron transfer from the thymine radical

$$[\text{thymine}]^- + \text{orotic acid} \rightarrow \text{thymine} + [\text{orotic acid}]^- \tag{65}$$

The bimolecular rate constant is 1.5×10^9 dm³ mole⁻¹ sec⁻¹. Values for transfer to orotic acid for other donor nucleic acid radicals are of the same order of magnitude, ranging from 6.5×10^8 dm³ mole⁻¹ sec⁻¹ for cytidylic acid to 3.5×10^9 dm³ mole⁻¹ sec⁻¹ for adenine.

4.3.3.1. Isomeric Electron Transfer. The effect of the carboxyl group on the relative electron affinity of the pyrimidine base depends upon its position in the ring, as was shown[123] by the pulse radiolysis of a dilute aqueous mixture of orotic acid and its isomer, isoorotic acid, 5-carboxyluracil. Similarly to orotic acid, isoorotic acid reacts rapidly with e_{aq}^- to form an electron adduct, although in this case the spectrum of the adduct is much weaker than that of orotic acid itself.

Figure 15 shows the electron adduct spectrum of isoorotic acid formed on pulse radiolysis of a 2 mM neutral solution of that acid. It is much weaker than that of the orotic acid electron adduct formed under isodose conditions.[101] When 5×10^{-3} M orotic acid is also present in the solution, the transient spectrum present after the pulse is similar to that of the electron adduct of the isoderivative. However, the spectrum changes and after 40 μsec resembles that of the orotic acid transient. The spectral change is

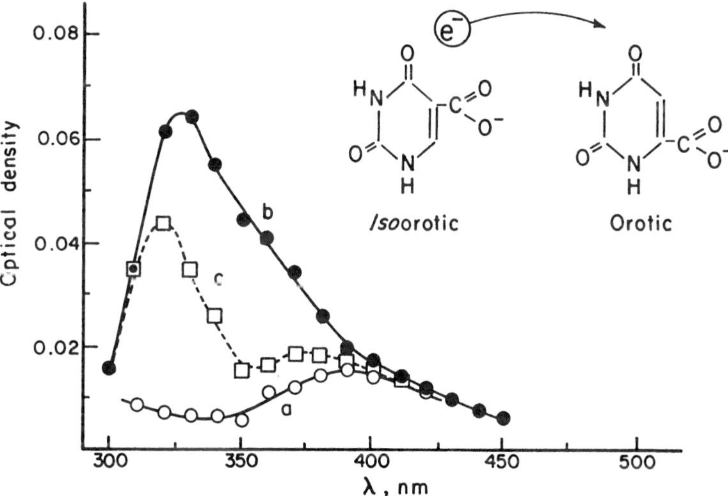

Fig. 15. Interisomer transfer in irradiated oxygen-free solution containing 2×10^{-3} M isoorotic acid, 5×10^{-5} M orotic acid, and 1 M *tert*-butanol. (*a*) Spectrum of electron adduct of isoorotic acid. (*b*) Spectrum of electron adduct of orotic acid. (*c*) Spectrum 40 μsec after pulse radiolysis of a mixture of isomers. Dose: 1 krad. (Adams et al.[123])

attributable to the electron transfer reaction

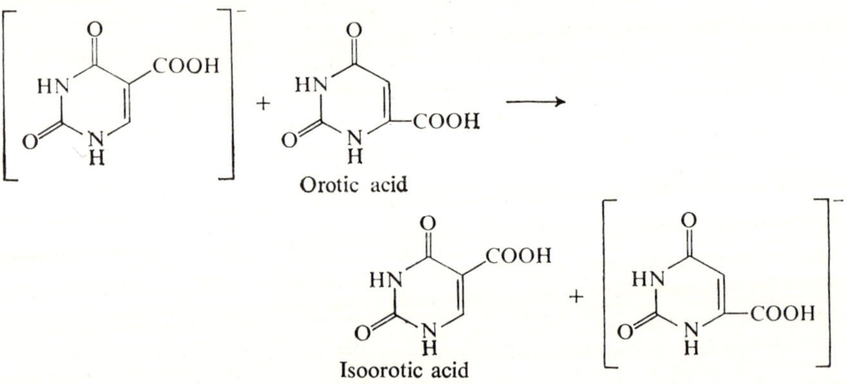

$$\text{(66)}$$

It is probable that, in such oxypyrimidines, after electron attachment some electron localization occurs in the vicinity of the 4-carbonyl group. However, location of the COO^- group in the 6-position would tend to favor electron delocalization by conjugation;

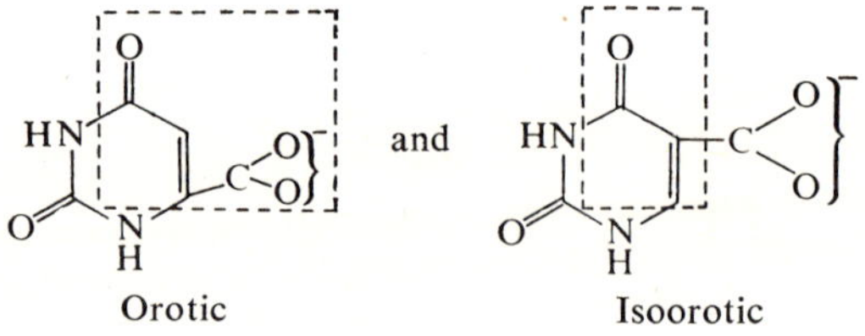

Orotic   and   Isoorotic

Substitution at the 5-position would result in less resonant interaction and would explain the displacement to higher energies of the absorption spectrum of the isotransient.

4.3.3.2. Bromouracil. The thymine analog, 5-bromouracil, owes its high electron affinity to the electronegative bromine atom. Reaction with the hydrated electron leads to dissociative electron capture in the sense that a bromide ion is eliminated

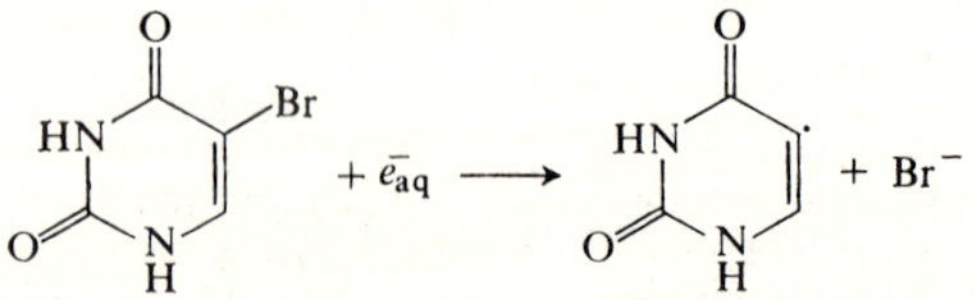

$$+ \, e_{aq}^- \longrightarrow \quad + \, Br^- \tag{67}$$

According to Adams and Willson,[125] a similar reaction occurs when the

electron donor is the electron adduct of thymine

$$(68)$$

The reaction, observed by a pulse radiolysis competition technique, is rapid ($k = 1 \times 10^9$ dm^3 mole^{-1} sec^{-1}) and may be of some relevance to the mechanism of radiosensitization by bromouridine incorporation (see Section 5).

4.3.4. *Radiation Chemistry of Bacteriophage in Dilute Aqueous Solution*

Damage to DNA in irradiated cells may result in part from the action of free radicals formed by radiolysis of water, although there is little information on the extent of the contribution from the indirect effect. Bacteriophage and the biologically active DNA, which can sometimes be extracted from the phage particles, are useful models for the study of such indirect damage.

In general, bacteriophage in aqueous suspension is much more sensitive to the damaging effects of ionizing radiation if the suspension is pure and relatively free of chemical contaminants. In the presence of nutrient broth, for example, doses of the order of megarads are often required to produce substantial inactivation while in contrast, pure suspensions are inactivated in the kilorad range. This vast difference in sensitivity to irradiation is clearly the result of an inactivating effect on the primary radicals produced by water radiolysis.

Bacteriophage consists of a protein shell, or coat, which envelops a central core of infective DNA. It would be reasonable to anticipate, therefore, that indirect action would involve mainly free-radical attack on this protein coat. According to Luthjens and Blok,[126] this process results in the ejection of DNA from damaged phage (T$_4$), although in the later stages of irradiation the efficiency falls off because the phage debris acts as a radical scavenger. There is some indication that in this phage another type of damage occurs which converts the phage to "subviral" particles which, although unable to infect the host cell, are still capable of infecting spheroplasts.

In irradiated suspensions of bacteriophage (T$_7$), the constituent DNA undergoes both single- and double-strand scission.[127, 128] The ratio of single- to double-strand breaks appears to depend, however, upon the irradiation conditions. Freifelder[127] reports that in broth less than half of the inactivated phage had double breaks; however, he found that in buffer, in which it is anticipated that indirect effects are not suppressed, almost all the inactivated phage had one or more such lesions.

There is now much evidence that double-strand breaks originating either by "direct" or "indirect" action are implicated in the inactivation process. However, according to Taylor and Ginoza,[129] the inactivation of double-stranded RF-DNA (i.e., "replicating-form DNA") extracted from ϕX-174 bacteriophage and irradiated in buffer at $-196°C$ appears to correlate more with single-strand breaks. As yet, there is no clear understanding of how these chain breaks occur.

Blok and co-workers[130] used the single-stranded DNA from ϕX-174 bacteriophage, irradiated in buffer solution, to investigate the relative in-activating efficiency of the water radicals. It was found from the effects of scavengers that in concentrated DNA solutions at least 95 percent of the inactivations are caused by OH radicals, although it was inferred that re-ducing radicals are not completely ineffective. The protective effect of oxygen indicates that the radical species O_2^- formed in the reaction

$$O_2 + e_{aq}^- \rightarrow O_2^- \tag{69}$$

does *not* inactivate DNA.

While both the oxidizing and the reducing radicals from water radiolysis are implicated in the inactivation of *protein-free* DNA in dilute aqueous solution, data for bacteriophage are conflicting. Dewey and Stein[131] in-vestigated the effects of atomic hydrogen produced in an electrodeless dis-charge, of hydrated electrons formed by photolysis of ferricyanide solution, and of the radicals produced by water radiolysis, on the activity of bacterio-phage T_7. The relative efficiency of each radical was obtained from a knowl-edge of the G values for the respective radicals produced under various defined conditions. It was concluded that (for a "standard" G value of 1.0) the inactivation efficiency of hydrogen atoms is higher than that of e_{aq}^- although this species appears to sensitize surviving phage to inactivation by hydrogen peroxide. It was also found that in this system also O_2^- radicals do not inactivate the phage. No inactivation was attributed to OH radicals, a result that is surprising in view of the sensitivity of numerous enzymes to OH inactivation (see Section 3).

Gampel and Powers[132] reinvestigated this system under conditions such that the indirect effect predominated. The respective contributions of the reducing radicals and of the OH radicals were also assessed by irradiation in N_2O, N_2, and O_2. An interesting observation was that the relative efficiencies in these three systems depend upon the extent of dilution of the growth medium. In a 1000-fold dilution, the relative efficiencies for N_2, O_2, and N_2O were, respectively, $18:5.5:10.5$, indicating that OH is responsible for about 30 percent of the inactivation in N_2. However, for a 10^5-fold dilution of medium, in which the sensitivity of the phage is increased, almost all the inactivation appears to be by OH. This effect of low solute concentration clearly demonstrates the problems encountered in the study of indirect action in irradiated suspensions of large macromolecules.

5. MODIFICATION OF CELLULAR RADIOSENSITIVITY

5.1. Introduction

Interest in the phenomenon of cellular radiosensitization (i.e., the increase in cellular radiosensitivity caused by change in the internal or external environment of the cell, before, during, or after irradiation) is illustrated by two fairly comprehensive reviews of the subject by Bridges[133] and by Emmerson.[134]

Of particular interest is the study of potential future applications of chemical radiosensitization in the radiation therapy of cancer. It is believed that in some tumors a proportion of the cells become hypoxic because the vascular system of the tumor decreases in efficiency as the tumor grows. Oxygen diffuses away from the blood capillaries in the tumors and is consumed by metabolizing cells. According to the Thomlinson–Gray model[135] there is therefore a decreasing oxygen gradient which is believed to extend about 100–200 μ away from the capillaries. Although beyond this limit the tumor cells die (necrosis) and constitute no further problem, in the boundary region there are some cells that are almost devoid of oxygen (hypoxic). These cells are radioresistant as a result of the hypoxia. They therefore constitute potential sites for regrowth of the tumor after the oxygen supply is reestablished following eradication of the well-oxygenated cells and shrinkage of the tumor after radiotherapy.

Because in clinical practice irradiation dose levels are limited by the tolerance limits of well-oxygenated healthy tissue (e.g., skin, connective tissue, or blood vessels) which are unavoidably irradiated, there is interest therefore in reducing the differential radiosensitivity of hypoxic and oxygenated cells. In principle, there are two ways of achieving this; first, by increasing the radiosensitivity of hypoxic cells without affecting well-oxygenated tissue (differential radiosensitization), and second, by reducing the radiosensitivity of healthy, well-oxygenated cells by chemical radioprotectors. It is the discussion of the molecular mechanisms of radiosensitization and radioprotection that forms the basis of this section.

5.2. The Oxygen Effect

The sensitization effect of oxygen on cells irradiated in the presence of oxygen compared with cells irradiated in anoxia is evident throughout all cellular radiobiology, both *in vitro* and *in vivo*. In many cellular systems, the lethal effect of irradiation is an exponential function of dose in both oxygen and oxygen-free systems, as is illustrated by the dose–response curve of the bacterium *Serratia marcescens* irradiated in suspensions of phosphate buffer;[136, 137] see Fig. 16. In some systems, however, survival curves exhibit a small shoulder (dotted line in Fig. 16). The simple exponential survival curve is given by the relationship

$$N = N_0 e^{-kD} \tag{5.1}$$

where N_0 is the initial number of cells and N is the number surviving a radiation dose D. The constant k is numerically equal to the reciprocal of the dose for 37 percent survival which, for a simple exponential relationship, is equal to the slope of the survival plot and is usually referred to as D_0. It is usual practice to express radiosensitivity of a given cellular system by D_0 expressed in rads. In current usage D_0 expresses the slope of the *linear* portion of a survival curve irrespective of the presence of any shoulder. In some respects it is more logical, however, to use the inactivation constant k, since it is directly proportional to radiosensitivity.[1]

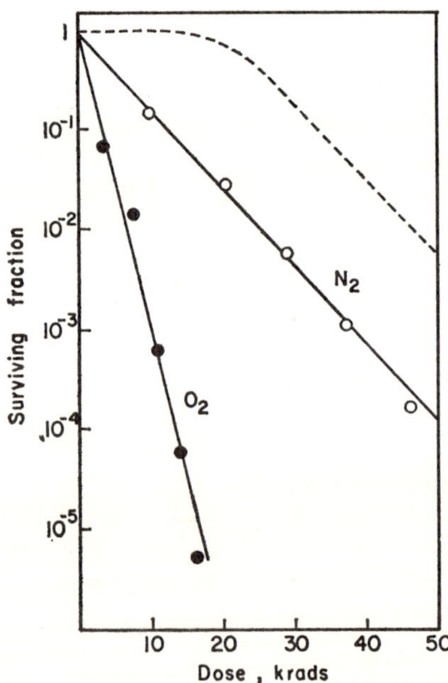

Fig. 16. Typical survival curves for bacteria irradiated in suspension showing the sensitizing effect of oxygen; data for *S. marcescens* grown in synthetic medium. (Dewey[136] and Adams and Cooke.[137])

In a given system the ratio of the slopes of the respective survival curves for oxygen and nitrogen is the "oxygen enhancement ratio" (OER) or the "dose modification factor" (DMF). Values of OER generally range from about 2 to 4, although there are instances (e.g., for neutron irradiation) in which the OER falls below 2.

Doubtless, as shown by early ESR studies of Powers, Ehret, and Smaller,[138] on irradiated bacterial spores, the oxygen effect involves fast free-radical processes. Application of fast-response techniques, in which oxygen is brought into contact with bacteria a short time before irradiation, has proved informative. Howard-Flanders and Moore[139] investigated the time scale of the oxygen effect on *Shigella flexneri* irradiated with 1.8-MeV electrons from a Van de Graaff generator. It was found that exposure of anoxic bacteria to oxygen about 20 msec before irradiation was sufficient to produce the oxygen effect, whereas addition of oxygen 5–10 msec *post-irradiation* produced no change in the anoxic response.

The available experimental time scale has been shortened (by Adams, Cooke, and Michael[140]) by mixing anoxic suspensions of bacteria with an oxygenated solution of buffer in a fast-flow, rapid-mix apparatus. It was shown that in *S. marcescens* the full oxygen effect operates within 4 msec, the limit of time resolution of the apparatus. Preliminary experiments[141] with mammalian cells, using the same apparatus, suggest a similarly fast response to oxygen contact.

Double-pulsing techniques have interesting possibilities. In a technique employed by Epp,[142, 143] two 550-KeV field emission electron sources are used to irradiate a thin layer of bacteria suspended on a millipore filter. Two separate pulses ($\sim$3 nsec) are delivered to the bacteria with separation times variable from the micro- to the millisecond region. The first pulse is designed to remove, by radiation chemical processes, all the oxygen from the bacteria. The time scale for rediffusion of oxygen back to the critical sites is obtained from the dependence of the radiation sensitivity of the cells to the second pulse as a function of the pulse separation time. It is found that the upper limit of the lifetime of the oxygen-sensitive species is about 200 μsec.

A plausible molecular mechanism for the cellular oxygen effect was offered in a "fixation hypothesis."[144–146] According to the mechanism proposed, free radicals are formed in the target molecule either by direct energy absorption or by an indirect effect. These radicals then react rapidly with oxygen to form peroxy radicals

$$R\cdot + O_2 \rightarrow RO_2 \quad \text{(fixation)} \qquad (70)$$

It was proposed that this reaction, considered to produce "irreversible damage," competes with a "repair" or "healing" reaction in which hydrogen atoms are transferred from SH compounds, XSH, occurring naturally within the cell

$$R\cdot + XSH \rightarrow RH + XS\cdot \quad \text{(repair)} \tag{71}$$

Although there is no direct evidence that this competiton applies in irradiated cells, there is no doubt that these reactions occur in model systems. Results from pulse radiolysis studies of such systems are described in Section 5.4.

5.3. Chemical Radiosensitization

5.3.1. *Potential Applications of Chemical Radiosensitizers*

The oxygen effect has a direct clinical application in the use of the hyperbaric oxygen tank in radiotherapy. In this technique, tumors are irradiated while the patient breathes oxygen at 3 atm pressure. The object is to oxygenate hypoxic cells by increasing the oxygen tension in the vascular system of the tumor, thereby extending the extracapillary diffusion gradient of oxygen.

During the decade 1961–1971, several clinical trials were initiated but the results were not immediately as promising as was originally anticipated from the results of radiobiological research. Discussion of possible reasons for this are outside the scope of this chapter. However, it has been suggested that the rate of oxygen consumption in poorly oxygenated regions may be such that the oxygen gradient is not extended in the hyperbaric oxygen chamber by an amount sufficient to oxygenate all hypoxic cells. For this reason alone there is interest in the search for chemical replacements for oxygen sensitization. The criteria that have been applied in the search for such compounds are as follows.

(1) *Differential Action against Hypoxic Cells* For obvious reasons the ideal sensitizers should be active only against hypoxic cells. The general sensitization of both hypoxic and well-oxygenated cells would be useless and probably detrimental to treatment.

(2) *Acceptable Toxicity* The concentration range over which a given compound is able to sensitize hypoxic cells in tumors must be considerably smaller than that which might cause unacceptable toxic responses elsewhere.

(3) *Acceptable Pharmacology* Apart from the toxicity problems, a suitable radiosensitizer must satisfy other pharmacological requirements. Included in the most important of these is resistance to metabolic breakdown. The poor vascular systems in tumors present a barrier to drug penetration and it would be unlikely that suitable concentrations could be built up in tumors over a very short period of time. A drug that would be rapidly inactivated by biochemical reaction would be useless. Further, the sensitizer would have to be resistant to inactivation by binding to protein and would

have to have suitable lipid-water distribution properties in order to diffuse the considerable distance away from capillaries that is required.

(4) *Cell-cycle Dependence* It is likely that the radioresistant hypoxic cells in tumors are arrested fairly early in the cell cycle, probably in G_1, the phase of the cycle before the onset of DNA synthesis. Clearly, suitable sensitizers should preferably operate throughout the cycle and certainly must be effective in the early phase.

Formidable as these requirements are, there are other problems in the search for potential radiosensitizers in radiotherapy. However, the large amount of published work in this field, admittedly most of it concerned with systems *in vitro*, has provided much justification for continuing this line of research. Of the various sensitizers available, most of them satisfy some of the requirements mentioned; toxicity varies sufficiently widely to offer hope that the problem can be overcome. Most sensitizers are active only against hypoxic cells, and there are recent indications that criteria (3) and (4) are met in some instances. Several interesting compounds have been characterized; the molecular mechanisms by which they are suggested to operate form the basis for the remainder of this section.

5.3.2. *Early Studies*

Oxygen is paramagnetic in its ground state. It is consequently not surprising that the structurally similar nitric oxide molecule has been examined for possible sensitizing ability. It was found by Howard-Flanders et al. to sensitize anoxic bacteria with an efficiency as great as that of oxygen;[147, 148] similar to oxygen, it had to be present at the time of irradiation. It was suggested therefore that, as in oxygen sensitization, an essential step in the mechanism is reaction with a radical formed in an essential target molecule, that is, "fixation"

$$\text{radical} + NO \longrightarrow \text{radical-NO} \tag{72}$$

Subsequently, it was found by Dale, Davies, and Russell[149] that the sensitizing effect is lost at much higher concentrations, a phenomenon observed more recently with other radiosensitizers.[150, 151]

An impetus to the field was provided by the discovery by Bridges[152, 153] that NEM sensitizes anoxic suspensions of bacteria. The investigation was prompted by the knowledge that NEM is an SH poison and has been used as such in some industrial processes. It was argued that if the natural radiosensitivity of cells is partly conditioned by SH compounds occurring naturally in the cell as radioprotectors, removal of these compounds by chemical reaction might increase the radiosensitivity. SH compounds react with NEM by addition to the 3-4 double bond in reactions that are generally quite fast[154]

$$RSH + \underset{\overset{\displaystyle CH-C}{\displaystyle CH-C}}{\big\|}\; NC_2H_5 \longrightarrow \underset{\overset{\displaystyle RS-CH-C}{\displaystyle CH_2-C}}{\big|}\; NC_2H_5 \qquad (73)$$

The sensitizing effect of NEM, usually confined to anoxic systems, has been demonstrated in many different bacterial cell systems. Naturally, this success led to the search for other radiosensitizers and a large number of widely differing compounds were found to be active, including the SH reagents, iodoacetamide[155] and *p*-chloromercuribenzoate.[156] Other compounds shown to be anoxic cell sensitizers are cupric ion,[157] the organic nitroxyl compounds[158, 159] which are stable free radicals of general formula

$$\overset{\displaystyle R}{\underset{\displaystyle R'}{{>}N-O}}$$

and numerous compounds of high electron affinity.[137, 160] The last group includes α-diketo derivatives—for example, biacetyl and the glyoxals, various quinones, and a range of substituted phenones—one of which, *p*-nitroacetophenone (PNAP), reveals several interesting properties from the viewpoint of potential studies *in vivo*.

It is now clear from the results of this work that the mechanisms by which all these compounds sensitize fall into different categories. Indeed, there is considerable evidence that "SH suppression" is not as important a mechanism of sensitization as was previously believed. As the following sections indicate, experiments with various model systems have done much to distinguish these categories and have led to improved understanding of the molecular processes involved.

5.3.3.   *Analysis of Hypothetical Mechanisms*

The flow diagram shown in Fig. 17 has been proposed as a convenient classification of mechanisms of radiosensitization.[161] Two main categories are suggested, namely, "radiation chemical" and "biochemical." The former includes all processes involving the *direct* participation of short-lived species (free radicals, excited states, and ions) in the sensitization processes. The biochemical category includes those processes in which the sensitization

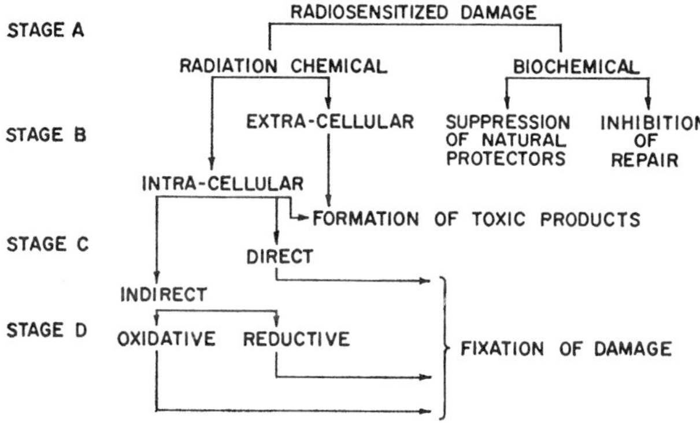

Fig. 17. Classification of mechanisms of radiosensitization.

results, either from interaction of the sensitizer with damaged cell components *after* all fast radiation chemical processes have ceased, or from modification of the environment or constitution of the cell *before* irradiation.

Although the distinction is somewhat arbitrary, in experimental practice it can be extremely precise. The relevant parameter is time of action. Reactions of free radicals and other transient species are very rapid and usually occur in the milli, micro, and submicro time range. Biochemical processes, however, occur usually over time scales that are longer by several orders of magnitude. Rapid mixing and other fast-response techniques exploit this vast difference in time scale in order to determine the general class in which a given sensitizer may be placed. These techniques, designed to reduce to a minimum the time interval between mixing of the sensitizer with the cells and subsequent irradiation of the mixed suspension, confirmed the role of free-radical processes in the oxygen effect. Other experiments have been carried out on the time scale of bacterial sensitization by NEM and TAN and on the protection of bacteria by glycerol;[162] these are discussed in Sections 5.3.4 and 5.4.2.

5.3.3.1. Radiation Chemical Processes—Extracellular. The general features of sensitization mechanisms involving interaction of the sensitizer with free radicals or other short-lived species produced by the radiation may be first considered.

Once it is established by fast-response experiments that such processes are involved, it is necessary to determine whether these initiating events occur *inside* or *outside* the cell membrane (stage B, Fig. 17). Although most *in vitro* experiments on cellular sensitization are conducted under conditions in which the growth medium is removed, either by centrifugation and resuspension or by extensive dilution, there is adequate evidence that sensitization *can* and *does* occur even when the extracellular fluid contains high concentrations of free-radical scavengers. Under these conditions sensitization produced by interaction of the sensitizer with free radicals outside, or on the surface of, the external membrane is very unlikely.

There are, however, some compounds that lead to apparent sensitization by radiation processes that occur *outside* the cell. These are compounds that react with radiation-induced free radicals produced in the extracellular fluid to form products that are extremely toxic to the cells; iodoacetamide (ICH_2CONH_2)[155] is a case in point. Dewey and Michael[163] showed, by a fast-injection technique, that cells added to an irradiated solution of iodoacetamide a short time after irradiation are extensively inactivated. Although it has been suggested[164] that the toxic product responsible is the iodine atom (or its complex, I_2^-), the duration of the posteffect observed[163] indicates that the lethal effect results from a nonradical toxic product. According to Moroson and Tenney,[165] the iodine molecule is an obvious possibility, particularly since similar sensitization has been observed with KI and KIO_3^-. As might be expected, iodoacetamide sensitization can occur in both nitrogen and oxygen.

Cramp[157] has shown that cupric ion sensitizes anoxic bacteria. In this instance, the effect is attributable to the highly toxic cuprous ion formed by reduction of Cu^{2+} either by e_{aq}^- or by other radicals

$$Cu^{2+} + e_{aq}^- \rightarrow Cu^+ \tag{74}$$

Absence of sensitization in the presence of oxygen or other oxidizing agents (e.g., $KMnO_4$) is the consequence of rapid reformation of the nontoxic cupric ion

$$Cu^+ + O_2 \rightarrow Cu^{2+} + O_2^- \tag{75}$$

It is not known, however, whether sensitization by cupric ion occurs intra- or extracellularly.

5.3.3.2. Radiation Chemical Processes—Intracellular. The large water content in bacterial and mammalian cells implies that about 70–80 percent of the energy deposition occurs in the aqueous component. In principle, therefore, molecular damage, which is a precursor to cell death, can occur, either by direct energy absorption in the relevant "target area," or by reaction of free radicals derived from the products of radiolysis of intracellular water. These damaging radicals may be either the primary radicals e_{aq}^-, OH, and H, or secondary radicals formed by their reaction with other cell constituents.

The determination of the relative importance of "direct" and "indirect" damage is a major and long-standing problem of cellular radiobiology; the origin of that part of the total damage that is susceptible to sensitization by chemical additives is still somewhat obscure. It is logical therefore to attempt to subclassify intracellular sensitizers into those that interact with "direct" damage and those that enhance the effects of the products of radiolysis of cell water (stage C, Fig. 17). Both types remain possibilities and indeed there is indirect evidence indicating that both general types of mechanism do in fact occur.

Sensitization of indirect damage may be subclassified further (stage D, Fig. 17). In many ionization phenomena (i.e., processes in which the ejected electron escapes from the electrostatic field of its parent ion), the subsequent chemistry of the system can be resolved into two distinct pathways—oxidative and reductive. In irradiated water, for example, reactions of OH radicals are generally oxidative, whereas both the hydrated electron and the hydrogen atom are powerful reducing agents. In the search for sensitization mechanisms, therefore, it is worthwhile examining the redox properties of known radiosensitizers; it is notable that radiosensitizers are generally strong oxidizing agents. Recognition of this fact led to the discovery of a large number of potent radiosensitizers, the "electron-affinic" group.[137, 160] However, by itself, recognition of an empirical relationship between the sensitizing ability of given compounds and their oxidizing properties is of little value, except that it narrows the search for additional radiosensitizers. The efficiency of sensitizers operating by intracellular binding of naturally occurring radioprotectors (e.g., SH) would also be expected to correlate with redox properties; SH compounds are indeed strong reducing agents. Further, if sensitizers were to act by interference with repair processes, it is not difficult to accept that here too a redox relationship might be apparent. However, it is when the redox property is considered in the light of results from rapid-mix experiments, or pulse radiolysis and ESR studies with model compounds, that an understanding of the mechanisms of sensitization begins to emerge. In the next section, the mechanism of action of known sensitizers is discussed in more detail.

5.3.4. *Specific Mechanisms of Known Radiosensitizers*

5.3.4.1. NEM. The belief that NEM (*N*-ethylmaleimide) sensitizes by intracellular SH binding led to much research, not only with this compound but with other compounds capable of binding SH. The development of this field has been reviewed in detail by both Bridges[133] and Emmerson[134] and there is therefore little point in reiteration. However, some of the more important experimental conclusions that have led to the opinion that SH binding is *not* the major cause of sensitization by this compound are worth repeating.

It was shown by Bridges[153] that when bacteria are incubated with NEM and then washed free of the sensitizer they are not sensitized to subsequent irradiation. It was anticipated that if SH binding is important to sensitization the effect should be observed even when free NEM is removed before irradiation. Dewey[166] confirmed that this procedure depresses SH levels considerably and that there is no regeneration within 2 hr. Further, Adams, Cooke, and Michael[140] showed that the time scale of sensitization of anoxic bacteria by NEM is extremely short, much shorter in fact than a reasonable estimate of the time required for binding of NEM with intracellular SH groups.

In these experiments anoxic suspensions of bacteria were mixed with a deaerated solution of NEM, in a fast-flow apparatus, 4 msec *before* irradiation with a train of electron pulses from a lineaɪ accelerator. When the sensitizer was added 4 msec *after* irradiation of the bacteria, very little sensitization was observed. Fast mixing of unirradiated cells with irradiated NEM produced no lethal response whatsoever, thus eliminating the possibility that the sensitization is a result of formation of toxic radiolysis products.

The extent of sensitization observed when NEM is added before irradiation depends upon both the concentration of the sensitizer and the pre-irradiation contact time (Fig. 18), effects that probably reflect the time required for diffusion of NEM to the site of its action. Although the rate of reaction of NEM with free SH groups is undoubtedly fast by nonradical standards (the half-life for reaction[154] of 1 mM NEM with 1 mM reduced glutathione at pH 7 is only $\sim$10 sec), Fig. 18 shows that the time scale for sensitization is far too short for such a mechanism to be relevant.

A suggestion originally made by Bridges[153] is that NEM reacts with a radiation-induced free radical in a vital target molecule. The implication in that suggestion is that, similar to the fixation hypothesis proposed for

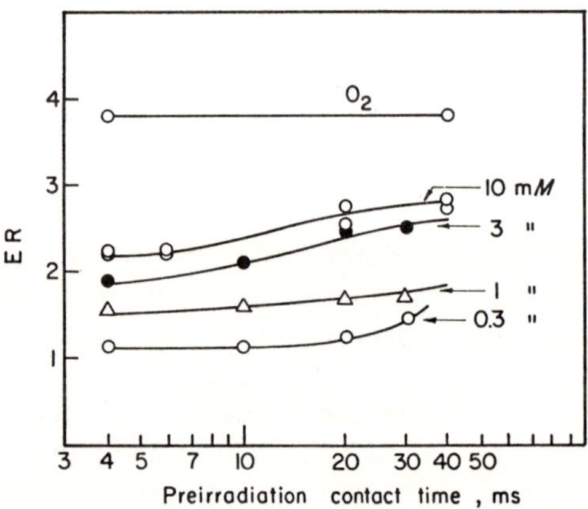

Fig. 18. Variation of the enhancement ratio (ER) of NEM with preirradiation contact time over the NEM concentration range 0.3–10 mM. (Adams, Cooke, and Michael.[140])

oxygen, NEM sensitizes by *competing* (not binding) with intracellular SH compounds which might otherwise repair some of the damage.

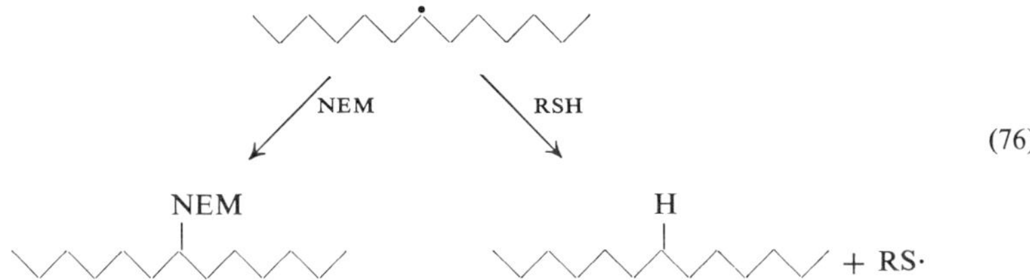

$$(76)$$

The general radiation chemistry of NEM in aqueous solution has been studied by both stationary-state[167] and pulse radiolysis methods,[168, 169] and it is now known that NEM is a very efficient radical scavenger, reacting with both e_{aq}^- and OH at rates that approach diffusion-limited values. In both cases radical adducts are formed

$$e_{aq}^- + NEM \rightarrow NEM^- \tag{77}$$

$$OH + NEM \rightarrow \cdot NEM-OH \tag{78}$$

at specific rates given by Ward, Johansen and Aasen[167] as $k_{77} = 3 \times 10^{10}$ dm^3 mole^{-1} sec^{-1} and $k_{78} = 6 \times 10^9$ dm^3 mole^{-1} sec^{-1}, respectively. In the absence of oxygen, the OH adduct reacts with a further molecule of NEM

$$\cdot NEM-OH + NEM \rightarrow products \tag{79}$$

Of considerable significance with respect to the mechanism of sensitization are the reactions of NEM with the electron and OH adducts of various simple nucleic acid derivatives, for example, thymine (Th)

$$Th-OH + NEM \rightarrow products \tag{80}$$

$$Th^- + NEM \rightarrow products \tag{81}$$

It has been suggested that reactions (80) and (81) proceed both by adduct formation and by electron transfer. As yet, few quantitative data are available concerning the rate constants for nucleic acid reactions of this type, although some preliminary results of Greenstock, Adams, and Willson[122] indicated that reactions of the electron adducts with NEM are very fast. However, as these pulse radiolysis experiments were carried out at high pH where NEM is less stable, the data are suspect and require to be repeated.

Pulse radiolysis has been used to study reactions of NEM with simple radicals derived from alcohols and other compounds.[168, 169] Both adduct formation and electron transfer occur, depending upon the nature of the radical involved. The formyl radical ion, CO_2^-, formed by hydrogen atom abstraction from formate ion, reacts exclusively by electron transfer[168, 169]

$$CO_2^- + NEM \rightarrow CO_2 + NEM^- \tag{82}$$

whereas the methanol radical CH_2OH forms a radical adduct

$$\cdot CH_2OH + \underset{\substack{CH-C \\ O}}{\overset{\substack{CH-C \\ O}}{\parallel}}NC_2H_5 \longrightarrow HO-CH_2\cdot CH-C \underset{\substack{\cdot CH-C \\ O}}{\overset{O}{\diagup}}NC_2H_5 \qquad (83)$$

Absolute confirmation of the ability of NEM to bind with DNA irradiated *in vitro* was provided by Johansen et al.[170] in an elegant tracer experiment in which the DNA was irradiated in an aqueous solution containing NEM labeled with ^{14}C. The ^{14}C activity was found to be incorporated in the DNA and showed the expected linear increase with radiation dose. The conditions of the experiment were such that the DNA, rather than the NEM, reacted with the free radicals from water, and therefore, the relevant reaction is

$$DNA\cdot + NEM \rightarrow adduct \qquad (84)$$

and *not*

$$DNA + NEM\cdot \rightarrow adduct \qquad (85)$$

5.3.4.2. The Stable Organic Nitroxyl Free Radicals. Following the demonstration that nitric oxide is a cellular radiosensitizer, it was natural that the related organic nitroxyl free radicals of general formula

$$\underset{R'}{\overset{R}{\diagdown}}NO$$

should be investigated (Table 18). The first compound tested was di-*tert*-butylnitroxide (DTBN), which Emmerson and Howard-Flanders[171] found to sensitize anoxic *Escherichea coli B/r* almost to the level of the oxygen effect. This study was followed by extensive work with the related TAN, which is effective at concentrations lower than that required for DTBN.[116] Thereafter, Emmerson, Fielden, and Johansen[119] showed that the double-ring compound norpseudopelletierine-*N*-oxyl (NPPN, Table IX) sensitizes at even lower concentrations.

TAN is very soluble in water, remarkably nontoxic to a wide range of bacteria, and generally sensitizes only under anoxic conditions. Unlike most sensitizers outside the *N*-oxyl group, which rarely mimic the full oxygen effect, TAN can sensitize some bacteria to a *greater* extent than oxygen. Unfortunately, however, the encouraging results from bacterial studies are not maintained in mammalian cell systems. For Chinese hamster cells in culture, the maximum degree of sensitization attained is small (enhancement

TABLE IX
Structures of Some *N*-Oxyl Sensitizers

Di-*tert*-butylnitroxide (DTBN)	
Triacetoneamine-*N*-oxyl (TAN)	
2,2,6,6-Tetramethyl-4-piperidonol-*N*-oxyl (TMPN)	
Norpseudopelletierine-*N*-oxyl (NPPN)	

ratio = 1.5), even when the TAN concentration is 10^{-2} *M* and near the toxic limit.[172] As of 1971 there were no authenticated reports of TAN sensitization of mammalian cells irradiated *in vivo*. However, these disappointing results may not be representative of the *N*-oxyls generally and, in view of their favorable properties, further work on mammalian cell sensitization by these compounds is certainly warranted.

Much indirect evidence from various model systems supports the suggestion[116, 173] that the mechanism of sensitization by the *N*-oxyls resembles the "fixation" mechanism proposed for oxygen, in the sense that these compounds form adducts with free-radical centers in DNA

$$\text{(radical)} + \text{TAN} \longrightarrow \text{(TAN adduct)} \tag{86}$$

The pulse radiolysis experiments referred to in Section 4.3.2 showed that TAN reacts rapidly with free radicals formed by OH attack on various nucleic acid derivatives including DNA itself. The residual absorption shown in the oscillogram in Fig. 14 is probably that of the TAN adduct of the DNA radical.

Absolute proof of radiation-induced TAN adduct formation was obtained from some isotopic-labeling experiments by Nakken, Sikkeland, and Brustad[174] in which calf thymus DNA was irradiated in aqueous solution containing ^{3}H-TAN. The activity thus became incorporated into the DNA and the adducts were extracted on a Sephadex column. The rapid-mix technique was used to study the reaction time scale for adduct formation.[175] The tritium-labeled TAN was added to the DNA solution at various periods after irradiation of the DNA in nitrogen, and the incorporated activity measured as a function of this postirradiation mixing time. The time profile indicated that at least two DNA transients are formed, one of which is surprisingly long-lived. The respective half-lives are ~ 70 msec and ~ 10 sec.

Emmerson, Fielden, and Johansen[119] correlated the rates of reaction of some N-oxyls with the hydroxythymine radical with their effectiveness as bacterial radiosensitizers

$$\text{(hydroxythymine radical)} + \text{TAN} \longrightarrow \text{(TAN adduct)} \tag{87}$$

Most of the N-oxyls normally used are analogs of the di-*tert*-butyl derivative DTBN, in which the NO group is sterically partially blocked by two large flanking groups. Such is not the case for the double-ringed compound NPPN (Table IX,) in which the NO group is exposed. It was argued that this compound would have a free-radical reactivity greater than that of the other N-oxyls and that, if adduct formation were relevant to sensitization, NPPN should show a corresponding increase in effectiveness as a radiosensitizer. Experimentally, this correlation is found, as is shown in Table X by the comparison of free-radical reactivities with the sensitization $C_{1/2}$ values,*

---

* Radiobiologists customarily refer to *these* concentrations as "k values."

TABLE X
Reaction Rates of Nitroxyls with the Hydroxythymine Radical Compared
with the Concentrations $(C_{1/2})$ Required to Give Half the Maximum
Sensitization in Anoxic AB2463 *recA* Bacteria[a]

Sensitizer	Reaction rate with hydroxy-thymine radical at 20°C, units of 10^8 dm^3 mole^{-1} sec^{-1}	$C_{1/2}$, μmoles/liter
TMPN	2.6	300
TAN	3.5	230
NPPN	6.4	60
O_2	19.0	8

[a] Emmerson, Fielden, and Johansen.[119]

that is, the respective concentrations of sensitizer required to produce 50 percent of the maximum sensitization effect.

5.3.4.3. PNAP and the Electron-Affinic Group. The suggestion that the property of radiosensitization is related to the electron affinity of the sensitizer[137, 160] led to the discovery of a wide range of active compounds (Table XI). The original hypothesis of Adams and Dewey[160] was that the sensitizers, by acting as electron transport agents, increase the efficiency of damage caused by reaction of hydrated electrons with sensitive sites within the cell. It was argued that the extent of such damage, arising from e_{aq}^- formed in the vicinity of these "target sites" would depend upon the diffusion distance of the hydrated electron. In the normal cell this distance is limited by the high concentrations of other electron-scavenging molecules likely to be present in the immediate neighborhood of the target sites. If, however, a foreign molecule (i.e., a sensitizer) is present whose electron affinity exceeds that of these electron scavengers, it can act as a competitive electron scavenger and also as an acceptor in electron transfer reactions involving radicals formed by reactions of e_{aq}^- elsewhere. The essential feature of the hypothesis is that the sensitizer–electron complex, stabilized by resonance, can diffuse farther than the hydrated electron and will react, finally, with a site of even higher electron affinity in the target molecule. The sensitizer acts, in this view, as an electron transport agent.

The hypothesis was proposed originally in explanation of the sensitizing effect of NEM because this compound was known to form a resonance-stabilized electron adduct and was expected to function as an acceptor in some electron transfer reactions [e.g., reaction (82)]. The proposal was "tested" by investigating the sensitizing properties of biacetyl, benzophenone, and benzoquinone, all of which form resonance-stabilized free radicals.

TABLE XI
Sensitizers of the Electron-Affinic Group

Compound	Structure
Biacetyl	$CH_3CO-COCH_3$
Glyoxals	$R-CO-CHO$
Indane-trione	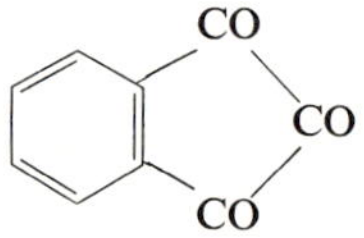 benzene ring fused to $CO-CO-CO$
Indane-dione	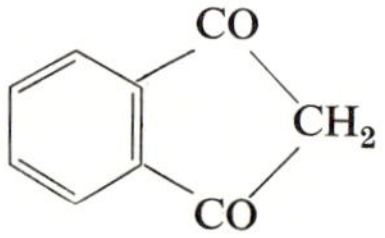 benzene ring fused to $CO-CH_2-CO$
Dimethylfumarate	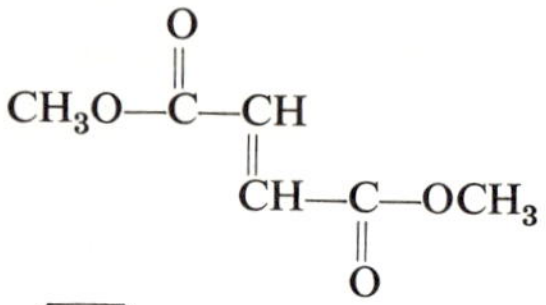 $CH_3O-\underset{O}{\overset{O}{C}}-CH=CH-C-OCH_3$
Phenylpyruvic acid	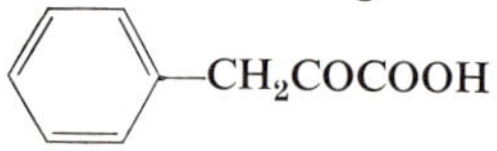 $C_6H_5-CH_2COCOOH$
Benzophenone	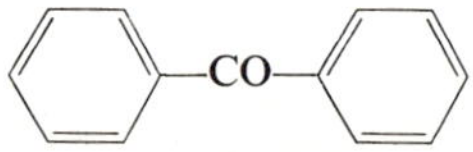 $C_6H_5-CO-C_6H_5$
Triacetylbenzene	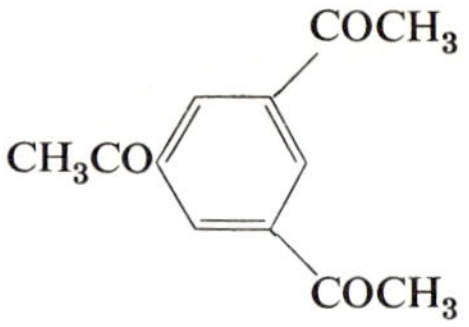 benzene ring with three $COCH_3$ groups
p-Nitroacetophenone	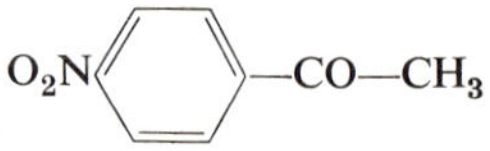 $O_2N-C_6H_4-CO-CH_3$
Methylnaphthaquinone	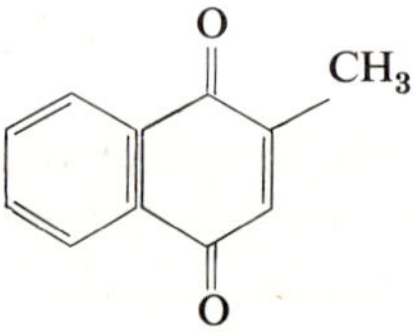 methyl-substituted naphthaquinone

Although this early model is now known to be incorrect as a working hypothesis, it was successful in that the demonstration of sensitization of anoxic bacteria by the first two compounds led to the search for others. In time, various other sensitizers were found, most of which were considerably more efficient than either biacetyl or benzophenone.[137] These included derivatives of pyruvic acid, dimethyl fumarate, keto derivatives of hydrindene (indane di- and triones) and some quinones. Compounds related to methyl glyoxal and phenyl glyoxal, shown to be sensitizers by Ashwood-Smith, Bridges, and co-workers,[176, 177] can be regarded as members of this class of sensitizers.

Sensitization by these compounds is confined generally to anoxic cell systems, although in indanetrione a slight effect is also observed in oxygenated suspensions.[137] The concentration range over which sensitization occurs varies widely; some compounds (e.g., menadione, 2-methylnaphthaquinone) are effective in the 10–100 μM range, whereas others (e.g., biacetyl), sensitize at concentrations of several millimolar. Evidence from both cellular[151] and pulse radiolysis studies[178] indicates that the higher the electron affinity the lower the concentration range required for sensitization.

Of particular interest is p-nitroacetophenone PNAP, a sensitizer that meets several of the requirements necessary for pilot studies *in vivo*.[179, 180] Both anoxic bacterial and mammalian cells are sensitized at concentrations well below the toxic limit, even in the presence of considerable quantities of extracellular protein. Further, PNAP sensitizes anoxic mammalian cells at all stages of the cell cycle, as is the case with oxygen.

Rapid-mix studies[181] have established that, similarly to NEM, PNAP sensitizes in the time range indicative of a fast radiation chemical process; thus SH suppression is not involved. The question therefore arises, "Do PNAP and other sensitizers in this group enhance damage produced by direct energy absorption in the target molecule, or do they enhance the damage attributable to indirect action within the cell?" The electron transport model (mentioned at the beginning of this section), an example of the latter type of mechanism, has serious limitations in its failure to account for the oxygen effect because there is ample evidence that O_2^- is relatively unreactive and certainly does not react with DNA. Further, although one-electron transfer processes do occur between nucleic acid derivatives and some typical sensitizers, the transfer occurs *to*, rather than *from*, the sensitizer. The electron affinity of the sensitizer is therefore higher than that of DNA, the likely target. This model has now been abandoned in favor of a direct-action mechanism in which it is postulated that the sensitizer enhances damage arising from absorption of energy directly in the target molecule.[137]

If the target molecule has electrical conducting properties (as is true for DNA), then electrons ejected in ionization processes in the molecule may migrate along the conduction bands to electron-deficient centers. This process can compete with charge recombination (i.e., neutralization) and the efficiency depends on the frequency and depth of electron traps in the molecule. Foreign molecules of higher electron affinity, present either as bound complexes or in loose association, increase the likelihood of *shallow* electron trapping and promote *thermal* electron migration. Such phenomena therefore reduce the probability of charge recombination, a process that does not necessarily lead to irreversible damage. Irreversible electron transfer to the sensitizer would lower the probability of this self-healing and would therefore lead to an increase in the number of free radicals at the ionization sites

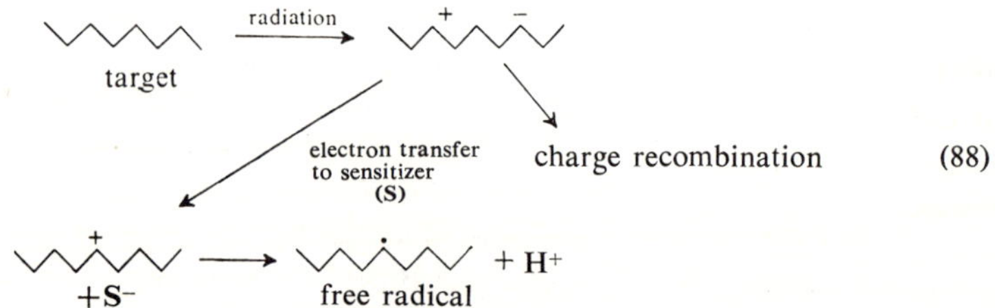

(88)

There is direct evidence in support of this mechanism. Several groups have used ESR spectroscopy to study free-radical formation in irradiated DNA containing various amounts of such sensitizers as menadione and other quinones,[182, 183] PNAP,[184] and NEM.[185] In all cases the total number of free spins produced was higher in the solid mixture than in the pure DNA control. For both the quinones and PNAP, the effect persists even when the sensitizer is present at levels of only a few mole percent. (In the NEM experiments, in which the DNA/NEM ratio was 1:1, the increase in the total number of spins is attributed, however, to interference with the efficiency of radical transfer from the sugar to the base during irradiation.[185])

As indicated in Section 4.3.3, it has been demonstrated that electron transfer reactions occur rapidly and quantitatively (1) between simple nucleic acid derivatives and (2) between such compounds and some of the radio-sensitizers. If these processes can occur between molecules diffusing freely in dilute aqueous solution at room temperatures, intramolecular electron transfer should occur in irradiated DNA, in which there is considerable resonant interaction between the heterocyclic bases. This transfer can only be aided by the presence of additional electron-trapping sites in or near the macromolecule.

5.3.4.4. Chemical Sensitization in Relation to the Oxygen Effect.   It is interesting that the sensitizers that act by fast intracellular radiation chemical processes generally appear to do so within the oxygen effect. That is, with the occasional exceptions referred to earlier, the maximum attainable degree of sensitization does not exceed the sensitizing effect of oxygen. It would be reasonable therefore to accept that these sensitizers act by a similar mechanism, although it is noteworthy that in almost all cases the maximum effect is significantly less than that of oxygen.

The systematic study of the radiobiology of the bacterial spore *Bacillus megaterium* (Powers and Tallentire[1] and others quoted therein) has provided strong evidence that in this system at least the oxygen effect contains two fast free-radical components, evidence that is supported by the double-pulsing bacterial experiments of Epp and co-workers.[142, 143] In the spore experiments the two components of the free-radical oxygen effect are in the ratio of 40:60. It is significant therefore that the maximum degree of sensitization of anoxic spore suspensions observed with biacetyl,[150] acetophenone,[151] and PNAP[151] is about 40 percent of the total oxygen effect. It was suggested by these investigators, therefore, that these compounds and probably others in the electron-affinity class mimic just one component of the total oxygen effect. The maximum degree of sensitization of anoxic mammalian cells by PNAP is also about 40 percent of the oxygen effect in this system[179, 180] although in bacteria the proportion may be a little higher.

As indicated in Sections 5.3.4.2 and 5.3.4.3, there is indirect evidence from model chemical systems in support of both the "adduct formation" and the "direct electron transfer" hypotheses. In cellular systems, however, there is as yet no unequivocal proof that either mechanism operates; the two models are not necessarily mutually exclusive and both remain possibilities. It is notable that several of these sensitizers react both by addition reactions and by electron transfer with the same radical.[168] The compound NEM reacts with the radical CH_2OH mainly by addition [reaction (83)], whereas the CO_2^- reacts exclusively by electron transfer. In the case of the ethanol radical, $CH_3\dot{C}HOH$, electron transfer accounts for about 10 percent of the total reaction. The latter processes appear to be more favored in reactions with PNAP, where the methanol radical reacts mainly by addition, presumably to the ring

$$NO_2\text{—}\langle\text{—}\rangle\text{—}COCH_3 + \dot{C}H_2OH \rightarrow NO_2\text{—}\langle\text{—}\rangle\text{—}CO\text{—}CH_3 \qquad (89)$$
$$\text{with } \dot{}CH_2OH \text{ substituent}$$

the isopropanol radical undergoes electron transfer

$$NO_2\text{—}\langle\text{—}\rangle\text{—}COCH_3 + \begin{array}{c}CH_3\\ \diagdown\\ \dot{C}\text{—}OH\\ \diagup\\ CH_3\end{array} \longrightarrow$$

$$NO_2\text{—}\langle\text{—}\rangle\text{—}\overset{O^-}{\underset{}{C}}\text{—}CH_3 + CH_3COCH_3 + H^+ \qquad (90)$$

For the ethanol radical both processes occur with about equal probability.

5.3.5. *The Bromodeoxyuridine Effect*

The halogenated pyrimidines in which a halogen atom is substituted for the 5-hydrogen atom in uracil or cytosine represent a class of sensitizing agents which has proved to be of much value in the development of molecular radiobiology. (See Kaplan[186] for a review of some of the relevant literature.) The interest stems from the observation by Djordjevic and Szybalski[187] that certain types of cells, when grown in a medium in which thymine (5-methyluracil) is replaced by 5-bromouracil or 5-bromodeoxyuridine (5-BUdR), incorporate the thymine analog into their DNA and become more radiosensitive. The bromouracil is incorporated without too much difficulty because the bromine atom has about the same van der Waals radius as the methyl group; consequently, incorporation does not produce much distortion of the DNA structure.

With respect to the mechanism of sensitization, certain possibilities were soon eliminated. A suggestion that the bromine atom increases the amount of energy absorbed was shown to be unfounded. It was also demonstrated that the bromine incorporation does not weaken hydrogen bonds, for the melting profile of the DNA was hardly affected. Although interference with repair processes has been implicated,[188] there is now much evidence that BUdR incorporation causes an increase in the yield of discrete radiation chemical "lesions," each of which may result in irreversible biological damage.

A mechanism has been suggested attributing the sensitization to the high electron negativity of the bromine atom substituted on the uracil ring.[189] It is known that various simple organic compounds containing substituent halogen atoms, when irradiated in solution, react rapidly with solvated electrons. In most cases the reaction involved is dissociative electron capture, that is,

$$RCl + e^-_{solv} \rightarrow R + Cl^- \tag{91}$$

Figure 19 represents simple stylized structures of normal and of 5-BUdR-substituted DNA (in both of which the two phosphodiester "backbones" are linked together by hydrogen-bonded base pairs). As mentioned in Section 5.3.4.3, the DNA molecule has conduction properties because there is considerable resonance energy associated with the π—π interaction of the pyrimidine and purine bases. It is reasonable to expect, therefore, that bromine atoms in DNA, into which 5-BUdR is incorporated, act as traps for electrons liberated in ionizations produced by *direct* absorption of energy at random sites elsewhere in the molecule. Provided appropriate conduction bands are available in the DNA, the bromine atoms may function as long-range traps for electrons liberated at considerable distances from the

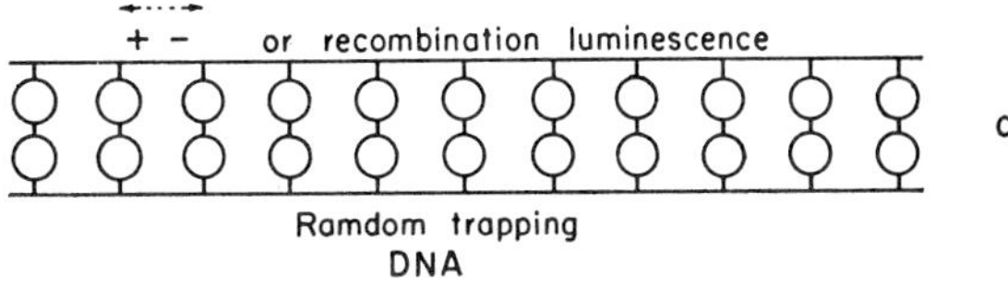

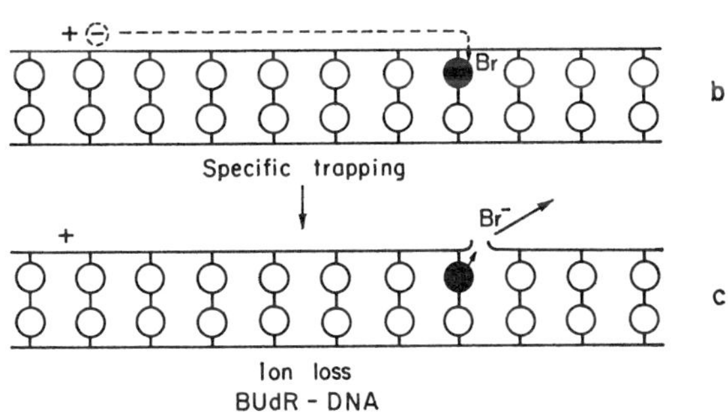

Fig. 19. Stylized model of the influence of 5-BUdR incorporation on electron trapping in irradiated DNA. (*a*) With DNA the initial charge separation process leads either to random trapping or recombination luminescence. (*b*) With 5-BUdR–DNA the initial charge separation leads most probably to electron capture at a Br atom and (*c*) free Br⁻ production.

5-BUdR bases. Such a process would occur in competition with charge neutralization, which in DNA is a process that need not necessarily lead to free-radical formation because there are adequate channels for dissipation of the excess energy liberated by charge recombination.

According to this mechanism, therefore, electron capture by the bromine atom would result in an increase in the total yield of free radicals per unit volume, because radical formation by this mechanism should be 100 percent efficient. Further, the damage would be channeled to specific sites in the substituted DNA, namely, the 5-position on the uridine ring.

As Section 4.3.3 indicates, bromouracil reacts in aqueous solution with e_{aq}^- by dissociative capture to form the uracil radical and the free bromide ion [Eq. (67)], a reaction that also occurs when the donor is the electron adduct of thymine [reaction (68)]. It is not unreasonable to anticipate, therefore, that *intra*molecular migration of electrons in irradiated 5-BUdR–DNA would also result in a similar dissociative electron capture reaction. Inspection of the 5-BUdR–DNA structure shows that the bromine position is close to the sugar backbone, and studies with dilute aqueous solution (by Zimbrick, Ward, and Myers[103]) show that the 5-uridyl radical can easily abstract a hydrogen atom from a ribose phosphate. A mechanism exists, therefore, for translation of the free radical on the pyrimidine base to a free radical on the sugar backbone; such a process would be a probable precursor of a strand break.

Some elegant experiments of Fielden and Lillicrap[190] on the luminescence from DNA irradiated with a single 1.6-μsec electron pulse have shown directly that the random replacement of a small number of thymidine subunits by 5-BUdR has a dramatic effect on the characteristics of the luminescence. The spectrum of the light emitted from irradiated DNA was observed over the time range 1 μsec to >1 sec; the luminescence decays with time and this time profile is also dependent on temperature.

Incorporation of 5-BUdR *increases* the total yield of luminescence, changes the spectrum of the emitted light, and alters the kinetics of the decay. The interesting fact that emerges is that, as the percentage of thymidine replacement increases, the luminescence profile of the substituted DNA approaches that observed from pure crystalline 5-BUdR. Some of the results are shown in Fig. 20. The decay curve of DNA with 25–40 percent of the thymidine replaced is similar to that of pure 5-BUdR. Indeed, the curves are almost superimposable except for a small component in the 0–100 μsec range. Where the curves superimpose, the emission spectra, both of which consist of a single phosphorescence peak at 500 nm, are identical. Significantly, the component that is characteristic of the 5-BUdR is absent if the DNA is denatured before irradiation. It is striking that the effect on the luminescence is detectable at the level of 0.03 percent incorporation.[191] At this level only 1 in 7500 of the base pairs contains a 5-BUdR group.

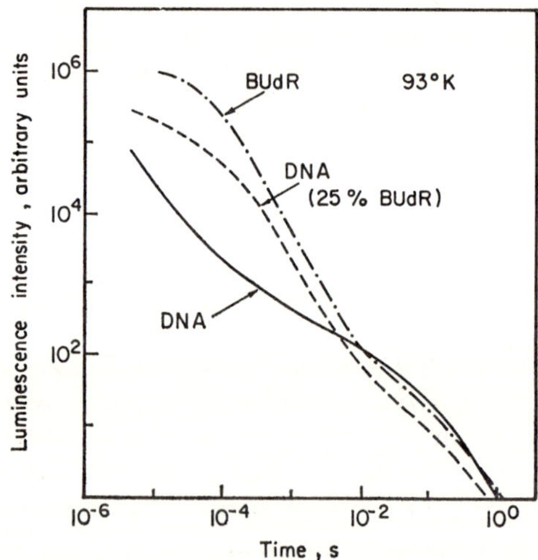

Fig. 20. Decay kinetics at 93 K for light emission (490 nm) from pulse-irradiated crystals of 5-BUdR, DNA with 25 percent thymine replaced by 5-BUdR, and normal DNA. (Fielden and Lillicrap.[190])

Although only a small amount of the energy initially absorbed is emitted as light, these experiments show directly that energy in some form can migrate considerable distances along the chain. Whether or not the migration involves excitation energy or direct thermal electron transfer remains to be established.

It is worth noting that the electron migration model for the 5-BUdR effect is similar in most respects to the mechanism proposed for sensitization by the electron-affinic group of compounds discussed in Section 5.3.4.3.

5.4. Radioprotection

5.4.1. *General*

The subject of protection of radiation damage in cells is an enormous field; so much so that probably more publications have been devoted to this type of work than to any other single area of radiation research. Clearly, it is not possible, within the confines of this section, to do other than mention a few of the proposals relevant to the molecular mechanisms of radiation protection. More detailed reviews[192–196] are available for those with a special interest in the subject.

It is convenient to classify mechanisms of radiation protection into those that involve fast radiation chemical processes and those in which the protection is the result of changes in the cell biochemistry or physiology. It is assumed by Pihl and Sanner[196] that in the "radiation chemical" processes the protective compounds act by direct participation in the free-radical reactions associated with damage to vital molecules. In the biochemical mechanisms however, protection is assumed to be a result of the action of the chemical compound either on the metabolism of the cell or on the biochemical processes that affect the response of the cell to the initial chemical damage after the fast radiation chemical reactions have been completed.

Although various types of simple organic compounds have been shown to act as radioprotective compounds, by far the most efficient of these are the organic mercaptans, in particular, the aminothiols,

$$NH_2 \cdot CH_2CH_2SH \qquad and \qquad \underset{\displaystyle NH_2}{\overset{\displaystyle SHCH_2CHCOOH}{|}}$$

Cysteamine   Cysteine

Studies on these two compounds constitute most of the published work on radioprotection of cellular systems, both *in vitro* and *in vivo*, and this section is concerned mainly with the mechanism, or mechanisms, by which such compounds are believed to act.

5.4.2. *Protection by Radical Scavenging*

On the assumption that some finite amount of the damage sustained by irradiated cells results from indirect action, it is reasonable to expect that some radioprotective compounds operate by scavenging of water radicals produced intracellularly. Although there is some doubt as to the extent of the contribution of the indirect effect to radiation damage in general, evidence from the irradiated bacteria, *E. coli* B/r[197] and *E. coli* B,[198] indicate that, in these systems at least, indirect action accounts for about 50 percent of the total lethal effect in anoxia. With regard to the identity of the radicals responsible for the indirect contribution, attempts have been made to establish a correlation between the protective ability of various agents and their absolute reactivities with the species e_{aq}^- and OH.[197, 198] There is no evidence of any such relationship so far as hydrated electrons are concerned, although there are some data implicating the hydroxyl radical.

The most efficient radioprotectors are the SH compounds, which are followed by the alcohols, particularly ethanol and glycerol, and some simple amino acids. A correlation between OH reactivity and protective efficiency shown in this series of compounds requires considerably more data for establishment beyond doubt.

5.4.3. *Mixed-Disulfide Formation*

Eldjarn and Pihl[199] noted that there appears to be a relationship between the protective efficiency of SH compounds and their ability to form mixed disulfides. The mixed-disulfide hypothesis states that the SH compound (e.g., cysteamine) reacts with SH groups of tissue protein (PSH) under oxidizing conditions to form a disulfide link

$$RSH + HS\text{---}protein \rightarrow RSS\text{---}protein \tag{91}$$

They suggested that when a mixed disulfide is attacked by a free radical one of the sulfur atoms is oxidized while the other is reduced

$$RSS\text{---}protein \rightarrow RSO_2H + protein\text{---}SH \tag{92}$$
$$\text{oxidized}$$

$$\rightarrow RSH + protein\text{---}SO_2H \tag{93}$$

According to the hypothesis, therefore, reactions (92) and (93) are equiprobable and, because the first reaction repairs the original protein damage, the probability of damage by indirect action is reduced by one-half. In model experiments it was established that most of the common SH protectors form stable mixed disulfides. For example, cysteine and cysteamine react *in vitro* with oxidized glutathione at physiological pH and temperature within minutes of mixing, while, in contrast, thiols that do not protect react slowly if at all.

Although the hypothesis remains, in several respects, a plausible mechanism of protection, it suffers from the obvious disadvantage that there is much indirect evidence that damage to SH-containing proteins is not the major cause of cell inactivation.

5.4.4. *The Hydrogen Transfer Mechanism of Radioprotection*

Of the various mechanisms proposed for the molecular basis of radioprotection by SH compounds, the hydrogen donation model has attracted the most attention. Proposed originally by Alexander and Charlesby[144] in the context of radiation polymer chemistry and applied subsequently to biological systems by Howard-Flanders,[146] it has provided a direct link between radiation chemistry and radiobiology.

The model postulates that a free radical formed in a cellular target molecule, either by direct or indirect action, is restored, or "repaired," by hydrogen transfer from the labile SH group of the protector

$$\mathord{\wedge\kern-2pt\wedge\kern-2pt\wedge}\; + \text{RSH} \longrightarrow \mathord{\wedge\kern-2pt\wedge\kern-2pt\wedge} + \text{RS}\cdot \tag{94}$$

It is supposed that in the absence of the protective agent the free radical is a precursor of molecular damage, expressed later at the cellular level. In the presence of oxygen, the radical can be converted into a peroxy radical, which is assumed to be irreparable, by a process occurring in competition with any SH-induced radical repair. Omerod and Alexander[201] have presented direct evidence of the kinetic competition in irradiated suspensions of bacteriophage and there is also much indirect evidence at the cellular level. Rapid-mix studies have shown, for example, that the interaction between the oxygen effect and cysteine protection in irradiated bacteria occurs in the millisecond time scale (cf. Adams[161]).

The repair reaction in simple model systems has been observed directly by various techniques, including ESR[200–203] and by pulse radiolysis of cysteamine[204, 205] and H_2S[206] solutions. The optical technique is useful because in solutions of excess RSH (e.g., cysteamine) the ion radical $RSSR^-$ absorbs strongly in the visible region of the spectrum

$$\text{RS}\cdot + \text{RS}^- \rightleftharpoons \text{RSSR}^- \tag{26}$$

In early studies the repair reaction was observed in pulse-irradiated dilute solution containing a simple alcohol and the SH compound (e.g., methanol and cysteamine). In N_2O-saturated solution, in which the OH radical is the main radical initiator, methanol scavenges OH to form the radical CH_2OH

$$OH + CH_3OH \rightarrow CH_2OH + H_2O \tag{95}$$

which is then repaired by the hydrogen transfer reaction

$$CH_2OH + RSH \rightarrow CH_3OH + RS\cdot \tag{96}$$

In pulse radiolysis the reaction rate can be measured by observation of the rate of formation of the complex $RSSR^-$ because reaction (26) is fast and is not rate determining. Subsequently, the repair reaction was also observed[205] in solutions containing simple polyethylene oxide (PEO) polymer of molecular weights ranging from 200 to 20,000. Table XII lists specific rates for radicals derived from several alcohols and PEOs. Although they are fairly large, the rate constants are much less than those for diffusion-controlled reactions and vary with the structure of the radical.

In Fig. 21, which shows the effect of polymer chain length on the rate constant for repair of PEO radicals by cysteamine, specific rates are plotted as a function of $2/n$ where $n = $ number of carbon atoms in the polymer molecule. The linear trend probably reflects the greater reactivity of polymer radicals produced by hydrogen abstraction at the terminal CH_2OH.[205]

Oxygen inhibits repair by competing for the free radical $X\cdot$ formed in the cellular target

$$X\cdot + O_2 \rightarrow XO_2\cdot \tag{97}$$

Because methods exist for accurate measurement of such reaction rates, it is possible to calculate the relative efficiency of radical "fixation" to radical repair by RSH. Values range from about 10 to several hundred but, within an order of magnitude or so, are consistent with the value of 30 indicated from experiments with bacteriophage by Howard-Flanders.[146]

TABLE XII
Relative Efficiency of Radical Repair

Solute radical (R)	$k_1(R\cdot + SH)$, in units of $10^7 \, dm^3 \, mole^{-1} sec^{-1}$	$k_2(R\cdot + O_2)$, in units of $10^7 \, dm^3 \, mole^{-1} sec^{-1}$	k_2/k_1
Isopropanol	42	410	10
Ethanol	14	460	33
Glucose	3.2	160	50
Methanol	8	480	60
Ethylene glycol	2.6	320	123
PEO 200	0.97	290	300
PEO 6000	0.5	190	380
PEO 20000	0.5	220	440

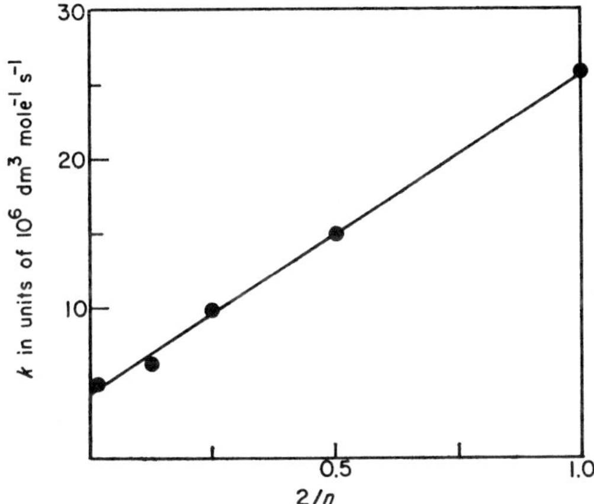

Fig. 21. Dependence of the rate constants for repair by cysteamine on chain length in PEO solutions, where n is the number of carbon atoms. The intercept on the left indicates the rate constant for repair of an infinitely long polymer. (Adams et al.[205])

Pulse radiolysis studies show that the efficiency of radical repair is pH-dependent, for the undissociated RSH molecule is the active agent.[204, 205] In Fig. 22, from Armstrong,[207] the solid line on the left shows the effect of pH on the rate constant for repair of PEO free radicals;[205] the dotted line is the pK_a curve for the SH group in cysteamine. The points on the right show the pH effect on the protective effect of cysteamine in a γ-irradiated solution of PEO, where the end point is the "gel dose"; that is, the minimum dose required for the onset of gelation. Protection occurs in both acid and alkaline media and, doubtless, general scavenging of OH radicals is partly responsible for the effect. However, an additional component, present at acid and neutral pH and amounting to about 40 percent of total protection, disappears in alkaline solution. The solid line in Fig. 22 (right) is the curve from Fig. 22 (left) normalized to the protection data.[207]

Although it remains to be demonstrated *directly* that hydrogen transfer is a mechanism of protection in cellular systems, it clearly is so in irradiated aqueous solution of model compounds.

5.4.5. *Thiourea and Selenourea*

According to Badiello et al.,[208] Thiourea, NH_2CSNH_2, and its selenium analogue, NH_2CSeNH_2, show radioprotective properties. It has been suggested that the protective effect of thiourea may be the result of its ability to act as a hydrogen donor in free-radical reactions in which the sulfhydryl

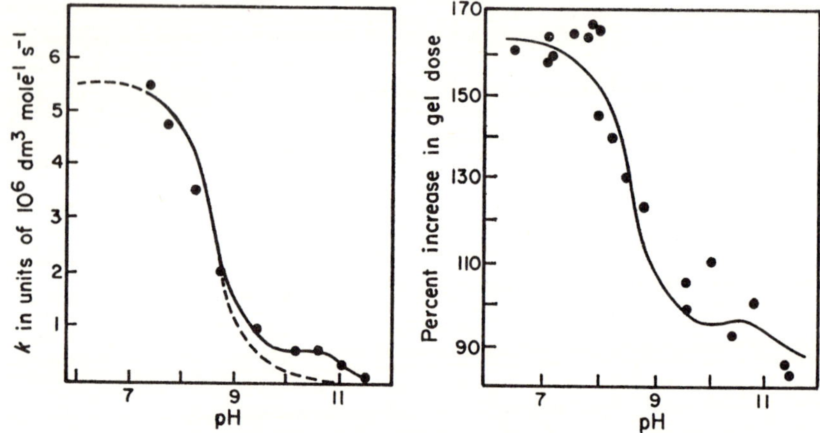

Fig. 22. Left: Effect of pH on the rate constant for repair by cysteamine of radicals from PEO 6000. Broken line, dissociation curve for SH in cysteamine. Right: pH-dependent component of cysteamine-induced protection of γ-irradiated PEO solutions. (Armstrong.[207])

is the active agent

$$NH_2CSNH_2 \rightleftharpoons NH{=}C(SH)NH_2 \tag{98}$$

$$X{\cdot} + NH{=}CSHNH_2 \rightarrow XH + NH{=}\overset{\cdot}{C}SNH_2 \tag{99}$$

Pulse radiolysis of thiourea solutions reveals a strong transient absorption with a maximum near 400 nm.[209] This absorption was originally attributed to the imino-RS radical, but pulse studies by Badiello and Fielden[210] with the seleno derivatives, in which a similar absorption band is observed, indicate that the band is attributable to the radical dimer anion, analogous to $RSSR^-$.

$$\left[\begin{array}{cc} \underset{\overset{|}{\underset{NH_2 \diagup \diagdown NH}{C}}}{Se} \text{\rule{3cm}{0.4pt}} \underset{\overset{|}{\underset{NH_2 \diagup \diagdown NH}{C}}}{Se} \end{array} \right]^-$$

However, although the experiments with free SH compounds, for example, cysteamine, demonstrate that free-radical repair by hydrogen transfer can and does occur *in vitro*, there is as yet no evidence that similar processes are indicated for either thiourea or selenourea. In the latter case the protective effects have been ascribed to general free-radical scavenging.[210]

References

1. E. L. Powers and A. Tallentire, in *Action Chimiques et Biologiques des Radiations*, M. Haissinsky, Ed., Masson, Paris, 1968, p. 3.
2. J. K. Thomas, *Ann. Rev. Phys. Chem.*, **21**, 17 (1970).

3. L. Augenstein, *Advan. Enzymol.* **24,** 359 (1962).

4. R. A. Luse, *Radiation Res. Suppl.* **4,** 192 (1964).

5. W. M. Garrison, in *"Current Topics in Radiation Research,"* Vol. IV, M. Ebert and A. Howard, Eds., Interscience, New York, 1968, p. 45.

6. B. M. Weeks and W. M. Garrison, *J. Chem. Phys.,* **25,** 585 (1956).

7. B. M. Weeks and W. M. Garrison, *J. Chem. Phys.,* **24,** 616 (1956).

8. B. M. Weeks, S. A. Cole, and W. M. Garrison, *J. Phys. Chem.,* **69,** 4131 (1965).

9. J. Holian and W. M. Garrison, *Chem. Commun.,* **14,** 676 (1967).

10. J. Kopoldova, J. Liebster and A. Babicky, *Intern. J. Appl. Radiation Isotopes* **14,** 489 (1963).

11. G. E. Adams and R. L. Willson, *Trans. Faraday Soc.,* **65,** 2981 (1969).

12. R. L. S. Willix and W. M. Garrison, *Radiation Res.,* **25,** 252 (1965).

13. R. L. S. Willix and W. M. Garrison, *Radiation Res.,* **32,** 452 (1967).

14. L. M. Dorfman, I. A. Taub, and R. E. Bühler, *J. Chem. Phys.,* **36,** 3051 (1962).

15. G. E. Adams and B. D. Michael, *Trans. Faraday Soc.,* **63,** 1171 (1967).

16. E. J. Land and M. Ebert, *Trans. Faraday Soc.,* **63,** 1181 (1967).

17. G. G. Jayson, G. Scholes, and J. Weiss, *Biochem. J.,* **57,** 386 (1954).

18. E. Eich and H. Rochelmeyer, *Pharm. Acta Helv.,* **41,** 109 (1966).

19. R. G. Armstrong and A. J. Swallow, *Radiation Res.,* **40,** 563 (1969).

20. A. J. Swallow, *J. Chem. Soc.,* 1334 (1952).

21. B. Shapiro and L. Eldjarn, *Radiation Res.,* **3,** 255 (1955).

22. P. Riesz and B. E. Burr, *Radiation Res.,* **16,** 661 (1962).

23. J. E. Packer, *J. Chem. Soc.,* 2320 (1963).

24. D. A. Armstrong and V. G. Wilkening, *Can. J. Chem.,* **42,** 2631 (1964).

25. D. El Samahy, H. L. White, and C. N. Trumbore, *J. Am. Chem. Soc.,* **86,** 3175 (1964).

26. A. Al-Thannon, R. M. Peterson, and C. N. Trumbore, *J. Am. Chem. Soc.,* **72,** 2395 (1968).

27. J. E. Packer and R. V. Winchester, *Chem. Commun.,* 826 (1968).

28. W. A. Armstrong and D. W. Grant, *Can. J. Chem.,* **41,** 1882 (1963).

29. R. Brdicka, Z. Spurny, and A. Fojtik, *Coll. Czech. Chem. Commun.,* **28,** 1491 (1963).

30. J. W. Purdie, *J. Am. Chem. Soc.,* **89,** 226 (1967).

31. G. E. Adams, G. S. McNaughton, and B. D. Michael, in *The Chemistry of Ionisation and Excitation,* G. R. A. Johnson and G. Scholes, Eds., Taylor and Francis, London, 1967, p. 281.

32. J. W. Purdie, *Can. J. Chem.,* **47,** 1029 (1969).

33. R. Braams, *Radiation Res.,* **27,** 319 (1966).

34. R. Braams, *Radiation Res.,* **31,** 8 (1967).

35. G. Scholes, P. Shaw, R. L. Willson, and M. Ebert, in *Pulse Radiolysis,* M. Ebert, J. P. Keene, A. J. Swallow, and J. H. Baxendale, Eds., Academic Press, New York, 1965, p. 151.

36. M. Simic, P. Neta, and E. Hayon, *J. Am. Chem. Soc.,* **92,** 4763 (1970).

37. E. S. G. Barron, S. Dickman, J. A. Muntz, and J. P. Singer, *J. Gen. Physiol.,* **32,** 537 (1949).

38. E. S. G. Barron and S. Dickman, *J. Gen. Physiol.,* **32,** 595 (1949).

39. E. S. G. Barron and P. Johnson, *Arch. Biochem. Biophys.,* **48,** 149 (1954).

40. S. Tanaka, H. Hatano, S. Ganno, and T. Okamoto, *Bull. Inst. Chem. Res., Kyoto Univ.,* **37,** 374 (1959).

41. R. Lange, A. Pihl, and L. Eldjarn, *Intern. J. Radiation Biol.,* **1,** 73 (1959).

42. R. Lange and A. Pihl, *Intern. J. Radiation Biol.,* **2,** 301 (1960).

43. R. Lange and A. Pihl, *Intern. J. Radiation Biol.,* **3,** 249 (1961).

44. S. J. Adelstein and C. A. Drakes, *Biochemistry*, **4,** 89 (1965).
45. R. J. Romanic and A. L. Tappel, *Arch. Biochem. Biophys.*, **79,** 323 (1959).
46. F. Friedberg and G. A. Hayden, *Arch. Biochem. Biophys.*, **98,** 485 (1962).
47. J. A. Winstead, *Radiation Res.*, **30,** 832 (1967).
48. G. Gorin, M. Quintiliani, and S. K. Airee, *Radiation Res.*, **32,** 671 (1967).
49. A. Pihl and T. Sanner, *Radiation Res.*, **19,** 27 (1963).
50. L. S. Myers and J. L. Abernethy, *Radiation Res.*, **22,** 334 (1964).
51. E. S. G. Barron, J. Ambrose, and P. Johnson, *Radiation Res.*, **2,** 145 (1955).
52. M. E. Jayko, B. M. Weeks, and W. M. Garrison, *Proc. 2nd Intern. Conf. Peaceful Uses Atomic Energy, Geneva*, **22,** 488 (1958).
53. W. D. Brown and J. H. A. Akounoglou, *Arch. Biochem. Biophys.*, **107,** 239 (1964).
54. J. Liebster and J. Kopoldova, *Radiation Res.*, **27,** 162 (1966).
55. B. S. Hartley, in *Structure and Activity of Enzymes*, T. W. Goodwin et al., Eds., Academic Press, New York, 1965.
56. D. K. Myers and M. C. Church, *Nature*, **213,** 636 (1967).
57. A. Fojtik and R. Brdicka, *Collection Czech. Chem. Commun.*, **33,** 462 (1968).
58. K. Dose, in *Physical Processes in Radiation Biology*, L. Augenstein et al., Eds., Academic Press, New York, 1964, p. 253.
59. L. K. Mee, G. Navon, and G. Stein, *Biochim. Biophys. Acta*, **104,** 151, 1965.
60. L. K. Mee, *Radiation Res.*, **21,** 501 (1964).
61. K. R. Lynn, *Radiation Res.*, **43,** 525 (1970).
62. L. K. Mee, G. Navon, and G. Stein, *Nature*, **204,** 1056 (1964).
63. R. Braams and M. Ebert, *Intern. J. Radiation Biol.*, **13,** 195 (1967).
64. J. V. Davies, M. Ebert, and R. J. Shalek, *Intern. J. Radiation Biol.*, **14,** 19 (1968).
65. G. E. Adams, R. L. Willson, J. E. Aldrich, and R. B. Cundall, *Intern. J. Radiation Biol.*, **16,** 343 (1969).
66. G. E. Adams, R. B. Cundall, and R. L. Willson, in *Chemical Reactivity and Biological Role of Functional Groups in Enzymes*, R. M. S. Smellie, Ed., Academic Press, New York, 1970, p. 71.
67. G. Gabor, G. E. Adams, R. Bisby, R. B. Cundall, J. L. Redpath, and R. L. Willson, to be published.
68. L. I. Grossweiner and Y. Usui, *Photochem. Photobiol.*, **13,** 195 (1971).
69. J. E. Aldrich, R. B. Cundall, G. E. Adams, and R. L. Willson, *Nature*, **221,** 1049 (1969).
70. C. C. F. Blake, *Proc. Roy. Soc. (London)*, **B167,** 435 (1967).
71. G. E. Adams, J. E. Aldrich, R. H. Bisby, R. B. Cundall, J. L. Redpath, and R. L. Willson, *Radiation Res.*, 1971, submitted for publication.
72. G. E. Adams, R. H. Bisby, R. B. Cundall, J. L. Redpath, and R. L. Willson, *Radiation Res.*, 1971, submitted for publication.
73. J. V. Davies, M. Ebert, and M. Quintiliani, in *Radioprotection and Radiosensitization*, H. L. Moroson and M. Quintiliani, Eds., Taylor and Francis, London, 1970, p. 87.
74. R. L. Willson, C. L. Greenstock, G. E. Adams, R. Wageman, and L. M. Dorfman, *Intern. J. Radiation Phys. Chem.*, **3,** 211 (1971).
75. G. Scholes, in *Radiation Chemistry of Aqueous Systems*, G. Stein, Ed., Weizmann Science Press of Israel, Jerusalem, 1968, p. 259.
76. P. Kieft, in *Biological Effects of Transmutation and Decay of Incorporated Radioisotopes*, I.A.E.A., Vienna, 1968, p. 65.
77. S. Person, in *Biological Effects of Transmutation and Decay of Incorporated Radioisotopes*, I.A.E.A., Vienna, 1968, p. 29.
78. E. Barrow, P. Johnson, and A. Cobure, *Radiation Res.*, **1,** 410 (1954).

79. G. Scholes, J. F. Ward, and J. J. Weiss, *J. Mol. Biol.*, **2**, 379 (1960).

80. G. Scholes, J. J. Weiss, and C. M. Wheeler, *Nature*, **178**, 157 (1956).

81. M. Daniels, G. Scholes, J. J. Weiss, and C. M. Wheeler, *J. Chem. Soc.*, 226 (1957).

82. B. Ekert and R. Monier, *Nature*, **184**, 58 (1959).

83. R. Latarjet, B. Ekert, S. Apelgot, and N. Reybeyrotte, *J. Chim. Phys.*, **58**, 1046 (1961).

84. G. Scholes and J. J. Weiss, *Nature*, **185**, 305 (1960).

85. B. Ekert and R. Monier, *Nature*, **188**, 309 (1960).

86. G. Scholes, J. F. Ward, and J. J. Weiss, *Science*, **133**, 2016 (1961).

87. G. Scholes and R. L. Willson, *Trans. Faraday Soc.*, **63**, 2983 (1967).

88. L. S. Myers, Jr., J. F. Ward, W. T. Tsukamoto, D. E. Holmes, and J. P. Julia, *Science*, **148**, 1234 (1965).

89. J. Holian and W. M. Garrison, *Nature*, **212**, 394 (1966).

90. G. Scholes and J. J. Weiss, *Exptl. Cell. Res. Suppl.* **2** (1952).

91. G. Scholes and J. J. Weiss, *Biochem. J.*, **53**, 567 (1963).

92. J. J. Conlay, *Nature*, **197**, 555 (1963).

93. J. Holian and W. M. Garrison, *J. Phys. Chem.*, **71**, 462 (1964).

94. J. Holian and W. M. Garrison, *Chem. Commun.*, **14**, 676 (1967).

95. B. Collyns, S. Okada, G. Scholes, J. J. Weiss, and C. M. Wheeler, *Radiation Res.*, **25**, 526 (1965).

96. A. R. Peacocke and B. Preston, *Proc. Roy. Soc. (London)*, **B153**, 90 (1960).

97. P. T. Emmerson, G. Scholes, D. Thompson, J. F. Ward, and J. J. Weiss, *Nature*, **187**, 139 (1960).

98. G. Scholes and M. Simic, *J. Phys. Chem.*, **68**, 1731 (1964).

99. C. L. Greenstock, J. W. Hunt, and M. Ng, *Advan. Chem. Ser.*, **81**, 397 (1968).

100. C. L. Greenstock, J. W. Hunt, and M. Ng, *Trans. Faraday Soc.*, **65**, 3279 (1969).

101. C. L. Greenstock, *Trans. Faraday Soc.*, **66**, 2541 (1970).

102. R. Danziger, E. Hayon, and M. Langmuir, *J. Phys. Chem.*, **72**, 3842 (1968).

103. J. D. Zimbrick, J. F. Ward, and L. S. Myers, Jr., *Intern. J. Radiation Biol.*, **16**, 505 (1969).

104. L. S. Myers, Jr., and L. M. Theard, *J. Am. Chem. Soc.*, **92**, 2868 (1970).

105. G. Scholes, R. L. Willson, and M. Ebert, *Chem. Commun.*, 17 (1969).

106. L. S. Myers, Jr., M. L. Hollis, and L. M. Theard, *Advan. Chem. Ser.*, **81**, 345 (1968).

107. H. Loman and M. Ebert, *Intern. J. Radiation Biol.*, **18**, 369 (1970).

108. L. S. Myers, Jr., A. Warnick, M. L. Hollis, J. D. Zimbrick, L. M. Theard, and F. C. Peterson, *J. Am. Chem. Soc.*, **92**, 2871 (1970).

109. L. S. Myers, Jr., M. L. Hollis, L. M. Theard, F. C. Peterson, and A. Warnick, *J. Am. Chem. Soc.*, **92**, 2875 (1970).

110. A. Kamal and W. M. Garrison, *Nature*, **206**, 1315 (1965).

111. H. Loman and J. Blok, *Radiation Res.*, **36**, 1 (1968).

112. P. Brown, M. Calvin, and J. Newmark, *Science*, **151**, 68 (1966).

113. E. M. Fielden and G. C. Stevens, 1971, private communication.

114. J. F. Ward and I. Kuo, *Intern. J. Radiation Biol.*, **18**, 381 (1970).

115. R. L. Willson, *Intern. J. Radiation Biol.*, **17**, 349 (1970).

116. P. T. Emmerson, *Radiation Res.*, **30**, 841 (1967).

117. P. T. Emmerson and R. L. Willson, *J. Phys. Chem.*, **72**, 3669 (1968).

118. R. L. Willson and P. T. Emmerson, in *Radiation Protection and Sensitization*, H. L. Moroson and N. Quintiliani, Eds., Taylor and Francis, London, 1970, p. 73.

119. P. T. Emmerson, E. M. Fielden, and I. Johansen, *Intern. J. Radiation, Biol.*, **19**, 229 (1971).

120. E. M. Fielden and P. Roberts, *Intern. J. Radiation Biol.*, **20**, 355 (1971).

121. G. E. Adams, B. D. Michael, J. T. Richards, and R. L. Willson, quoted in *Advan. Chem. Ser.*, **81**, 289 (1968).

122. C. L. Greenstock, G. E. Adams, and R. L. Willson, in *Radioprotection and Sensitization*, Moroson, H. L. and M. Quintiliani, Eds., Taylor and Francis, London, 1970, p. 65.

123. G. E. Adams, C. L. Greenstock, J. J. van Hemmen, and R. L. Willson, *Radiation Res.*, **49**, 85 (1972).

124. L. M. Dorfman, N. E. Shank, and S. Arai, *Advan. Chem. Ser.*, **82**, 75 (1968).

125. G. E. Adams and R. L. Willson, in *Symposia, Proc. 4th Intern. Congr. Radiation Res. (Evian), 1970*, Gordon and Breach New York, 1971, in press.

126. L. H. Luthjens and J. Blok, *Intern. J. Radiation Biol.* **16**, 101 (1969).

127. D. Freifelder, *Proc. Natl. Acad. Sci., U.S.* **54**, 128 (1965).

128. D. L. Dewey, *Intern. J. Radiation Biol.*, **12**, 497 (1967).

129. W. D. Taylor and W. Ginoza, *Proc. Natl. Acad. Sci., U.S.*, **58**, 1753 (1967).

130a J. Blok, L. H. Luthjens, and A. L. M. Roos, *Radiation Res.*, **30**, 468 (1967).
   b J. Blok and W. S. D. Verhey, *Radiation Res.*, **34**, 689 (1968).

131. D. L. Dewey and G. Stein, *Radiation Res.*, **44**, 345 (1970).

132. Z. Gampel and E. L. Powers, 1971, private communication.

133. B. A. Bridges, in *Advances in Radiation Biology*, Vol. 3, L. G. Augenstein, R. Mason, and M. Zelle, Eds. Academic Press, New York, 1969, p. 123.

134. P. T. Emmerson, 1971, in press.

135. R. H. Thomlinson and L. H. Gray, *Brit. J. Cancer*, **14**, 555 (1960).

136. D. L. Dewey, *Radiation Res.*, **19**, 64 (1963).

137. G. E. Adams and M. S. Cooke, *Intern. J. Radiation Biol.*, **15**, 457 (1969).

138. E. L. Powers, C. F. Ehret, and B. Smaller, in *Free Radicals in Biological Systems*, M. S. Blois, Jr., H. W. Brown, R. M. Lemmon, R. O. Lindblom, and M. Weissbluth, Eds., Academic Press, New York, 1961, p. 351.

139. P. Howard-Flanders and D. Moore, *Radiation Res.*, **9**, 422 (1958).

140. G. E. Adams, M. S. Cooke, and B. D. Michael, *Nature*, **219**, 1368 (1968).

141. G. E. Adams, J. C. Asquith, and B. D. Michael, 1971, unpublished.

142. E. R. Epp, *Trans. N.Y. Acad. Sci., Ser. 2*, **32**, 253 (1970).

143. E. R. Epp, H. Weiss, A. Santo-Masso, and J. M. Heslin, *Abstr. 19th Ann. Meeting Radiation Res. Soc.*, Academic Press, New York, 1971, p. 63.

144. P. Alexander and A. Charlesby, *Radiobiology Symposium*, Butterworth's, London, 1954, p. 49.

145. T. Alper, *Radiation Res.*, **5**, 573 (1956).

146. P. Howard-Flanders, *Nature*, **186**, 485 (1960).

147. P. Howard-Flanders, *Nature*, **80**, 1191 (1957).

148. P. Howard-Flanders and P. Jockey, *Radiation Res.*, **13**, 466 (1966).

149. W. M. Dale, J. V. Davies, and C. Russell, *Intern. J. Radiation Biol.*, **4**, 1 (1961).

150. A. Tallentire, N. L. Schiller, and E. L. Powers, *Intern. J. Radiation Biol.*, **14**, 397 (1968).

151. A. Tallentire and G. P. Jacobs, 1971, private communication.

152. B. A. Bridges, *Nature*, **108**, 415 (1960).

153. B. A. Bridges, *J. Gen. Microbiol.*, **26**, 467 (1961).

154. J. D. Gregory, *J. Am. Chem. Soc.*, **77**, 3922 (1955).

155. C. J. Dean and P. Alexander, *Nature*, **196**, 1324 (1962).

156. A. K. Bruce and W. H. Malchman, *Radiation Res.*, **24**, 473 (1965).

157. W. A. Cramp, *Radiation Res.*, **30**, 221 (1967).

158. P. T. Emmerson and P. Howard-Flanders, *Nature*, **204**, 1005 (1964).

159. P. T. Emmerson, *Radiation Res.*, **26**, 54 (1965).

160. G. E. Adams and D. L. Dewey, *Biochem. Biophys. Res. Commun.*, **12**, 473 (1963).

161. G. E. Adams, in *Radiation Protection and Sensitization*, H. L. Moroson and M. Quintiliani, Eds., Taylor and Francis, London, 1969, p. 3.

162. T. Brustad and B. Singsaas, *Radiation Res.*, **45**, 94 (1971).

163. D. L. Dewey and B. D. Michael, *Biophys. Biochem. Res. Commun.*, **21**, 397 (1965).

164. L. Mullenger, B. B. Singh, M. G. Ormerod, and C. J. Dean, *Nature*, **216**, 372 (1967).

165. H. L. Moroson and D. Tenney, *Radiation Res.*, **36**, 418 (1968).

166. D. L. Dewey, *Progr. Biochem. Pharmacol.*, **1**, 59 (1965).

167. J. F. Ward, I. Johansen, and J. Aasen, *Intern. J. Radiation Biol.*, **15**, 163 (1969).

168. P. Higgins, E. M. Fielden, G. E. Adams, and R. L. Willson, 1971, unpublished.

169. M. Simic and E. Hayon, 1971, private communication.

170. I. Johansen, J. F. Ward, K. Siegel, and A. Sletton, *Biochem. Biophys. Res. Commun.*, **33**, 349 (1968).

171. P. T. Emmerson and P. Howard-Flanders, *Nature*, **204**, 1005 (1964).

172. L. Parker, L. D. Skarsgard, and P. T. Emmerson, *Radiation Res.*, **38**, 493 (1969).

173. P. T. Emmerson and P. Howard-Flanders, *Radiation Res.*, **26**, 54 (1965).

174. K. F. Nakken, T. Sikkeland, and T. Brustad, *FEBS Letters*, **8**, 33 (1970).

175. T. Brustad, W. B. G. Jones, and K. I. Nakken, *Intern. Radiation Phys. Chem.*, 1971, in press.

176. M. J. Ashwood-Smith, D. M. Robinson, J. H. Barnes, and B. A. Bridges, *Nature*, **216**, 137 (1967).

177. M. J. Ashwood-Smith, J. H. Barnes, J. Huckle, and B. A. Bridges, in *Radiation Protection and Sensitisation*, H. L. Moroson and M. Quintiliani, Eds., Taylor and Francis, London, 1970, p. 183.

178. G. E. Adams and R. L. Willson, 1971, unpublished.

179. G. E. Adams, J. C. Asquith, D. L. Dewey, J. L. Foster, B. D. Michael, and R. L. Willson, *Intern. J. Radiation Biol.*, **19**, 575 (1971).

180. J. D. Chapman, R. G. Webb, and J. Borsa, *Intern. J. Radiation Biol.*, **19**, 561 (1971).

181. G. E. Adams, B. D. Michael, and M. E. Watts, 1971, unpublished.

182. C. Nicholau, O. Korner, and A. Cristea, Studia. Biophys., Berlin, **1**, 59 (1966).

183. T. Morley, 1971, to be published.

184. J. Freeman, 1971, private communication.

185. M. Farcasiv, R. Istratoiv, and P. Milvy, *Radiation Res.*, **46**, 415 (1971).

186. H. S. Kaplan, in *Radiation Protection and Sensitisation*, H. L. Moroson and M. Quintiliani, Eds., Taylor and Francis, London, 1970, p. 35.

187. B. Djordjevic and W. Szybalski, *J. Exptl. Med.*, **112**, 509 (1960).

188. J. T. Lett, G. Parkins, P. Alexander, and M. G. Ormerod, *Nature*, **203**, 593 (1964).

189. G. E. Adams, in *Current Topics in Radiation Research*, M. Ebert and A. Howard Eds., Vol. III, Interscience, New York, 1967, p. 37.

190. E. M. Fielden and S. C. Lillicrap, in *Radiation Protection and Sensitization*, H. L. Moroson and M. Quintiliani, Eds., Taylor and Francis, London, 1970, p. 81.

191. E. M. Fielden, S. C. Lillicrap, and A. B. Robins, 1971, private communication.

192. Z. M. Bacq, *Chemical Protection against Ionizing Radiation*, Thomas, Springfield, Ill., 1967.

193. K. F. Nakken, in *Current Topics in Radiation Research*, M. Ebert and A. Howard, Eds., Vol. I, Interscience, New York, p. 49, 1965.

194. *Brookhaven Symp. Biol.*, **20**, *Recovery and Repair Mechanisms in Radiobiology*, 1967.

195. *Panel on Radiation Damage to the Biological Molecular Information System with Special Regard to the Role of SH Groups*, I.A.E.A., Vienna, 1969.
196. A. Pihl and T. Sanner, in *Radiation Protection and Sensitisation*, H. L. Moroson and M. Quintiliani, Eds., Taylor and Francis, London, 1970, p. 43.
197. I. Johansen and P. Howard-Flanders, *Radiation Res.*, **24**, 184 (1965).
198. T. Sanner and A. Pihl, *Radiation Res.*, **37**, 216 (1969).
199. L. Eldjarn and A. Pihl, in *Mechanisms in Radiobiology*, Academic Press, New York, 1960, p. 23.
200. B. Smaller and E. C. Avery, *Nature*, **183**, 539 (1959).
201. M. G. Ormerod and P. Alexander, *Radiation Res.*, **18**, 495 (1963).
202. B. B. Singh and M. G. Ormerod, *Biochem. Biophys. Acta.*, **109**, 204 (1965).
203. A. Pihl, T. Henriksen, and T. Sanner, *Radiation Res.*, **35**, 235 (1968).
204. G. E. Adams, G. S. McNaughton, and B. D. Michael, *Trans. Faraday Soc.*, **64**, 902 (1968).
205. G. E. Adams, R. C. Armstrong, A. Charlesby, B. D. Michael, and R. L. Willson, *Trans. Faraday Soc.*, **65**, 732 (1969).
206. W. Karmann and A. Henglein, *Ber. Bunsenges. Physik. Chem.*, **71**, 421 (1967).
207. R. C. Armstrong, *Intern. J. Radiation Biol.*, **16**, 197 (1969).
208. R. Badiello, A. Trenta, M. Mattu, and S. Muzetti, *Med. Nucl. Radiobiol. Latina*, **10**, 57 (1967).
209. A. Charlesby, P. J. Fydelor, P. M. Kopp, J. P. Keene, and A. J. Swallow, in *Pulse Radiolysis*, M. Ebert, J. P. Keene, A. J. Swallow, and J. H. Baxendale, Eds., Academic Press, New York, 1965, p. 193.
210. R. Badiello and E. M. Fielden, *Intern. J. Radiation Biol.*, **17**, 1 (1970).
211. L. Michaelis and M. L. Menton, *Biochem. Z.*, **49**, 333 (1963).
212. H. Lineweaver and B. Burke, *J. Am. Chem. Soc.*, **56**, 658 (1934).

X-Ray Damage to DNA and Loss of Biological Function: Effect of Sensitizing Agents

PETER T. EMMERSON, *Department of Biochemistry, The University, Newcastle upon Tyne, England*

Contents

1. INTRODUCTION

If biological effects of radiation could be understood simply in terms of the yield and nature of the primary radiation-induced radicals, their reaction rates with intracellular substances, and the identity of the products of these reactions, then radiation chemists could, in theory, write down most of the chemistry involved. In practice, however, the situation is more complex, for the response of a living organism to ionizing radiation is determined not only by the initial radiation chemistry but also by the secondary biochemical events which follow. The living cell, it seems, does not stand idly by while some of its essential molecules are violated, but instead mobilizes various enzyme systems to repair some of the most crucial damage.

A radiosensitizing agent may be defined as a substance that increases the effect of radiation on any given irradiated system. One of the most widely studied effects of radiation is the decrease in colony- or clone-forming ability of irradiated cell populations, in which case the sensitizing agent increases the number of cells that fail to produce colonies or clones following a given dose of radiation. Research into the mechanisms by which sensitizing agents affect the radiation products within the cell may help in the elucidation of the processes normally involved in cell death, including those to do with repairing radiation injury. When these processes are better understood, it is hoped that it may be possible to find a sensitizing agent that can be used to sensitize a tumor selectively during radiation therapy.

The literature on radiosensitization is vast, covering biological systems ranging in complexity from viruses to man. To keep this review reasonably short and to preserve some continuity, the discussion is confined mainly to work on simple systems such as bacteria and cultured mammalian cells, and to those sensitizing agents the mechanisms of action of which are at least partially understood.

There are reviews that cover the sulfhydryl-binding sensitizing agents[1] and the radiosensitization of bacterial spores.[2] Unless required to illustrate a particular point, these topics are not covered here, nor is the sensitization caused by incorporation of base analogs, such as 5-bromouracil (BU) into cellular deoxyribonucleic acid (DNA). Sensitization by BU incorporation has already aroused much interest and has been used in clinical trials in radiation therapy. However, because BU is effective only when incorporated into DNA, the mechanism of sensitization is different from the mechanisms considered in this review.

Absorption of ionizing radiation by a cell is based only on the electron density of the absorbing species, and no constituent of the cell can be expected to escape injury, although, of course, the smaller it is the better are its chances. Given that every component of the cell can be damaged, the question arises whether all damage is potentially lethal, or whether, statistically, cell death is more likely to arise from damage to some critical structure such as the DNA or perhaps the cell membrane. In consideration of the mechanism of radiosensitization, this question is important and is discussed first.

1.1. Importance of the Integrity of DNA

The dry weight of a typical bacterium is made up of approximately 4 percent phospholipid, 5 percent polysaccharide, 5 percent DNA, 6 percent lipid, 10 percent RNA, and 70 percent protein.[3] Although, on this basis, damage to DNA would occur slightly less frequently than damage to, say, RNA or proteins, DNA damage when it does occur can be more troublesome to the cell because of the unique role DNA plays as the repository of the cell's genetic information. DNA molecules are very long; *Escherichia coli* DNA has a molecular weight of approximately 2.5×10^9 daltons,[4] which corresponds to approximately 10^7 nucleotides. Because these long molecules must act as templates for their own replication, the whole DNA content of a cell can be prevented from replicating by only one irreparable blockage or discontinuity per DNA molecule. The genetic information needed for specifying the structures of cell components is stored in DNA, and the flow of genetic information within the cell proceeds in general from DNA to RNA and thence to proteins and other cellular constituents. It is to be expected on general grounds that such a system would be most vulnerable at its source, because damaged RNA and protein can be replaced by newly synthesized molecules.

This intrinsic vulnerability of the cell in its DNA is presumably reflected by the fact that cells have evolved sophisticated, genetically controlled, enzymic processes to repair damaged DNA. No such mechanisms are known for the repair of damage to other cellular constituents. There is little doubt that DNA damage would be a major cause of cell death if such damage were not frequently repaired. However, assuming that much of the DNA damage can be repaired, the question arises whether the efficiency of repair is sufficiently high to bring non-DNA damage into contention for the role of lethal damage. Although this chapter is concerned only with the effects of ionizing radiation, it may be useful at this stage to digress briefly and to consider one aspect of ultraviolet irradiation. Absorption of ultraviolet is more specific than is that of ionizing radiation, and the chromophores of the heterocyclic bases in DNA and ribonucleic acid (RNA) are responsible for most of the absorption, by the cell, of the 254-nm wavelength emitted by a typical low-pressure mercury discharge germicidal ultraviolet source. Therefore, most ultraviolet damage occurs in nucleic acids, where the main photoproducts are pyrimidine dimers formed between adjacent pyrimidine bases in the same DNA strand.[5, 6] Extremely sensitive mutants of the bacterium *E. coli* that lack two repair systems have been obtained. These mutants are killed by the formation of only one or two pyrimidine dimers in their DNA; by contrast, wild-type cells can survive an ultraviolet dose that produces several thousand dimers in their DNA.[7] However, even at the very high ultraviolet doses that kill the wild-type cells, the action spectrum for this killing corresponds to that of nucleic acids. Thus despite two very efficient repair systems the lethal effect of ultraviolet irradiation on *E. coli* seems to be predominantly on the DNA.

In the case of ionizing radiation, the situation is more complex because of both the less specific nature of the absorption and the lack of precise information about the yield and nature of radiation products and mechanisms by which they can be repaired. However, on the basis of the level of the X-ray sensitivity of mutants that are apparently defective in repairing X-ray-damaged DNA, it appears that X-ray damage may be less efficiently repaired than is ultraviolet damage. This fact, together with the general observation that X-ray sensitivity seems to correlate with defective DNA repair processes, suggests that DNA damage can be implicated in cell killing by ionizing radiation.

Direct evidence or other proof that cell killing involves DNA damage is, however, lacking—although several reviewers have presented indirect evidence in support of this notion.[8–10] For example, incorporation of halogenated pyrimidines such as BU, an analog of thymine, into the DNA of many viruses and cells sensitizes them to ionizing radiation. Szybalski and Lorkiewicz[8] showed that the degree of sensitization by certain bacteria resultant from BU incorporation is similar to that of the biological activity of the DNA extracted from the same cells when tested in a transforming system.* No sensitization is found if 5-fluorouracil is incorporated into the RNA of bacteria.[11] These results suggest that the greater sensitivity is a reflection of radiation damage in the DNA. Also, the X-ray sensitivity of various organisms can be roughly correlated with their total DNA content[12,13] and their DNA base composition.[14, 15] The correlation between the X-ray sensitivity of a wide variety of organisms and their total DNA content is

---

*In a transforming system DNA from a drug-resistant strain of the bacterium *Bacillus subtilis* is used to treat a drug-sensitive strain and converts a small fraction of recipient cells to drug resistance; cf. "transforming DNA."

particularly striking and is discussed in more detail later in the light of present knowledge about DNA repair processes.

The ideal experiment to determine the contribution that DNA damage makes to cell killing would be to remove the DNA from the cell, irradiate either the DNA or the remainder of the cell separately, replace the DNA, and compare the survival with suitable controls. Unfortunately, this experiment cannot be done with the techniques of contemporary molecular biology. However, a similar although less direct procedure is to irradiate bacteria heavily and then to introduce into them some unirradiated DNA by infection with a suitable bacterial virus. Because viruses inject only their DNA into the host,[16] and because in order to develop into a viable virus this DNA requires the host's protein-synthesizing machinery, it is possible in this way to determine the radiosensitivity of some of the intracellular enzyme systems. Figure 1 shows the result of an experiment by Howard-Flanders[10] with the T2 bacteriophage system. Apparently, an X-ray dose of 600 krads to the bacterium *E. coli*, which massively decreases its survival, has no detectable effect on its capacity to act as host and produce one or more viable viruses, whereas 100 krads to the virus reduces their survival by about 90 percent. A dose of only 10 krads is sufficient to kill 90 percent of the bacteria. Thus, there is a striking difference between the radiation sensitivity of the ability

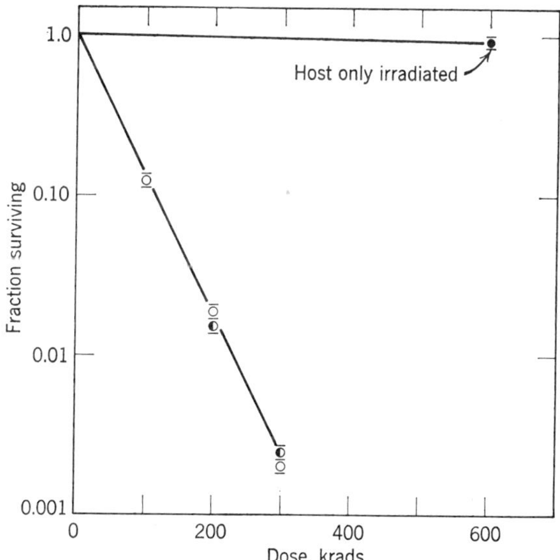

Fig. 1. Fraction of bacteriophage T2 suspended in nutrient broth surviving irradiation, and retaining the ability to multiply, as a function of the dose when only the phage is irradiated (○), when both phage and the host bacterium (*E. coli*) are irradiated before absorption (◐), and when only the host is irradiated (●). (Howard-Flanders.[10])

of the virus to multiply or the colony-forming ability of the bacterium on the one hand, and the capacity of the bacterium to act as host for virus growth on the other. The greater radiation sensitivity of the viral DNA compared with that of the protein-synthesizing machinery of the host cell indicates the overwhelming importance of damage to DNA in this system.

While there is much indirect evidence that indicates the critical importance of DNA damage in cell killing, there is little evidence for the importance of damage to other structures within the cell. There have been suggestions by Alper[17] and by Mehrishi[18] that damage to the cell membrane may be particularly important, perhaps especially at the site at which the DNA is attached.[17] These speculations seem to lack experimental support.

Throughout this chapter it is assumed that the survival of a cell following exposure to ionizing radiation may be largely determined by its ability to preserve the integrity of its DNA so that the full complement of genetic information can be transmitted from one generation to the next. Results pertinent to the mechanisms by which oxygen, N-ethylmaleimide (NEM), nitric oxide, and N-oxyl radicals sensitize cells to ionizing radiation are considered in some detail and attention is drawn to the fact that these sensitizing agents appear to have at least one property in common, the ability to form adducts with radiation-induced DNA free radicals.

2. OXYGEN AS A SENSITIZING AGENT

Oxygen, a stable biradical, is by far the best known and most extensively studied sensitizing agent. To be effective it must be present during irradiation.[19] Such diverse systems as bacteria,[20] cultured mammalian cells,[21] broad-bean roots,[22] and mouse tails[23] are sensitized by oxygen; but viruses[24] and transforming DNA[25] are not sensitized unless compounds possessing sulfhydryl groups are present.[26, 27]

2.1. Damage to DNA Bases: Formation of Organic Peroxides

One of the keys to our present understanding of the role of oxygen in increasing radiation lethality was the demonstration that many organic molecules irradiated in the presence of oxygen produce relatively stable hydroperoxides. For example, aqueous DNA solutions irradiated in the presence of oxygen give a high yield of hydroperoxides ($G = 1$).[28, 29] The presence of oxygen during irradiation of aqueous thymine solutions leads to the formation of hydroperoxides[30–32] and increases the destruction of thymine.[30–35] This is, therefore, one chemical system in which the effect of oxygen is positive in that it *sensitizes* the solute to the effect of ionizing radiation and therefore parallels the biological oxygen effect. Not all chemical systems behave in this way. Oxygen has the reverse effect in the radiation-induced decomposition of formic acid, apparently because it interferes with a chain reaction.[36]

Measurement of the extent of destruction of the bases in irradiated solutions of calf thymus DNA, determined by acid hydrolysis and paper chromatography, indicate that thymine is destroyed at a somewhat higher rate than the other bases for irradiation both in the presence[31, 37] and in the absence[38] of oxygen. Because thymine is an important constituent of DNA, and because its radiation chemistry has been extensively studied, it provides a useful model chemical system in which to study the effect of various sensitizing agents. Accordingly, the reaction schemes for its radiation-induced degradation in the presence and absence of oxygen are considered in some detail.

When dilute, oxygenated, aqueous thymine solutions are exposed to ionizing radiation, the main reactions appear to be[34, 39]

$$H_2O \rightsquigarrow OH, H, e_{aq}^- \tag{1}$$

$$H + O_2 \longrightarrow HO_2 \tag{2}$$

$$e_{aq}^- + O_2 \longrightarrow O_2^- \tag{3}$$

$$\text{Thymine} + OH \longrightarrow (I) \tag{4}$$

$$(I) + O_2 \longrightarrow (II) \tag{5}$$

$$(II) \xrightarrow{[H]\ \text{donor}} \tag{6}$$

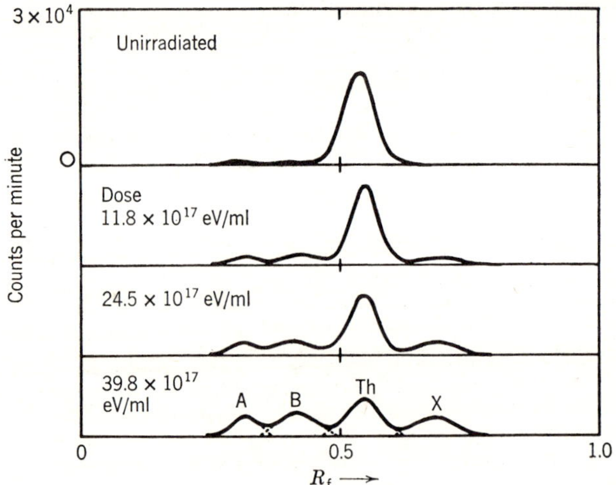

Fig. 2. Radiochromatograms of 2×10^{-4} M [^{14}C]thymine in n-propanol–HCl (85:15) irradiated at 2°C in the presence of oxygen. A and B were found to cochromatograph in several different solvent systems with authentic *cis*- and *trans*-5,6-dihydroxythymines (thymine glycols). X was not identified. (Emmerson, unpublished work.)

The 5-hydroperoxy-6-hydroxythymine produced in reaction (6) decomposes slowly at room temperature and neutral pH by an unknown reaction mechanism to give *cis*- and *trans*-5,6-dihydroxythymine (thymine glycols). Saturation of the 5-6 double bond in pyrimidines renders them vulnerable to alkaline hydrolysis; compounds such as thymine glycols and hydroxy hydroperoxides can be detected on paper chromatograms by treatment with alkali followed by spraying with p-dimethylaminobenzaldehyde,[40] with which they produce characteristic colors. Figure 2 shows the result of radiochromatography of [^{14}C]thymine irradiated in aqueous solution in the presence of oxygen.

Under anoxic conditions the absence of reactions (2), (3), and (5) results in a buildup of radical intermediates according to reaction (4) and also, presumably, according to reactions such as

$$\text{Thymine} + \text{H} \longrightarrow \qquad\qquad (7)$$

(III)

$$\text{Thymine} + e_{\text{aq}}^- \longrightarrow \left[\text{(IV)} \right]^{-} \qquad (8)$$

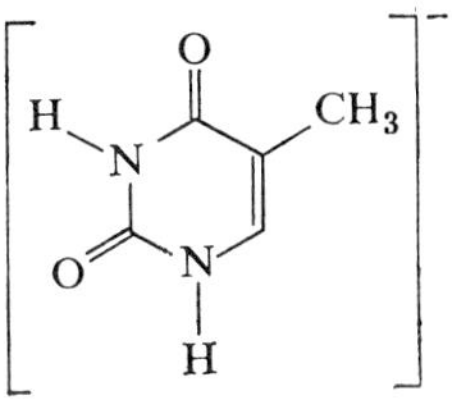

(IV)

In Eq. (7) the hydrogen atom attachment site appears to be the C-6 position,[41] as is the case with OH radical attack in reaction (4). The product of reaction (8) is shown without specifying the position of the electron because there is evidence that e_{aq}^- does not destroy the chromophoric group in thymine.[35, 42]

The yields of the radical intermediates produced in reactions (4), (7), and (8) are determined by the yields of the primary species; that is, approximately, $G(e_{\text{aq}}^-) = 2.6$, $G(\text{OH}) = 2.6$, $G(\text{H}) = 0.5$ for ^{60}Co γ-rays.[43] Consequently, the initial extent of attack on thymine by the primary species is higher under anoxia than in the presence of oxygen, for oxygen scavenges hydrogen atoms and hydrated electrons according to reactions (2) and (3), and there is no evidence that the products of these reactions can react with thymine. The maximum extent of destruction of thymine irradiated with ^{60}Co γ rays in dilute aqueous solution in the presence of oxygen would be approximately $G = 2.6$. At the other extreme, a G value of nearly 6 could reasonably be expected under anoxia if attack by all of the three species e_{aq}^-, OH, and H leads to destruction of thymine. In order to explain the observed *diminished* $G(-\text{thymine})$ in the absence of oxygen, it is suggested that various biradical disproportionation reactions regenerate thymine; for example,

$$\text{(I) or (III) or (IV)} + \text{(I) or (III) or (IV)} \rightarrow \text{Thymine} + \text{products} \qquad (9)$$

Examples of plausible back-reactions are

$$\text{(I)} + \text{(III)} \longrightarrow \text{thymine} + \qquad (10)$$

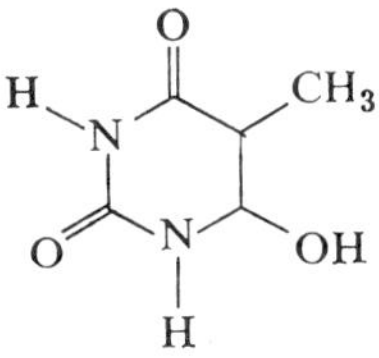

$$\text{(I)} + \text{(I)} \longrightarrow \text{thymine} + \qquad (11)$$

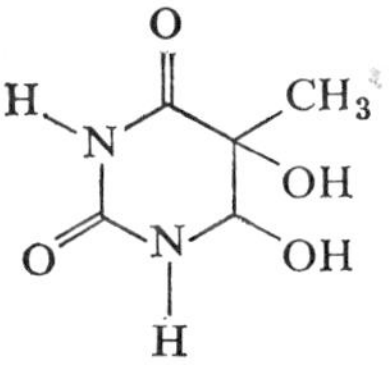

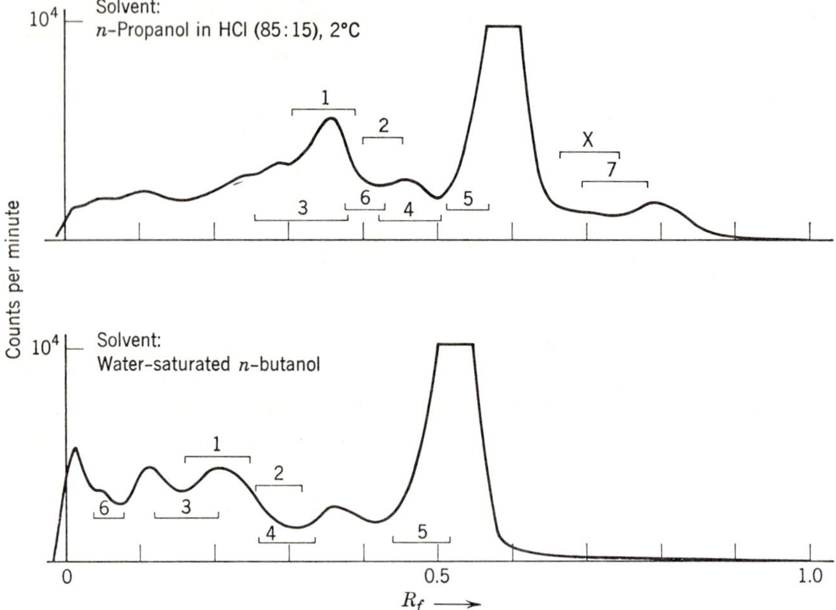

Fig. 3. Radiochromatograms of [¹⁴C]thymine irradiated at 2°C under anoxia. The positions of several substances that are possible radiation products are shown. The positions (indicated by braces) of known compounds added as tracers are as follows: 1, hydroxymethyluracil; 2, urea; 3 and 4, *cis-* and *trans-*5,6-dyhydroxythymines (i.e., thymine glycols); 5, dihydrothymine; 6 and 7, β-amino, and β-ureido isobutyric acids. The product X is unidentified. (Emmerson, unpublished work.)

Because both of the hydration products, 5-hydroxy-6-hydrothymine and 5-hydro-6-hydroxythymine are unstable and break down at room temperature to give thymine and water,[41] reaction (10) would ultimately result in the conversion of two thymine radical intermediates to thymine. Reactions such as (11) partially regenerate thymine and partially give rise to thymine irradiation products. Comparison of Figs. 2 and 3 shows the greater diversity of product on irradiation of thymine solutions under anoxic conditions as compared with the products of irradiation in the presence of oxygen. By determining the area under the peaks in radiochromatograms such as these, some idea can be obtained of the G values involved, although this method always suffers the disadvantage that there may be more than one substance contributing to the peak. The destruction of thymine in the presence and absence of oxygen measured in this way is illustrated in Fig. 4. Similar results were obtained when the destruction of thymine was measured by a decrease in absorption of the chromaphoric group. In these experiments the presence of oxygen increased $G(-$ thymine$)$ from $\sim$1.0 to $\sim$2.0. The product X at R_f

0.7 in Fig. 2 was not identified. It is neither a hydroperoxide nor a glycol. The products *A* and *B* are thymine glycols. However, because the yield of organic peroxides immediately after irradiation in the presence of oxygen is nearly as high ($G = 1.7$) as the extent of destruction of thymine ($G = 2.0$ for X-rays),[33] it follows that most of the observed radiation products must be decomposition products of the peroxides produced in reactions (5) and (6).

Many of the radiation products of thymine irradiated under anoxia remain to be identified. Those that have been identified so far include *cis-* and *trans*-5,6-dihydroxythymine (thymine glycols), 5,6-dihydrothymine, 5-hydroxy-6-hydrothymine, 5-hydro-6-hydroxythymine, 5-hydroxymethyl-uracil, and urea.[32, 41, 44] In a similar study with ^{14}C-labeled uracil, a wide range of similar products was identified.[45] However, their yields were not determined and it is not known whether in these experiments the overall radiation-induced destruction of uracil was increased by oxygen.

The extent of destruction of thymine irradiated in aqueous solution in the presence of oxygen appears to be determined by competition between the forward reaction (5) and the backreaction (9). For explanation of the increased cell killing by irradiation in the presence of oxygen, back-reactions under anoxia must also be invoked. However, these are not likely to be simply of the type shown in reaction (9), for it appears improbable that

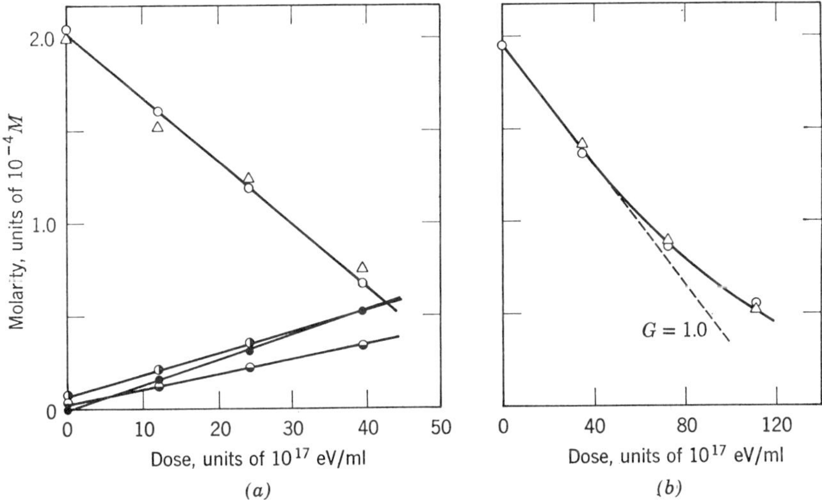

Fig. 4. Destruction of thymine irradiated in aqueous solution in the presence of either oxygen or nitrogen. (*a*) Irradiation in oxygen; this figure derives from Fig. 2. (*b*) Irradiation in nitrogen. ○ and △, thymine concentrations established by optical density and gas chromatography, respectively; *cis-* (◖) and *trans-* (◗) 5,6-dihydroxythymines (i.e., glycols); unidentified product. (●). $G(-\text{thymine}) = 2.0 \pm 0.1$. $G(cis) = 0.47$; $G(trans) = 0.68$; $G(X) = 0.80$.

there would be a buildup in the cell cytoplasm of the species (I), (III), and (IV) to significant concentrations. Instead, substances such as those containing sulfhydryl groups, which are capable of donating hydrogen atoms, are thought to participate in back-reactions. These reactions may be written in the general form[29, 46, 47]

$$R\cdot + X\text{—}SH \rightarrow RH + XS\cdot \tag{12}$$

where R· is a radiation-produced free radical from a molecule essential for cell survival and X—SH is a sulfhydryl. It is for this reason that sulfhydryl compounds figure so prominently in discussion of mechanisms of sensitization. Strong experimental support for reactions such as (12) is presented in Section 3.1 in connection with proposed mechanisms of sensitization of sulfhydryl-binding agents.

Reactions such as (5) have been studied in detail by Willson,[48] using a pulse radiolysis technique. Irradiation of nitrous oxide-saturated solutions of thymine or thymidylic acid with a pulse of electrons produces a broad absorption in the region 300–400 nm, which has been attributed by Scholes, Willson, and Ebert[49] to the product of reaction (4) because the solvated electrons primarily formed are converted to OH radicals by nitrous oxide[50, 51] by the reaction

$$e_{aq}^- + N_2O \rightarrow OH + OH^- + N_2 \tag{13}$$

and product (IV) is thus prevented from interfering. This absorption was found to decay rapidly in the presence of oxygen.[48] Figure 5 shows the results of an experiment with thymidylic acid. Pseudo-first-order rate constants for the disappearance of the absorption as functions of oxygen concentration, shown in Fig. 6, are compatible with reactions such as that

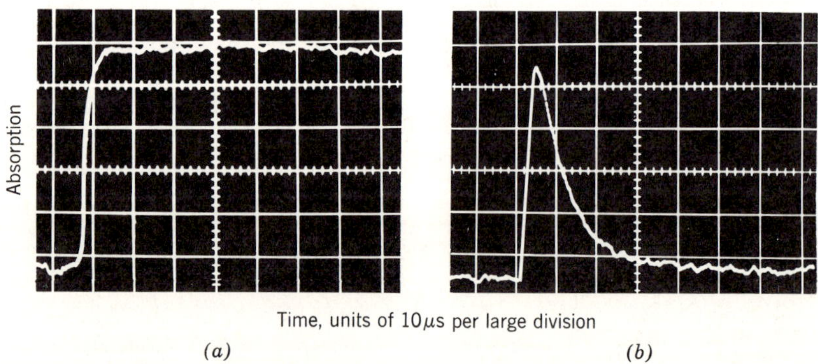

Fig. 5. Oscillograms showing the effect of oxygen on the decay of the absorption produced at 400 nm by pulse irradiation of aqueous neutral solutions of thymidylic acid in the presence of nitrous oxide. Nitrous oxide converts e_{aq}^- to OH. (a) No oxygen; (b) 1.3×10^{-4} M oxygen. (From Willson.[48])

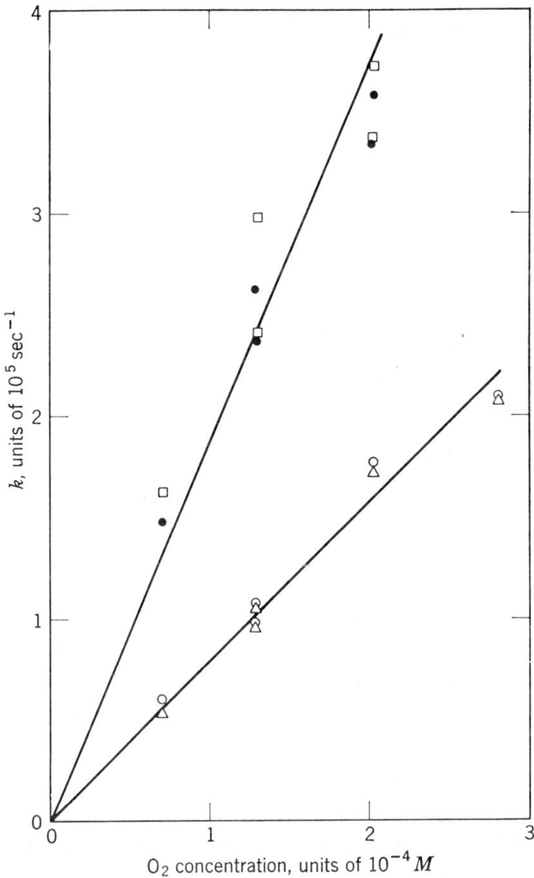

Fig. 6. First-order rate constant as a function of oxygen concentration. Thymine, 400 nm; □, pH, 5.5; ●, pH 7. Thymidylic acid, pH 7; ○, 350 nm; △, 400 nm. (From Willson.[48])

given for thymine in Eq. (5) and show that the reactions are very fast. Some of the absolute rate constants for reactions of oxygen with radicals formed from various bases, nucleosides, and nucleotides are presented in Table I. Adenine, deoxyadenylic acid, and calf thymus DNA gave results that were difficult to interpret. However, overall, these experiments demonstrate a very prominent effect of oxygen in the radiolysis of many of the components of DNA.

Saturation of the 5-6 double bond in thymine such as in reactions (4) to (6) may interfere with its ability to pair with adenine in the complementary strand of DNA because it would result in a stereochemical change from a

TABLE I

Absolute Rate Constants for the Reactions of Oxygen
with the OH Adducts of Various Bases, Nucleosides, and
Nucleotides[a]

Solute	Rate constant, units of $10^9\,\mathrm{dm^3\,mole^{-1}\,sec^{-1}}$
Thymine	1.9
Cytosine	2.0
Thymidine	1.0
Cytidine	1.2
Thymidylic acid	0.8
Deoxycytidylic acid	1.1
Deoxyadenylic acid	0.9

[a] From Willson.[48]

planar to a nonplanar conformation, which might produce some local distortion of the DNA.[41, 52] However, the distortion may be only slight because the N-1, C-2, N-3, and C-4 atoms of thymine would be expected to remain fairly planar and it is the imino and carbonyl groups at N-3 and C-4 that are involved in the hydrogen bonding with adenine.

An elegant method of determination of the yield of the hydroxyhydroperoxides produced in the pyrimidine bases in ³H-labeled DNA on irradiation in the presence of oxygen has been described by Cerutti and his co-workers.[53, 54] These radiation products can be selectively reduced by sodium borohydride, and their cleavage products estimated by paper chromatography. This circumvents the difficulty that radiation products such as the hydroperoxides are extremely unstable and do not withstand the usual hydrolysis procedures required for isolation of bases from DNA. This technique may well find wide application in the estimation and identification of base damage by ionizing radiation and its repair, investigations of which hitherto have been plagued by the instability of the initial radiation products.

Apparently, cupric ion is able to oxidize the radicals produced by OH attack at the 5–6 bond in pyrimidines to give a high yield of glycols.[55] For example, with uracil

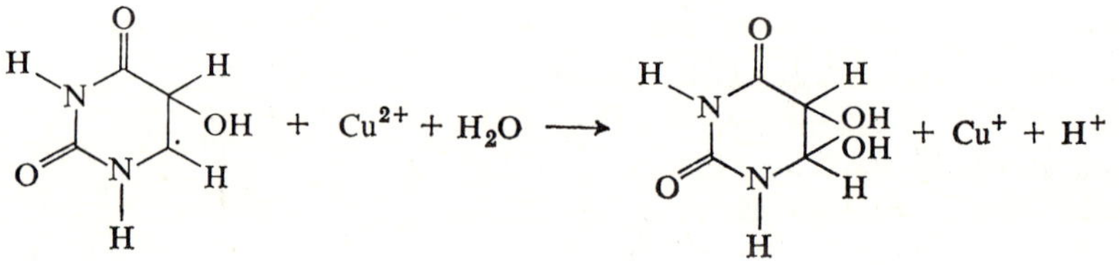

Such a reaction is analogous to reaction (5) in that it would be expected to compete with back-reactions such as (9). On this basis cupric ion might be expected to be a sensitizing agent for anoxic cells. Unfortunately, however, it is very toxic and cannot be tested.

The mechanisms considered above for the radiation-induced destruction of pyrimidines are also applicable to the purines. Thus the degradation of adenine in the presence of oxygen is presumed by Scholes, Ward, and Weiss[31] to proceed via a peroxyl radical intermediate

$$\text{(15)}$$

although Weiss and his co-workers[28] previously reported that in this case no hydroperoxides are detectable. Glycols or hydroxy hydroperoxides produced at the 4-5 double bond in purines would be expected to liberate carbonyl compounds on mild hydrolysis. These have been detected by Holian and Garrison.[56–58] For example, irradiation of xanthine solution in the presence of oxygen produces alloxan[57]

$$+ \; 4H_2O \; + \; O_2 \;\longrightarrow\;$$

$$+ \; 2NH_3 \; + \; HCOOH \; + \; H_2O_2 \qquad \text{(16)}$$

and $G(- \text{base}) = G(\text{alloxan}) = 2.0$ for Co γ rays.[57] In the case of adenine the reaction is strongly influenced by change in pH. $G(- \text{adenine})$ decreases from 2.1 at pH 1 to 1.2 at pH 7,[57] with the result that destruction of adenine in neutral solutions is only about one-half that of thymine in solutions of the same concentration. Because the reaction rates of OH radicals with thymine and adenine are very similar (4.6×10^9 and 4.3×10^9 dm^3 mole^{-1} sec^{-1}, respectively), it appears likely that the low value for the destruction of adenine is attributable to the occurrence of back-reactions.[34]

Hems[59] reports that irradiation of purines under anoxia produces a low yield of 4-amino-5-formamidopyrimidines, presumably as the result of an initial OH reaction at C-8. However, most of the radiation products remain to be identified, and it has not yet been conclusively demonstrated that oxygen increases the overall destruction of purines in solution.

2.2. Production of Single-Strand Breaks in DNA *in Vivo*

The older physico-chemical methods of determination of molecular weight of DNA (and thereby the number of breaks per molecule), such as light scattering and viscometry, have now largely given way to sedimentation studies in density gradients, which permit the measurement of a wider range of molecular weights.[60]

McGrath and Williams introduced a very useful technique[61] in which degradation of the DNA as the result of inadvertent shearing during manipulation is minimized by gentle lysis of the cells directly on top of an alkaline sucrose gradient which is then centrifuged so that the sedimentation of the DNA can be followed. This technique has been extensively used to study the production of single-strand breaks in several different bacteria[61–63] and cultured mammalian cells.[64–71] Table II presents yields of single-strand breaks in the presence or absence of oxygen for several different systems, as determined by this and other techniques. The results are here expressed in terms of a quantity, G_d, the number of single-strand breaks produced in the course of estimated absorption of 100 eV of energy by the DNA (calculated on the basis of mass). It has no theoretical significance and is quantitatively imprecise—but it is useful here for comparison purposes.

Boyce and Tepper[72] developed a sensitive method of determining the number of radiation-induced single-strand breaks which exploits the fact that a certain bacterial virus (bacteriophage λ) forms covalently closed double-stranded circular DNA molecules of uniform molecular weight (32×10^6 daltons) in suitable host bacteria. When these covalently closed circular DNA molecules are exposed to alkaline conditions, the hydrogen bonding between the two strands is broken but the strands cannot become disentangled. Young and Sinsheimer[73] state that the resultant, collapsed but still circular (i.e., "supercoiled"), molecule sediments several times faster than would be expected from its molecular weight. The presence of a single-strand break in this structure permits complete separation of the two strands into a single-strand, rod-shaped molecule and a single-strand circle both of which sediment about three to four times more slowly than does intact, covalently closed circular DNA. Therefore production of the first break in the covalently closed circular DNA molecules can be readily determined from the rate of conversion of rapidly to slowly sedimenting DNA (Fig. 7).

TABLE II

Production of Single-Strand Breaks in the DNA of Various Organisms Irradiated with Ionizing Radiation in Presence or Absence of Oxygen

Organism	Yield of Single-Strand Breaks					
	Breaks $\times$ rad^{-1} (10^6 daltons)$^{-1}$		$G_d{}^a$			
	O_2	N_2	O_2	N_2	OER[b]	Reference[c]
Escherichia coli B/r	1.8×10^{-7}		0.17			McGrath . . .[61]
Escherichia coli K12		7×10^{-7}		0.7		Rupp . . .[76]
Escherichia coli K12 heated at 52°C		$2\text{–}6 \times 10^{-6}$		2–6		Rupp . . .[76]
λ in *E. coli* (λ)	1.4×10^{-6}	5×10^{-7}	1.4	0.5	2.8	Boyce . . .[72]
λ in *E. coli* (λ)	4.2×10^{-6}	1.3×10^{-6}	4.1	1.3	3.3	Johansen . . .[74]
F′ episone in *E. coli*	1.1×10^{-6}		1.1			Freifelder[90]
Murine lymphoma	1.4×10^{-6}	6.9×10^{-7}	1.43	>0.67	<2.1	Lett . . .[64]
Murine lymphoma	1.6×10^{-6}	1.6×10^{-6}	1.5	1.5	1	Dean . . .[62]
Murine lymphoma	6×10^{-7}		0.6			Veatch . . .[91]
Micrococcus radiodurans	2.1×10^{-6}	6.9×10^{-7}	2.0	0.67	3	Dean . . .[62]
Micrococcus radiodurans + EDTA	2.1×10^{-6}	2.1×10^{-6}	2.0	2.0	1	Dean . . .[62]
Phage B3	4.5×10^{-6}	4.5×10^{-6}	4.3	4.3	1	Freifelder[77]
Phage B3 + histidine ($10^{-3} M$)	2.6×10^{-6}	2.6×10^{-6}	2.5	2.5	1	Freifelder[77]
Tobacco mosaic virus (frozen)	2.5×10^{-6}		2.4			Englander . . .[92]

[a] G_d is the number of breaks per 100 eV absorbed *directly* by the DNA. By numerical coincidence, G_d is approximately equal to the number of events $\times$ rad^{-1} (10^{12} daltons)$^{-1}$.

[b] OER, oxygen enhancement ratio.

[c] Only the first author is listed. A series of dots indicates multiple authorship.

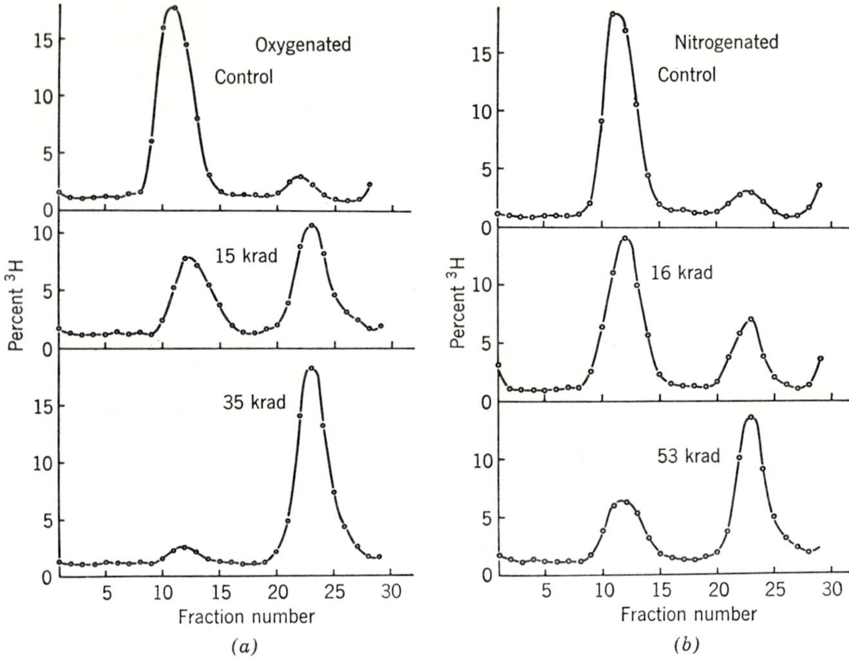

Fig. 7. Effect of X-irradiation on covalent "circular DNA" in cells of *E. coli* lysogenic for phage λ. The covalent "DNA circles" sediment more rapidly than do the "nicked DNA circles" in the alkaline sucrose gradient and thus appear in the peaks on the left. The gradient decreases from 20-percent sucrose at the left to 5 percent at the right. *Escherichia coli* lysogenic for phage λ contain repressor which prevents the λDNA from replicating. (From Boyce and Tepper[72].)

Boyce and Tepper[72] found that the rate of production of X-ray-induced single-strand breaks in the circular DNA of phage λ within the immune host bacterium was 2.8 times higher for irradiation in the presence of oxygen than in its absence (Fig. 8, Table II). They also showed that the breaks produced in the presence or absence of oxygen were repaired by the cell on incubation for 10 min following irradiation. There is some evidence that the number of breaks is underestimated in these experiments, because some rapid repair processes apparently can operate even at ice temperature. Johansen, Gurvin, and Rupp[74] repeated these experiments using a more rapid lysis technique and found a higher yield of breaks, although the sensitizing effect of oxygen was approximately the same. Their values of $G_d = 4.1$ for the production of single-strand breaks in the presence of oxygen and 3.3 for the oxygen enhancement ratio are among the highest so far determined.

Pollard and Weller[75] found that, if bacteria are heated to 52°C for 10 min before irradiation, growth is delayed but the viability of the cells is unaffected.

This treatment has been shown to result in a fivefold increase in the yield of single-strand breaks, as measured by sedimentation studies, possibly because of an effect on the cellular repair process.[76]

Freifelder[77] measured the yield of strand breaks in the DNA from phage B3 which provides a homogeneous population of molecules of molecular weight about 19 million daltons. When such a homogeneous DNA sample is irradiated, the broken molecules sediment more slowly than the main body of intact molecules; their yield can be estimated by analysis of the sedimentation pattern in an analytical ultracentrifuge. Table II includes data obtained on irradiation of phage B3 suspended in buffer and in buffer together with the radical scavenger, histidine. There is no oxygen effect in either case. Because in intact phage the DNA is partially protected by the protein coat, and because phage probably contains rather little water, the DNA is possibly inactivated predominantly by direct action if the irradiation is conducted in the presence of a radical scavenger. Moreover, there is no evidence to suggest that phage carry their own repair enzymes, and it is perhaps not surprising that the observed yield of breaks in phage B3 is high.

The data presented in Table II leave unanswered the question of whether there is an oxygen effect in the *initial* production of single-strand breaks. The difficulty stems partly from the fact that some single-strand breaks are rapidly repaired, even at 0°C in bacteria. Moreover, damage in the presence of oxygen can be expected to produce strand breaks with end groups different from those formed in nitrogen and these may be repaired at different rates.

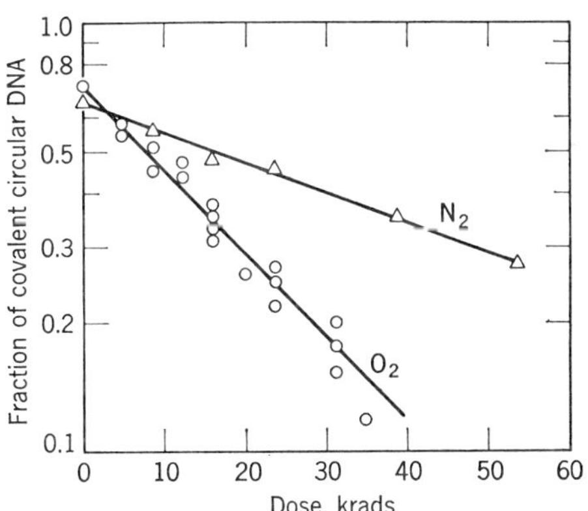

Fig. 8. Plot of fraction of intact "DNA circles" against X-ray dose delivered to ³H-labeled phage λ superinfecting a λ-lysogen, in the presence of either oxygen or nitrogen. Note that breaks produced in the presence of either oxygen or nitrogen are easily repaired by incubating the irradiated cells at 37°C prior to lysis on the sucrose concentration gradient. (From Boyce and Tepper[72].)

Lett et al.[64] reported that oxygen increases the yield of radiation-induced, single-strand breaks in the bacterium *Micrococcus radiodurans* and mouse lymphoma cells by a factor between 2 and 3. This conclusion was later withdrawn, in the case of the lymphoma cells, after reinterpretation of the data and, in the case of the bacterium, after the discovery that the effect is not observed if the irradiation is carried out in the presence of ethylenediamine tetraacetic acid (EDTA).[62, 78] It was suggested that EDTA inhibits the enzymes responsible for repairing the damage produced in nitrogen, but not the damage produced in the presence of oxygen. Thus the yield of breaks in oxygen remains unaffected by EDTA at a G_d value of 2.0, and the observed yield in nitrogen increases from 0.67 to 2.0.

2.3. Production of Single-Strand Breaks *in Vitro*

In addition to his work with intact phage, Freifelder[77] also determined the yields of single-strand and double-strand breaks in free bacteriophage B3 DNA (i.e., *in vitro*) under a variety of conditions. The rate of strand breakage *in vitro* was found to be independent of the presence of oxygen in all cases. There was no oxygen effect on irradiation of free DNA, DNA in intact phage, or DNA in intact phage suspended in 10^{-3} M histidine plus 10^{-3} M cysteine. Freifelder concluded that the oxygen effect in most biological systems probably represents a killing mechanism which is independent of strand breakage. This conclusion is, however, at variance with the results of Boyce and Tepper[72] and Johansen, Rupp, and Gurvin.[74]

Table III presents the yields of single- and double-strand breaks found for DNA of phage B3 under a variety of conditions. Also included are some data obtained by Kapp and Smith[79] for calf thymus DNA X-irradiated in the presence of oxygen and sedimented in alkaline sucrose gradients. These data agree surprisingly well, considering the differences in experimental technique and concentrations of the irradiated DNA samples. Both groups noted the reduction in yield of breaks in the presence of 10^{-3} M histidine. This protective effect, presumably the result of radical scavenging, is small, especially in Freifelder's experiment in which the DNA concentration was relatively low (20 $\mu g/ml$). Taken at face value, the fact that the protective effect of 10^{-3} M histidine is approximately the same for a 20 $\mu g/ml$ solution as it is for a 2.5 mg/ml solution indicates either that histidine is a very poor scavenger, or that indirect effects are of little importance. Histidine appears to be a good scavenger for OH radicals ($k = 3 \times 10^9$ dm^3 mole^{-1} sec^{-1});[80] but fairly poor for the hydrated electron ($k = 6 \times 10^7$ dm^3 mole^1 sec^{-1}).[81] It might be interesting to study the protective effect in these systems of higher concentrations of scavengers. In this way the damage exclusively attributable to direct absorption of radiation by the DNA could be estimated.

TABLE III
X-Ray Induced Single-Strand and Double-Strand Breaks in DNA in Aqueous Solution (i.e., *in Vitro*)

| Source | Irradiation medium | DNA concentration | Breaks × rad^{-1} (10^6 daltons)$^{-1}$ | | | | Reference[a] |
| | | | Single | | Double | | |
			O_2	N_2	O_2	N_2	
Phage B3	Buffer	20 μg/ml	3.3×10^{-4}	3.3×10^{-4}			Freifelder[77]
Phage B3	Buffer + 10^{-3} M histidine	20 μg/ml	1.25×10^{-4}	1.25×10^{-4}	2.7×10^{-6}		Freifelder[77]
Calf thymus	Saline	2.5 mg/ml	2.4×10^{-4}				Kapp . . .[79]
Calf thymus	Saline + 10^{-3} M histidine	2.5 mg/ml	9.2×10^{-5}				Kapp . . .[79]
Sarcina lutea	Buffer	40 μg/ml	5.2×10^{-5}				Summers . . .[82]
Escherichia coli B	Buffer	40 μg/ml	5.4×10^{-5}				Summers . . .[82]
Phage ϕ29	Saline	10 μg/ml			1.9×10^{-6b}		Summers . . .[84]

[a] Only the first author is listed. A series of dots indicates multiple authorship.
[b] Based on the assumption[84] of molecular weight = 10^7 daltons.

Summers and Szybalski[82] have described a novel technique for the determination of molecular weight which does not involve density gradient analysis. They introduce a cross-link between the double strands of DNA and then measure the number of radiation-induced single-strand breaks by their effect on the ability of the DNA to renature after it has been denatured by heating. In this way some single-strand portions of DNA are lost and can be separated from perfectly renatured DNA by fractionation in a polyethylene glycol–dextran two-phase system, or in a cesium chloride density gradient. Data obtained by this technique are included in Table III.

2.4. Double-Strand Breaks

The inactivation of phage T7 by X-irradiation, in both the presence and absence of oxygen, can be correlated with the number of double-strand, but not with the number of single-strand, breaks.[83] Approximately 12 single-strand breaks are produced for every double-strand break. Inactivation of phage in this connection can be caused by a failure to inject the fragmented phage genome into the host cell, for it would certainly be expected that injection would be interrupted by a double-strand break. According to Summers and Szybalski,[84] the number of double-strand breaks produced in phage ϕ28 can likewise be correlated with inactivation.

Kaplan[63] studied the production of double-strand breaks in the DNA of bacteria exposed to X-rays in the presence of oxygen and demonstrated an approximate correlation between the degree of sensitization caused by incorporation of the base analog BU and the degree to which this incorporation increases the yield of double-strand breaks. He found no evidence that double-strand breaks could be repaired in this system.

A similar technique was used by Veatch and Okada[68] to determine the yield of double-strand breaks on irradiation of aerobic cultured mouse lymphoma cells. They found that the ratio of double-strand breaks to single-strand breaks was approximately 10. Their estimate of the absolute yield of single-strand breaks is given in Table II.

Very little attention has been addressed to the question whether or not oxygen increases the production of double-strand breaks.

2.5. Damage to Deoxyribose in DNA

By comparison of the extent of destruction of the chromophoric groups in the free bases, nucleosides, and nucleotides in oxygenated aqueous solution, it can be concluded that approximately 80 percent of the free radicals attack the base moiety in a nucleotide and the remainder attack the sugar.[31, 34] It is not known whether this conclusion also applies to polynucleotides, such as DNA.

Early studies by Scholes, Weiss, and their colleagues[85, 86] on the irradiation of aqueous solutions of DNA demonstrated production of phosphomonoester groups and liberation of inorganic phosphate, indicating breakage of the DNA chain at the phosphodiester bond. Damage to the sugar component could also be inferred from postirradiation liberation of phosphate from nucleotides, which was thought to be the consequence of production of carbonyl groups at the C-3' position of the deoxyribose and the resultant labilization of the phosphate group.[87, 88]

Krushinskaya and Shal'nov[89] provided direct evidence that the deoxyribose residues are attacked in a demonstration that irradiation of aerobic aqueous DNA solutions produces malonic aldehyde by rupture of the C-3'–C-4' bond (Fig. 9). This observation was confirmed and extended by Kapp and Smith[79] in a more detailed study which relates the yield of these malonic aldehyde-like products to the yield of single-strand breaks. Figure 9 illustrates the reaction thought to produce the malonic aldehyde. It was

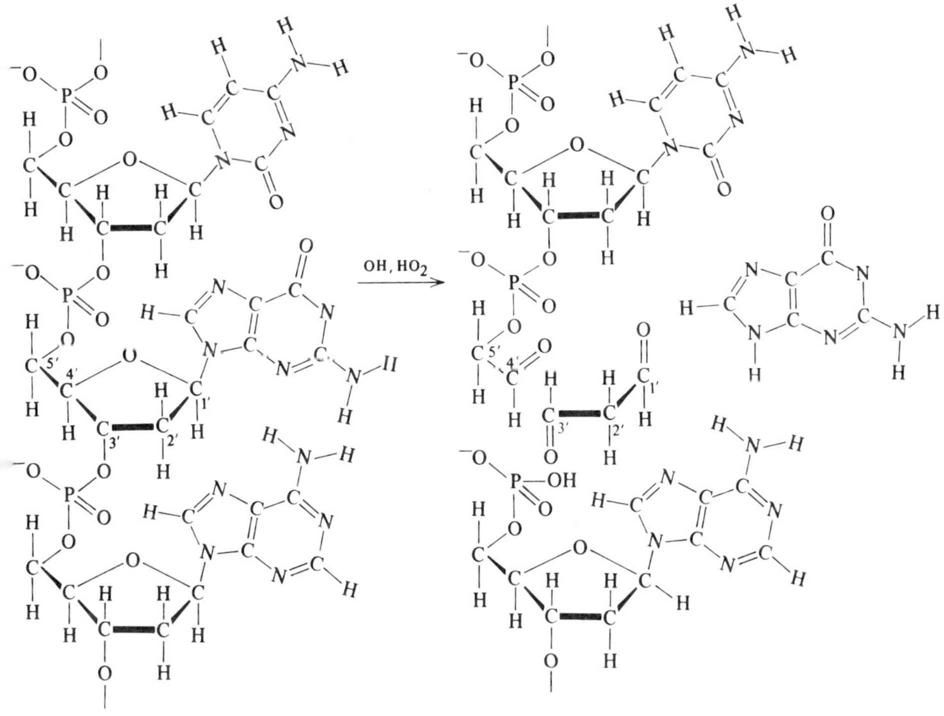

Fig. 9. Proposed reaction scheme for breakage of DNA chains with formation of malonic aldehyde. (From Krushinskaya and Shal'nov[89] and Kapp and Smith.[79])

found that, for every single-strand break produced in the DNA by X-irradiation in aqueous aerated solution, 0.60 molecules of the malonic aldehyde-like product, 0.33 molecules of inorganic phosphate, and 1.35 molecules of phosphomonoester groups are produced. By use of the enzyme polynucleotide kinase to catalyze the transfer of a phosphate group from γ[32P]ATP to the 5'-hydroxyl termini of the irradiated DNA, it was demonstrated that 29 percent of the liberated phosphomonoester groups are 5'-phosphoryl groups which outnumber 5'-hydroxyl termini by 32 to 1.[79] No experiments were reported for irradiation under anoxia. The results of such experiments will be awaited with considerable interest because this system appears to be a good one for studying the differences produced by irradiation under anoxic and aerobic conditions.

Some information regarding the relative importance of various types of damage can be obtained from a comparison of the sedimentation properties of the irradiated replicative form of the DNA of the phage ϕX174 with its reproductive ability. When this DNA is irradiated in oxygenated buffer solution, production of single-strand breaks can be distinguished from that of double-strand breaks by different rates of sedimentation in sucrose gradients. By testing the biological activity of each fraction, it is possible to determine to what extent single-strand breaks and double-strand breaks are lethal. Lethal events that do not appear as strand breaks are ascribed to base damage. Braams[81] found that in this system, all double-strand breaks are lethal but constitute only 3 percent of the total damage, only a very few of the single-strand breaks, if any, are lethal, and base damage is the main cause of inactivation (75–97 percent). It is not known what fraction of the nucleotide damage is associated with the sugar. This is a phage system and therefore should be compared with cells with considerable caution. The same reservation applies to the observation by Hariharan and Cerutti[54] that 30 percent of the lethal effects of γ rays on the single-stranded RNA phage R17 can be accounted for by base damage.

Bopp and Hagen[188] found that the number of observable single-strand breaks in DNA irradiated in aqueous solution is increased by a factor of 1.5. They concluded that this factor is a measure of latent radiation-induced damage to the deoxyribose moiety.

2.6. Mechanism

Although oxygen is probably the most widely studied radiosensitizing agent, the mechanism by which it operates—although understood in the broad sense—remains to be elucidated in detail, particularly with regard to the initial radiation products. There is little doubt that the most important general mechanism for the action of ionizing radiation in oxygenated solutions of organic molecules is hydroperoxyl radical formation, followed probably by reduction to a hydroperoxide. What is obscure is whether this process plays a part in the production of single- and double-strand breaks and sugar damage in DNA.

The number of single-strand breaks produced in DNA strands by X-rays is probably affected by the surrounding material. In the cytoplasm of bacteria, strand breakage appears to be reduced by anoxia as in the Boyce–Tepper experiments.[72] This behavior may be connected with the composition of the cytoplasm, for Freifelder[77] did not observe anoxic protection in phage DNA irradiated in histidine solutions.

Double-strand breaks, sugar damage, and base damage appear to be the main contenders for the role of lethal damage. Whether the yield of double-strand breaks in DNA *in vivo* is increased by oxygen remains to be determined, but many experiments lend support to the idea that base damage is increased by oxygen. It should be a simple matter to determine whether the yield of malonic acid derivatives from deoxyribose in DNA is increased by oxygen.

From the restricted viewpoint of this review, the most important conclusion is that oxygen can increase the effect of radiation by the production of peroxyl radical adducts

$$R\cdot + O_2 \rightarrow RO_2\cdot$$

3. *N*-ETHYLMALEIMIDE AS A SENSITIZING AGENT

Several theories have been postulated to explain the radiosensitizing properties of NEM, first discovered by Bridges[93, 94] in 1960 in experiments with *E. coli*. One centers around the fact that this compound is a sulfhydryl-binding agent

$$X\text{---}SH + \begin{array}{c} CH\text{---}C \\ \| \\ CH\text{---}C \end{array} N\text{---}C_2H_5 \rightarrow \begin{array}{c} X\text{---}S\text{---}CH \quad C \\ | \\ CH_2\text{---}C \end{array} N\text{---}C_2H_5 \qquad (17)$$

Although there is some evidence that in certain systems NEM slightly sensitizes oxygenated cells,[1, 93, 95] its most pronounced effect is observed under anoxia.

3.1. The Sulfhydryl-Binding Theory

The S—H bond is relatively weak, and the hydrogen atom is readily abstracted by free radicals. Therefore a sulfhydryl compound may be capable of donating a hydrogen atom to free radicals produced in reactions such as (18), with reconstruction of the original molecule.[29, 46, 47]

$$RH + OH \rightarrow R\cdot + H_2O \tag{18}$$

$$R\cdot + XSH \rightarrow RH + XS\cdot \tag{19}$$

In many cases in which the free radical is not produced by hydrogen abstraction, reaction with sulfhydryls could still restore the original molecule. For example, because 5-hydro-6-hydroxythymine is unstable with respect to dehydration, the initial addition of a hydroxy group to thymine is essentially reversed by sulfhydryl groups.

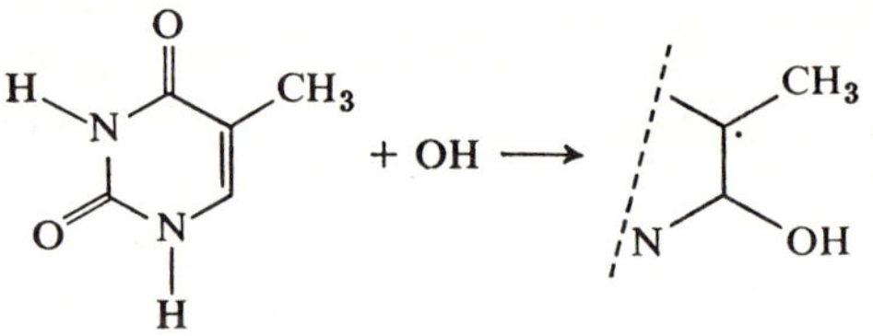

$$\text{(20)}$$

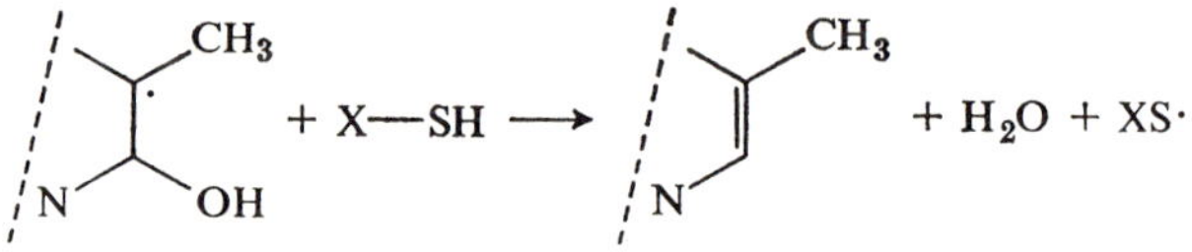

$$\text{(21)}$$

It was proposed that sulfhydryl-binding agents might sensitize cells by removing sulfhydryls and inhibiting reactions such as (19) and (21), in which case other, unspecified reactions of $R\cdot$ were thought to be lethal to the cell. Lynch and Howard-Flanders[47] demonstrated a relationship between the sensitization of the bacteria *Shigella sonnei* by NEM and the removal of cellular sulfhydryl groups. Later, Johansen and Howard-Flanders[96] studied the radiation protection afforded to *E. coli* B/r by mercaptoethanol in the presence of small amounts of oxygen and concluded that, at low oxygen concentrations, there is evidence for competition between oxygen and mercaptoethanol, for the more mercaptoethanol added the higher was the concentration of oxygen required to create a given degree of sensitization. They found that 40 μM oxygen competed about equally well with 10 mM mercaptoethanol and concluded that this result is the consequence of competition by the reaction

$$R\cdot + O_2 \rightarrow \text{inactive product} \tag{22}$$

with reaction (19). They also showed that NEM reduces the level of sulfhydryl groups in the cell by 90 percent, an effect they postulated to be responsible for the sensitizing properties of this compound.

Evidence for reaction (19) has been provided by Adams, NcNaughton, and Michael[97, 98] who demonstrated that the methanol radical reacts with the sulfhydryl group in cysteamine to form methanol

$$OH + CH_3OH \rightarrow H_2O + \overset{\cdot}{C}H_2OH \tag{23}$$

$$\overset{\cdot}{C}H_2OH + X\text{—}SH \rightarrow CH_3OH + XS\cdot \tag{24}$$

$k_{24} = 6.8 \times 10^7$ dm^3 mole^{-1} sec^{-1}. However, it is not known whether or not reactions analogous to (24) occur with DNA free radicals within the cell. The value of k_{24} appears to be sufficiently low to permit competition from other reactions.

On the strength of the sulfhydryl-binding theory, many sulfhydryl-binding agents have been tested for ability to sensitize cells to radiation. A considerable number have been found to be effective, but a large fraction of these have been found to sensitize by virtue of some other property, such as by the production of toxic radiation products. Sensitization by the formation of toxic radiation products in a suspension of cells in buffer can be eliminated by the addition of efficient radical scavengers and is perhaps less interesting or potentially less useful than sensitization mediated wholly within the cell.

3.2. Nonformation of Toxic Radiation Products

One test for production of toxic radiation products involves irradiation in a suspension medium containing a large concentration of radical scavenger.[99] If the scavenger can enter the cell, so much the better. Deoxythymidine (TdR) is quite useful in this connection because it reacts rapidly with OH ($k = 3.9 \times 10^9$ dm^3 mole^{-1} sec^{-1}),[34] H ($k = 2.5 \times 10^8$ dm^3 mole^{-1} sec^{-1}),[34] and probably e_{aq}^- ($k = 1.7 \times 10^{10}$ dm^3 mole^{-1} sec^{-1} for thymine and 1.5×10^9 dm^3 mole^{-1} sec^{-1} for thymidylic acid),[34] it is very soluble in water, and it can enter the cell. (The solubility of TdR has a high temperature coefficient. It can be used at 10^{-1} M at ice temperature as a supersaturated solution.) Figure 10 shows that NEM still sensitizes *E. coli* B/r in the presence of 10^{-1} M TdR. Many reported sensitizing agents do not sensitize bacterial suspensions in the presence of high concentrations of TdR (e.g., cupric ion,[100] *p*-hydroxymercuribenzoate,[101] iodoacetic acid,[102] and iodoacetamide[103]) – author's unpublished observation.

Another test for the production of toxic radiation products is afforded by the "rapid-mix" technique. In this technique, pioneered by Dewey and Michael,[104] bacteria are irradiated with a pulse of electrons and then rapidly mixed with unirradiated sensitizer, or the sensitizer is irradiated and then rapidly mixed with unirradiated bacteria. If irradiated NEM solution is rapidly mixed with unirradiated bacteria *Serratia marcescens*, there is no effect on their viability, showing that toxic products are not formed.

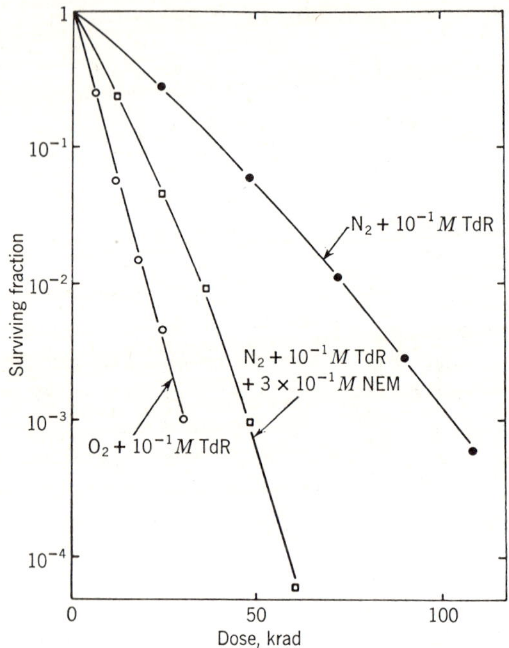

Fig. 10. Survival of colony-forming ability of the bacterium *E. coli* B/r as a function of the X-ray dose delivered in the presence of both NEM and TdR under anoxia. (Emmerson, unpublished results.)

3.3. Effect of Removal of NEM before Irradiation

Some doubt was cast on the validity of the sulfhydryl-binding theory in a critical experiment by Bridges,[94] who showed that bacteria incubated with NEM and then washed free of it were not sensitive to radiation. If the sensitization had been caused by removal of sulfhydryl groups, some sensitization would have been expected. The lack of observed sensitization is apparently not attributable to any rapid replacement of sulfhydryl groups after removal of NEM. Dewey[106] showed that NEM reduces the sulfhydryl content by about 70 percent, and that there is no regeneration within 2 hr. It could, however, always be argued that NEM binds irreversibly to some critical sulfhydryl groups within the cell and cannot therefore be washed out. This possibility cannot be entirely discounted, but the observation that removal of NEM immediately prior to irradiation abolishes sensitization must weaken the theory that the radiosensitizing properties of this substance are related exclusively to its sulfhydryl-binding properties.

Partial sensitization of the bacteria *Micrococcus sodonensis* has been reported in experiments in which the cells were washed free of NEM prior to irradiation[95] but, again, it is difficult to establish that no NEM remained in the cell. There may be some variation from one species to another with regard to the effect of NEM. However, it is striking that careful removal of NEM prior to irradiation has been reported to abolish sensitization completely in at least three different bacterial species, *E. coli*,[94] *Pseudomonas* species[1] and *S. marcescens*.[106]

3.4. Sensitization Rate

The argument for the sulfhydryl-binding theory is further weakened by the fact that rapid-mix experiments show that NEM is effective when added even only a few microseconds before irradiation (Fig. 11). Adams, Cooke and Michael[105] have pointed out that the sulfhydryl-binding reaction [reaction (17)] is not likely to occur in this time period.

3.5. Electron Affinity

One of the first theories concerning the mechanism of sensitization of NEM that was independent of its sulfhydryl-binding properties was that of Adams and Dewey.[107] Noting that NEM reacts with the solvated electron to produce a resonance-stabilized radical ion, they proposed that the sensitizing properties could be the result of increase in the effective lifetime

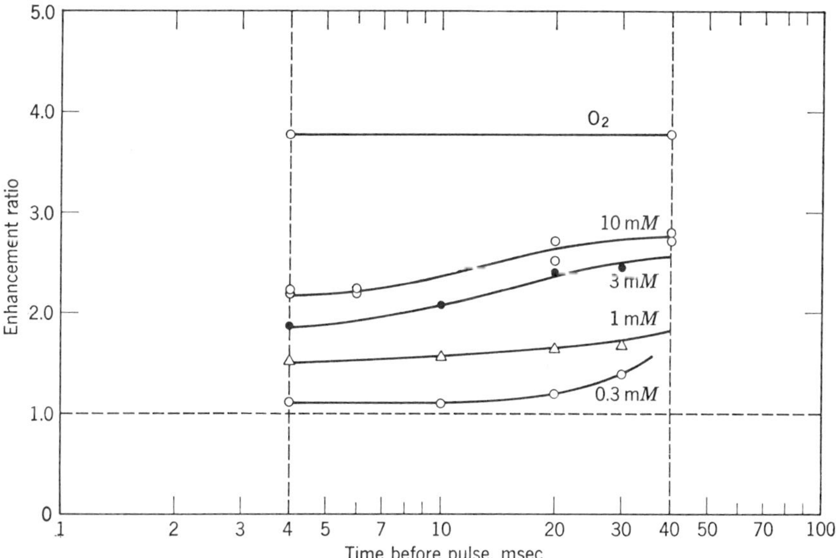

Fig. 11. Variation of the enhancement ratio afforded by NEM to the anoxic bacteria, *S. marcescens*, when added at various times before the irradiation pulse in the rapid-mix technique. (From Adams, Cooke, and Michael.[105])

of the electron, with consequent increase in the probability of its reaction with a molecule essential to cellular survival

$$e_{aq}^- + \begin{array}{c} CH-C \\ \parallel \\ CH-C \end{array}\!\!\!\!\begin{array}{c} O \\ \diagup \\ N-C_2H_5 \\ \diagdown \\ O \end{array} \rightarrow \begin{array}{c} CH-C \\ \parallel \\ CH-C \end{array}\!\!\!\!\begin{array}{c} O^- \\ \diagup \\ N-C_2H_5 \\ \diagdown \\ O \end{array} \rightleftharpoons \begin{array}{c} CH-C \\ \parallel \\ CH-C \end{array}\!\!\!\!\begin{array}{c} O \\ \diagup \\ N-C_2H_5 \\ \diagdown \\ O^- \end{array} \qquad (25)$$

Among the known sensitizing agents oxygen is one of the most electron affinic. However, Blok, Luthjens, and Roos[108] showed that O_2^- does not react with DNA in aqueous solution. Moreover, molecular oxygen is thought to sensitize by reacting with radiation-induced radicals to form hydroperoxyl radicals, as discussed in Section 2.1. Thus largely as the result of its failure to encompass the radiosensitizing properties of oxygen, this theory has been practically abandoned, although not before it had provided the impetus for a successful search for other electron-affinic radiation sensitizing agents of general properties similar to those of NEM. Many sensitizing agents, most of which contain carbonyl groups, have been found in this way. They include various aromatic keto compounds, biacetyl, quinones, indane-trione,[109, 110] methyl glyoxal,[111] phenyl glyoxal,[112] phenylpyruvic acid,[110] and *p*-nitroacetophenone.[98, 113]

Another hypothetical mechanism that involves the electron-affinic properties of compounds such as NEM was proposed by Adams.[109, 110] It derives from the observation that electron transfer processes occur from nucleotides to the sensitizers and not in the other direction, as would have been required by the model we have just considered. The new theory assumes that the absorption of radiation energy by a target molecule results in the formation of an ion pair. The ejected electron becomes thermalized and is trapped (1) at some suitable site on the molecule. There is a certain probability that the electron finds its way back to its original site (2) and thereby nullifies any radiation effect. The role of an electron-affinic sensitizer, as proposed by Adams, is to react with the electron (3) and to compete with the charge recombination. It is suggested that thereafter the remaining positive ion would then decay to a neutral free radical; see Adams, this volume.

Greenstock, Adams, and Willson[114] showed that in aqueous alkaline solution electron transfer reactions do occur from negatively charged nucleotide radical anions to various electron-affinic sensitizing agents in aqueous alkaline solution. In such case, however, the situation is somewhat different, for there is no reason to believe that the nucleotide is damaged in any way by the transfer process.

3.6. Reaction with DNA Radicals to Produce Adducts

Oxidizing agents may be defined as substances having a high electron affinity. They are able to participate in redox reactions either by accepting an electron in an electron transfer reaction or by formation of a covalent bond. Oxygen oxidizes a radical $R\cdot$ by formation of a covalent bond in the hydroperoxyl radical

$$R\cdot + O_2 \rightarrow ROO\cdot \tag{27}$$

The possibility that NEM might react in similar fashion was first suggested in 1961 by Bridges,[94] who proposed that NEM "could react across its double bond with a carbon-free radical center." This proposal seems to have attracted remarkably little attention, possibly partly because Bridges went on to consider an alternative possibility that NEM was perhaps sensitizing vital proteins in the cell by reacting with radiation-induced sulfhydryl groups formed from S—S bonds.

Bridges' suggestion that NEM can react directly with radiation-induced radicals is supported by Johansen et al.,[115] who showed that such reactions can occur with DNA radicals *in vitro*. On irradiation of solutions of ^{3}H-labeled DNA in the presence of ^{14}C-labeled NEM in deoxygenated solutions, the ^{14}C label becomes bound to the DNA. Radiation-induced binding does not occur in the presence of oxygen. Examination of the kinetics of the system showed that adduct formation is the result of the reaction

$$DNA\cdot + NEM \rightarrow products \tag{28}$$

and not of

$$DNA + NEM\cdot \rightarrow products \tag{29}$$

The production of adducts between DNA and NEM was confirmed independently by Buu and Emmerson (unpublished results). However, such adduct formation has not yet been demonstrated in intact cells.

The formation of NEM adducts with DNA· radicals suggests the study of similar reactions by the pulse radiolysis technique with model substances such as thymine. Unfortunately, the results of such experiments are not as clear-cut as those from analogous experiments with oxygen (see Section 2) and with *N*-oxyl radicals (see Section 5). If the hydroxythymine radical does react with NEM, the product of the reaction apparently has absorption characteristics similar to those of the reactants so that the reaction cannot be studied by following decay of absorption by the hydroxythymine radical.[116] However, NEM can be shown to react with the radical produced by OH attack on deoxyribose. When nitrous oxide-saturated solutions of deoxyribose are pulse-irradiated in the presence of NEM, there is an *increase* in absorption at 400 nm not found in its absence. Analysis of the kinetics of this increase reveals the occurrence of a reaction

$$R \cdot + NEM \rightarrow products \tag{30}$$

This reaction was found to be moderately fast ($k_{30} \approx 3.2 \times 10^8$ dm^3 mole^{-1} sec^{-1}) but the product, or products, were not identified (Emmerson and Fielden, unpublished).

3.7. Alternative Mechanisms

Mullenger and Ormerod[95] have recently concluded that NEM sensitizes the bacteria *M. sodonensis* by three different mechanisms. Because pre-treatment of the cells with NEM, followed by removal of sensitizer, resulted in partial sensitization to cells irradiated both in the presence and in the absence of oxygen, they concluded that some of the sensitization was the result of binding of NEM with sulfhydryls. However, they noted that, according to the usual concepts of the sulfhydryl-binding theory, such sensitization should be observed only in absence of oxygen, for it is thought to involve the competition

$$R \cdot + XSH \rightarrow RH + XS \cdot \tag{31}$$

$$R \cdot + O_2 \rightarrow RO_2 \cdot \tag{32}$$

A further sensitization was found with stationary-phase cells, when NEM was present during irradiation, an effect that could be eliminated by N_2O or nitrate ion. Both these species react rapidly with e_{aq}^-. Mullenger and Ormerod consequently concluded that this sensitization is the result of prolongation, by NEM, of the lifetime of the electron, as suggested by Adams' theory.[107] However, as noted in Section 3.5, the NEM$^-$ radical anion is not likely to be a very reactive species. The third mechanism for which these investigators reported evidence was based on their observation of a small degree of sensitization when NEM was added after irradiation, an effect they ascribed to inhibition of the enzymic repair processes. They did not discuss the possibility that NEM might react directly with radicals produced in vital molecules.

The results with these bacteria are complex and difficult to interpret. Large differences were found between stationary- and exponential-phase cells. With exponential-phase cells the full degree of sensitization was observed in cells treated with and washed free of NEM before irradiation. NEM was found also to sensitize stationary-phase cells in the presence of air. None of the theories discussed in this chapter (except toxic product formation) can account for sensitization in the presence of oxygen unless the concentration of oxygen is sufficiently low to allow significant competition from other reactions.

Attempts to associate the radiosensitizing properties of NEM with its capacity for binding sulfhydryl compounds (recently reviewed by Bridges[1]) have not met with much success, partly because, according to the sulfhydryl-binding theory, the role of the sensitizing agent is indirect and therefore perhaps more difficult to prove or disprove. Formation of DNA–NEM adducts on the other hand is a direct mechanism and it may be possible to demonstrate that it operates within the cell. If adducts are produced within the cell, their yields should be determinable and comparable with the degree of inactivation of the cell. Experiments along these lines could help to establish to what extent, if any, adduct formation is important in the sensitization process.

3.8. Sensitization of Higher Cells

Despite formidable problems associated with toxicity, some attempts have been made to study possible radiosensitizing effects of NEM on higher cells.

Klimek[117] has reported a dose-modifying factor of 1.5 in anoxic human (HeLa) cells in tissue culture treated with 2×10^{-6} M NEM. However, this rather small degree of sensitization was found at concentrations so near the toxicity limit (3×10^{-6} M) that the effect may possibly be the result of some trivial effect, such as a radiation-induced increase in permeability of the cell. It is striking that NEM is much more toxic to cultured mammalian cells than it is to bacteria. Mammalian cells seem to be able to tolerate only about 10^{-3} times as much NEM as the typical bacterium.

Moroson and Furlan[118] studied the effect of injecting (5 mg/kg; 4×10^{-5} M overall) NEM on the growth rate of mouse ascites tumor cells irradiated in the abdomen of the mouse. They concluded that NEM does sensitize the tumor cells but that the sensitization is not confined specifically to anoxic cells.

4. NITRIC OXIDE AS A SENSITIZING AGENT

Nitric oxide (NO) is a free radical with an unpaired electron in a π molecular orbital and is thus, to a degree, electronically similar to oxygen, which has two such unpaired electrons.

4.1. Anoxic Bacteria

In 1957, Howard-Flanders[119] showed that nitric oxide sensitizes anoxic bacteria to X-rays. Indeed, at low concentrations, nitric oxide was found to be as effective as oxygen in sensitizing the cells. Figure 12 shows that the sensitizing effect (expressed as the reciprocal of the 10-percent survival dose in rads) is almost identical for equal oxygen and nitric oxide concentrations so that Howard-Flanders and Jockey[120] concluded that these substances are equivalent molecule for molecule in their action. They also showed that nitric oxide is effective at low concentration only if present during irradiation and proposed that both oxygen and nitric oxide produce their effect by reacting with a radical formed in a molecule essential for the multiplication of the cell. Presumably, the initial product of the reaction between an organic free radical and nitric oxide would be a nitroso compound

$$R\cdot + NO \rightarrow RNO \tag{33}$$

Such reactions are well known to occur in the gas phase[121] and have been employed to establish reaction mechanisms in aqueous solution.[122] To what extent such intracellular radiation products would be lethal to the cell is uncertain.

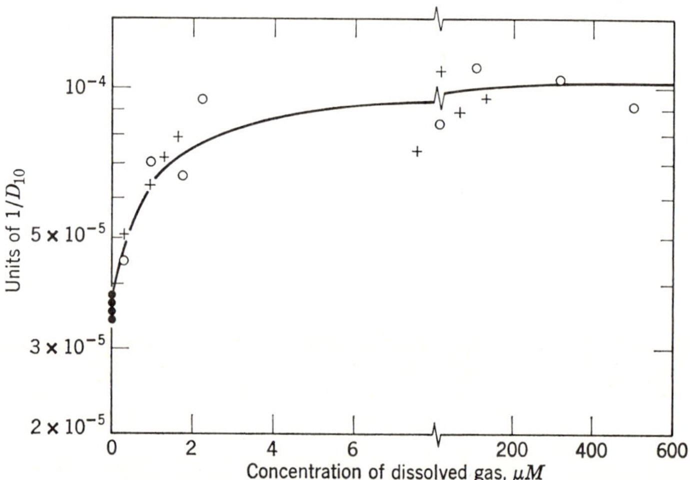

Fig. 12. Radiation sensitivity of bacteria *S. sonnei* (expressed as the reciprocal of the 10-percent survival dose D_{10} in rads) as a function of the concentrations of oxygen ($\bigcirc$) and of nitric oxide ($+$). The anoxic condition (under nitrogen) is shown by $\bullet$. (From Howard-Flanders and Jockey.[120])

At higher nitric oxide concentrations, the situation is more complex, a not surprising effect of a compound known to be much more reactive than oxygen. Dale, Davies, and Russell[123, 124] found that millimolar amounts of nitric oxide can sensitize the bacteria *Shigella flexneri* when present before irradiation and then subsequently removed. However, it is hard to demonstrate conclusively in such a case that a significant amount of the gas does not remain in the cell; a $1\text{-}\mu M$ residue would be enough to give one-half of the maximum sensitization at $2°C$. Dale, Davies, and Russell[124] also showed that higher concentrations of nitric oxide protect the bacteria, as would be expected of an efficient radical scavenger.[105, 124, 125]

Interpretation of results obtained at such higher concentrations has been further complicated by the fact that treatment of the bacteria *S. sonnei* with $10^{-3}\ M$ nitric oxide reduces the sulfhydryl concentration in the cell.[126] However, there is no evidence that this effect contributes to the sensitization.

4.2. Higher Organisms

Kihlman[127] showed that nitric oxide sensitizes the radiation production of chromosomal aberrations in anoxic bean roots to the same extent as does oxygen. It also sensitizes mouse ascites tumor cells[128] and insects[129] but is protective toward dried seeds[130] and spores.[2, 131]

4.3. Effect of Nitrogen Dioxide Production

In strict anoxia nitric oxide is not normally toxic to microorganisms if the temperature is kept below $15°C$.[120, 132] However, it cannot be used even under hypoxic (i.e., low oxygen concentration) conditions because it oxidizes to toxic nitrogen dioxide. Such behavior appears to preclude its application as an adjuvant in radiation therapy and explains why relatively little attention has been given to the study of this very efficient sensitizing agent.

4.4. Mechanism

Although results obtained in experiments with high concentrations of nitric oxide are complex, those at low nitric oxide concentrations are simpler and show a striking similarity to experiments at low concentrations of oxygen. Perhaps there are several different mechanisms by which nitric oxide sensitizes cells, but it appears likely that the most important mechanism, and that responsible for the rapid increase in sensitization at low concentrations, can well be adduct formation between DNA· free radicals and nitric oxide; see Eq. (33).

5. *N*-OXYL RADICALS AS SENSITIZING AGENTS

N-Oxyls are stable organic free radicals which are usually soluble in water. They are relatively new compounds, synthesized for the first time in 1959 by Lebedev and Kazarnovski.[133] Prior to that time the stability of the then-known organic free radicals was generally ascribed to delocalization of the unpaired electron over a conjugated system. Because the unpaired electron in the organic *N*-oxyl radicals is associated with the N—O bond, and is not delocalized further, their stability was at first attributed to steric hindrance.[134–136] The structures of several *N*-oxyl compounds are presented in Fig. 13, in which it can be seen that the N—O group is usually flanked by *tert*-butyl or similar bulky groups. Dupeyre and Rassat[137] have noted that the *N*-oxyls are similar to nitric oxide, in which stability is ascribed to the properties of the three-electron N—O bond.[134, 138] Unhindered *N*-oxyl radicals, such as dimethyl nitroxide[139] and diethyl nitroxide,[140, 141] can also be prepared but, although they do not dimerize, they are unstable and decompose.

N-Oxyl radicals are readily prepared by oxidation of the corresponding amine by hydrogen peroxide in the presence of sodium tungstate.[133, 142, 143]

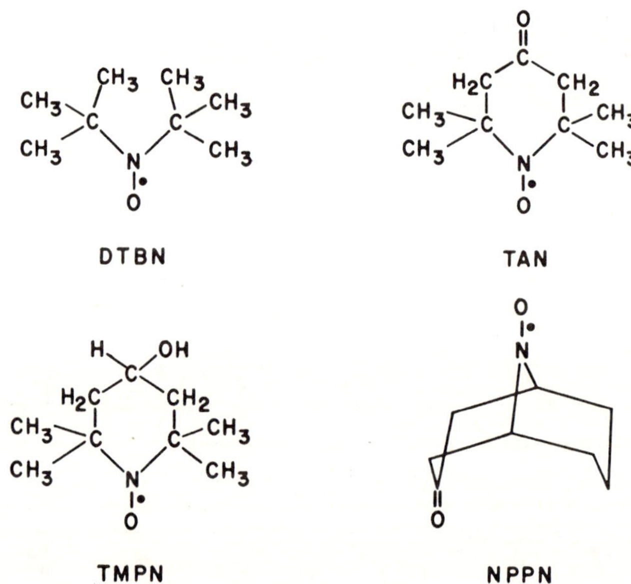

Fig. 13. Structures of some *N*-oxyl radicals: DTBN, di-*tert*-butyl nitroxide; TAN, triacetoneamine *N*-oxyl; TMPN, 2,2,6,6-tetramethyl piperidinol *N*-oxyl; NPPN, norpseudopelletierine *N*-oxyl.

5.1. Preferential Sensitization of Anoxic Bacteria to X-Rays by *N*-Oxyls

Di-*tert*-butyl nitroxide (DTBN), the first *N*-oxyl to be tested as a sensitizing agent, was found to raise the sensitivity of anoxic *E. coli* B/r almost to the same level as that when fully oxygenated.[144] It does not sensitize aerobic bacteria and therefore fulfils one of the criteria for a suitable adjuvant in radiation therapy, where it has long been thought that an agent that preferentially sensitizes anoxic tissue would have some therapeutic advantage. Subsequently, several other *N*-oxyls were tested and all were found to sensitize only anoxic cells. The *N*-oxyls were not detectably toxic to *E. coli* up to the maximum concentrations tested (almost 10^{-1} M), and *E. coli* B/r were even able to grow in broth containing 10^{-2} M DTBN. Relatively high concentrations of DTBN (10^{-2} M) were necessary to produce maximum sensitization.[145]

Most of the work done so far on sensitization by *N*-oxyls has been with triacetoneamine *N*-oxyl (TAN), which was found to sensitize *E. coli* B/r at significantly lower concentrations, with the maximum effect reached at approximately millimolar levels.[99] Because it was hoped eventually to conduct sensitization experiments on cancer cells *in vivo*, TAN had the additional advantage that it is considerably less toxic to mice when injected than the previously studied *N*-oxyls. Adult mice weighing approximately 20 g were able to survive at least 13 mg (0.7 ml of 10^{-1} M TAN in buffered saline) injected intraperitoneally.[99]

The possibility that the observed sensitization by TAN could be attributable to production of toxic radiation products of the *N*-oxyl was discounted by the fact that TAN still sensitizes bacteria in the presence of 10^{-1} M deoxythymidine.[99] Emmerson and Howard-Flanders[144, 145] proposed that the *N*-oxyls sensitize anoxic cells by the same mechanism as that involving oxygen, namely, by reacting with free radicals produced in vital molecules.

TAN is less effective than is oxygen in sensitizing the bacteria *E. coli* B/r. Such is not the case, however, with certain radiation-sensitive mutants of *E. coli*. In particular, mutant strains that have lost the capacity to form genetic recombinants by virtue of a mutation in a gene designated *recA* were found to be sensitized approximately twice as much by 10^{-1} M TAN as they are by oxygen. Figure 14 shows that the dose reduction factor is about 4 in TAN, and only about 2 in oxygen.[147] It is notable that TAN does not sensitize the *recA* strain in the presence of oxygen; oxygen apparently protects this strain from TAN damage. This situation can be understood from a consideration of the kinetics because, as is shown in Table IV, TAN reacts more slowly with organic free radicals that does oxygen (see Table I) but apparently yields a product that is more damaging to this particular bacterial strain.

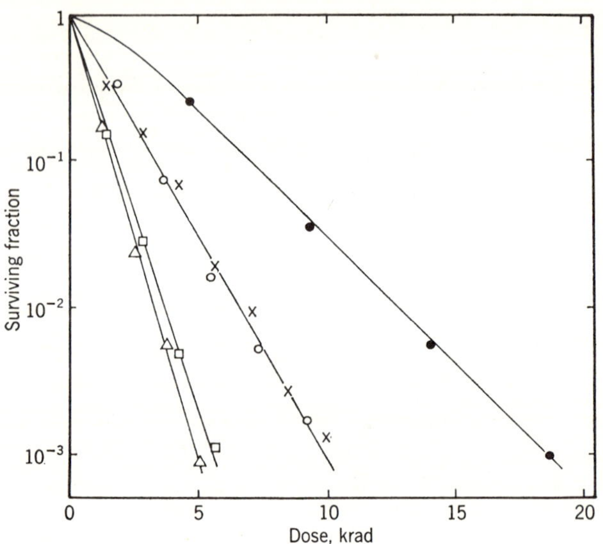

Fig. 14. The fraction of the bacterium *E. coli* K12 *recA* retaining the capacity for colony formation when plated on nutrient agar as a function of the X-ray dose delivered in the presence and absence of TAN under anoxia and oxia. ●, nitrogen alone; ○, oxygen alone; ×, oxygen plus 10^{-2} *M* TAN; □, nitrogen plus 10^{-3} *M* TAN; △, nitrogen plus 10^{-2} *M* TAN. (From Emmerson.[147])

TABLE IV

Specific Rates of Reaction of TAN with Free Radicals Produced by Hydroxyl Radical Attack on DNA and on Some of Its Components[a]

Solute	Wavelength, nm	Rate constant, units of 10^8 dm^3 mole^{-1} sec^{-1}
DNA	310	2.1
Thymine	400	3.5
Deoxythymidylic acid	350	1.0
Deoxycytidylic acid	350	1.2
	310	2.0
Deoxyguanylic acid	400	2.3
	550	2.4

[a] From Emmerson and Willson,[156] and Willson and Emmerson.[116]

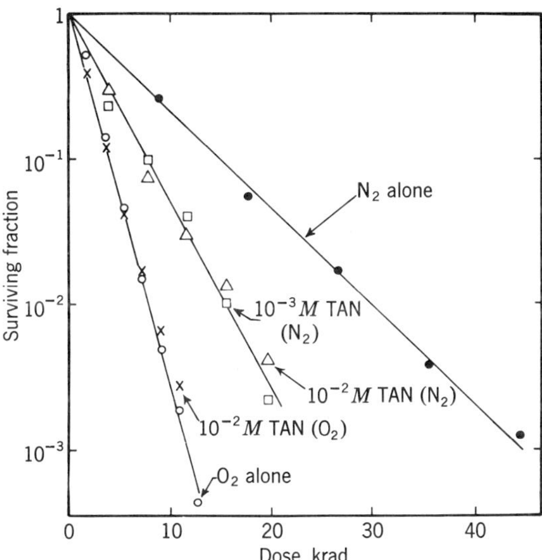

Fig. 15. Surviving fraction of *E. coli* K12 *recC* as a function of X-ray dose delivered in the presence and in the absence of TAN under oxia and anoxia. Legend as in Fig. 14. (From Emmerson.[147])

In contrast to the result with the *recA* strains, Fig. 15 shows that oxygen sensitizes strains carrying a mutation in another gene affecting recombination (i.e., *recC*) about twice as effectively as does TAN. The result shows that the presence or absence of genes concerned with DNA repair and recombination can affect the relative effectiveness of two sensitizing agents by a factor of 4.[147] (As with *E. coli* B/r, TAN does not sensitize in the presence of oxygen.) The large enhancement caused by oxygen in the *recC* strain does not fit an apparent trend reported by Alper[186] that radiation-sensitive strains are sensitized by oxygen to a lesser extent than are normal strains.

TAN is even more effective in sensitizing strains of the bacterium *E. coli* K12 carrying both *uvr* and *recA* mutations. Howard-Flanders and his colleagues[76] showed that 10^{-3} M TAN sensitizes this strain, which can neither excise pyrimidine dimers nor form genetic recombinants, with a dose modifying factor of nearly 6, while oxygen, in contrast, sensitizes by a factor of only 1.3.

Bacteria other than *E. coli* that have been shown to be sensitized by *N*-oxyls in anoxia include *S. marcescens*.[148–150]

Norpseudopelletierine *N*-oxyl (NPPN), a relatively unhindered free radical, was found to sensitize the *recA* mutant of *E. coli* when present at a significantly lower concentration than is required for TAN or for TMPN for the same degree of sensitization.[151] This correlates with the rate of reaction of these *N*-oxyls with organic free radicals and is discussed in Section 5.3.

5.2. Effect on Radiosensitivity of Anoxic Cultured Mammalian Cells

Parker, Skarsgard, and Emmerson[152] showed that TAN sensitizes anoxic suspensions of cultured Chinese hamster cells to X-rays, but the degree of sensitization is small. The maximum dose reduction factor (ca. 1.5) required 10^{-2} *M* TAN, at which concentration slight toxic effects could be detected. Revesz and Littbrand,[153] using a somewhat different technique, reported an enhancement ratio with Chinese hamster cells of only 1.1 for 2.5×10^{-4} *M* TAN, and no sensitization at all for 2×10^{-2} *M* TAN.

TAN does not sensitize mouse lymphoma cells irradiated under anoxia *in vitro* and then assayed by counting the number of clones produced on incubation in nutrient agar.[76]

Hewitt and Blake[154] were unable to demonstrate any sensitization by TAN in experiments in which anoxic leukemia cells were irradiated in mice. Foster[155] was unable to detect any sensitization of anoxic fern spores by TMPN.

5.3. Adduct Formation with Organic Free Radicals

As noted in Section 2.1, irradiation of nitrous oxide-saturated solutions of thymine with a pulse of electrons yields a species with a broad transient absorption in the region 300–400 nm, which decays relatively slowly over tens of microseconds. This absorption, attributed[49] to the 6-hydroxythymine radical (**I**) produced in reaction (4) decays rapidly in the presence of TAN,[156] as shown in Fig. 16. A graph of the first-order rate constant against concentration of TAN, shown in Fig. 17, is linear, indicating a bimolecular reaction which by analogy with reaction (5) would be expected to be

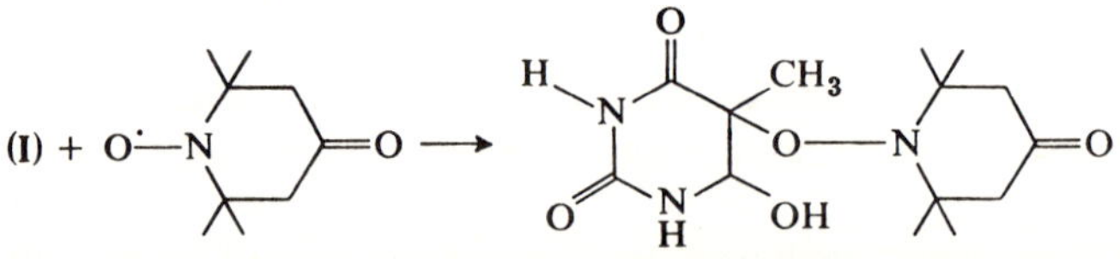

$$(\text{V}) \tag{34}$$

rather than

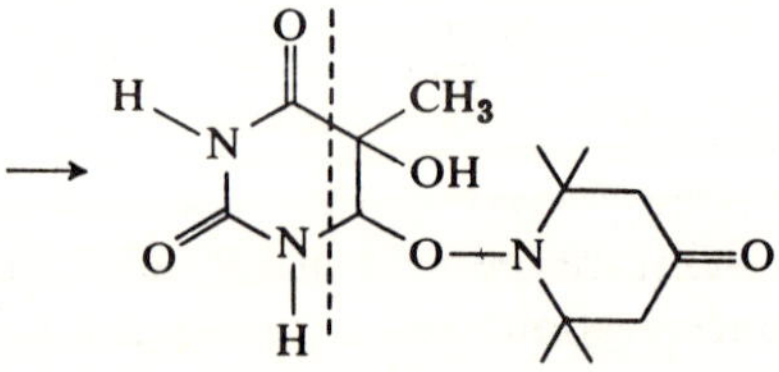

$$(\text{VI})$$

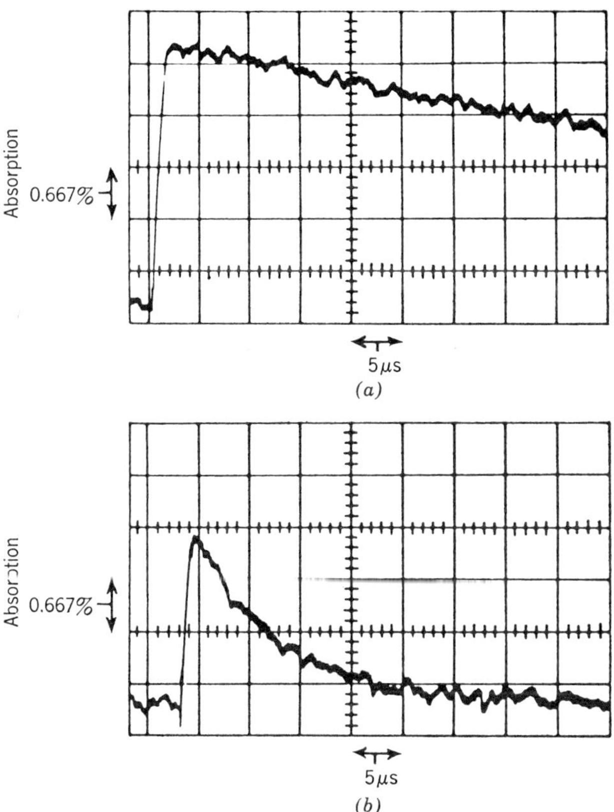

Fig. 16. Oscilloscope traces showing the effect of TAN on the decay of the absorption at 400 nm produced by pulse-irradiating aqueous N_2O-saturated solutions of thymine (6×10^{-4} M, pH 5.5). (a) No TAN. (b) 3.75×10^{-4} M TAN. (From Emmerson and Willson.[156])

or

$$\text{(VII)}$$

However, it is apparently impossible to build molecular models of product (V) of reaction (34) because of steric hindrance between the methyl group of thymine and the methyl groups of the N-oxyl. However, it is possible to

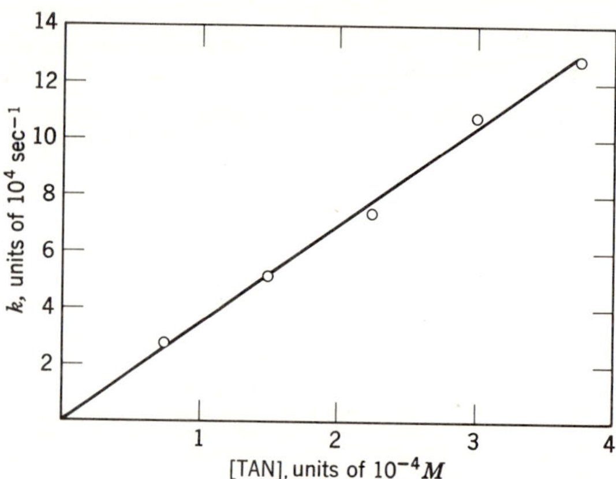

Fig. 17. Plot of first-order rate constant against concentration of TAN. (From Emmerson and Willson.[156])

build a model in which the positions of the 6-OH group and the N-oxyl moiety are reversed (**VI**), or the addition is to the methyl carbon (**VII**). A similar model can also be built of the adduct produced by attachment of the N-oxyl to the C-6 position of thymine in deoxythymidine, and although model building was not carried further the relatively exposed position of the 5-6 bond of thymine in double-helical DNA suggests that such adducts can also be formed in DNA.

When ^{14}C-labeled thymine solutions are irradiated under anoxic conditions in the presence of TAN, the main radiation product appears at $R_f = 0.8$ on subsequent paper chromatography, using the solvent n-propanol–1 N HCl (85:15). This product is completely absent if [^{14}C]thymine is irradiated without TAN and may be the thymine–TAN adduct (Fig. 18; Emmerson, unpublished results). TAN appears to increase the G value for destruction of thymine. Pulse radiolysis studies have shown that TAN reacts quite readily with free radicals produced by OH attack on nucleic acids and various nucleosides and nucleotides.[116] The reaction rates are presented in Table IV.

Unlike the equivalent reaction with oxygen, the reaction between the DNA free radicals and TAN is complex but the kinetics are measurable by pulse radiolysis in terms of decay of absorption at 310 mμ by DNA(OH)· in the presence and the absence of TAN.[116] Willson and Emmerson[116] attribute the initial rapid decay at 310 mμ to the reaction

$$\text{DNA(OH)·} + \text{TAN} \rightarrow \text{products} \qquad (35)$$

with specific rate ca. 2.1×10^8 dm^3 mole^{-1} sec^{-1}. They suggested that the

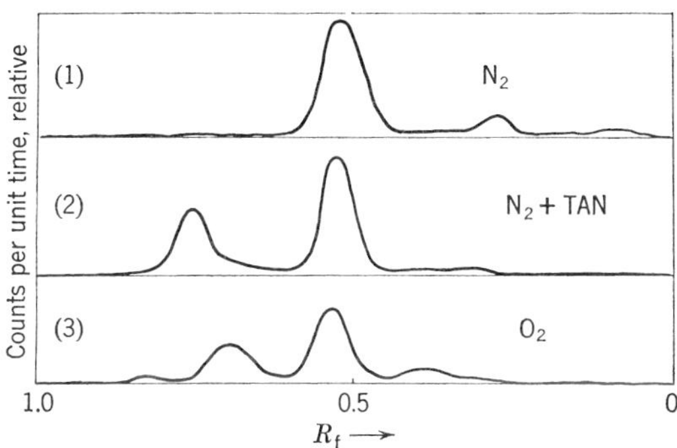

Fig. 18. Radiochromatogram of [^{14}C]thymine developed in *n*-propanol – 1 *N* HCl (85:15), after irradiation with a 120-krad dose of X-rays: (*a*) under nitrogen (*b*) under nitrogen and containing 5 × 10^{-4} *M* TAN, (*c*) under O$_2$. (Previously unpublished.)

reaction product is an adduct between DNA and TAN, analogous to the product of reaction (34).

Evidence for such adduct formation has been obtained in experiments with radioactively labeled *N*-oxyl radicals. Nakken, Sikkeland, and Brustad[157, 158] irradiated solutions of DNA and [^{3}H]TAN under anoxia and separated the ^{3}H-labeled DNA–TAN adducts from [^{3}H]TAN on Sephadex columns. By measuring the yield of adducts as a function of the relative concentration of TAN and DNA, they demonstrated that the reaction was between a DNA· radical and TAN and not vice-versa. Other evidence for adduct formation has been obtained in unpublished experiments by Buu and Emmerson[116] in which excess calf thymus DNA was irradiated under anoxia in the presence of ^{14}C-labeled 2,2,6,6-tetramethyl-4-acetate-[^{14}C]-1-oxyl.

5.4. Comparisons of Reaction Rates

The concentration of either *N*-oxyl radicals or of oxygen required to produce any given degree of radiosensitization appears to be roughly correlated with the reaction rates of these sensitizers with the hydroxythymine radical.[156] Because the *N*-oxyls react more slowly than oxygen, a higher concentration is required for maximum sensitization.

NPPN is less hindered sterically than are *N*-oxyls such as TAN. It is found to react approximately twice as fast as TAN with the hydroxythymine radical.[151] This increased speed of reaction, presumably associated with the steric factors involved, correlates well with the concentrations of the various sensitizers required to give one-half of the maximum sensitization (κ value), as shown in Table V. A concentration of only 60 μM of NPPN was found to give half the maximum sensitization. This is approximately one-quarter of the concentration of TAN required for the same effect.

TABLE V
Specific Rates of Reactions between Various Sensitizing Agents and the Hydroxythymine Radical Compared with Concentrations of those Sensitizers Required to Give Half-Maximum Sensitizing Effect (κ Value) to Anoxic Bacteria *E. coli* K12 *recA*[a]

Sensitizer	Specific rates of reaction with hydroxythymine radical, $dm^3\,mole^{-1}\,sec^{-1}$	κ-value μM
TMPN	2.6×10^8	300
TAN	3.5×10^8	230
NPPN	$6.4\text{–}9.2 \times 10^8$	60
O_2	1.9×10^{9}[b]	8[c]

[a] From Emmerson, Fielden, and Johansen.[151]
[b] Willson.[48]
[c] Johansen.[161]

5.5. Effect of DNA–TAN Adducts on the Functioning of DNA in Bacteria

An opportunity to distinguish the effects of radiation on DNA from those on the rest of the cell is afforded by experiments in which male and female strains of *E. coli* are mated and DNA is thereby transferred from the male to the female. This transferred DNA can be distinguished from the female chromosomal DNA by suitable genetic tests. The sex difference in *E. coli* is determined by the presence in the male of an extra length of DNA called the sex factor or F (for fertility) factor. Some sex factors also contain segments of the bacterial chromosome. For example, an F*lac*+ strain is a male, or donor, strain that possesses a sex factor and also a segment of DNA that codes for an enzyme or enzymes which enable the cell to grow in a medium containing lactose in place of the usual glucose. If such an F*lac*+ donor strain is irradiated, and then mated with a *lac*− recipient, the effect of irradiation on the sex factor DNA can be studied by measuring the number of lactose-fermenting colonies produced. It can be shown in this way that the presence of TAN during irradiation has a very large effect on the DNA.[159] The results are particularly striking if the recipient strain carries the *recA* mutation. Figure 19 shows the results of an experiment in which the donor strain was irradiated under anoxia in the presence or absence of $10^{-3}\,M$ TAN, mated with either normal or *recA* female bacteria, and plated on media on which only those recipient strains that have received a functional *lac*+ gene

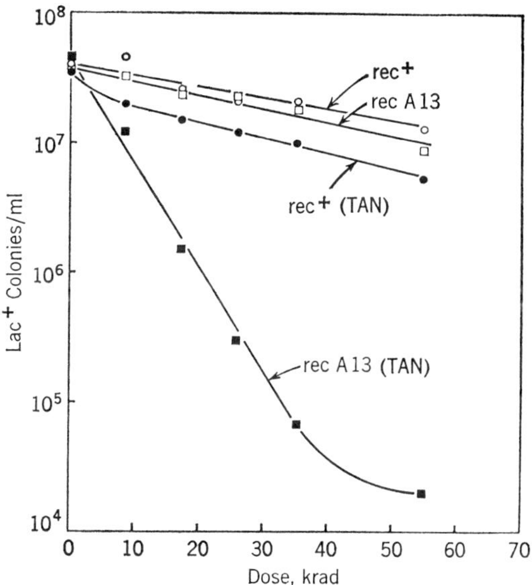

Fig. 19. The number of *lac⁺ ura⁺* bacterial colonies produced on selective agar as a function of X-ray dose delivered to anoxic suspensions of an *E. coli* strain carrying F*lac⁺* (*ura*) in the presence and absence of 10^{-3} *M* TAN before mating with the F⁻*lac* (*ura⁺*) strains *rec⁺* and *recA*. The results show that the episomal DNA is greatly modified by the presence of TAN during irradiation. *Editorial note*: *lac⁺ura⁺* means "ability to utilize lactose as a carbon (i.e., energy) source; *Ura⁺* means ability to grow without uracil."

from the incoming DNA can form colonies. In the experiments with the *recA* recipient, the presence of TAN during irradiation of the donor sensitizes the sex factor DNA by a factor of 7.5. This sensitization does not appear to result from any appreciable decrease in ability of the irradiated cell to conjugate or to transfer the episome, because the same irradiated donor when crossed with the *recA⁺* recipient gives a much higher yield of *lac⁺* colonies, and the survival curve is very similar in slope to the control irradiated in the absence of TAN. It is possible to conclude from these experiments that TAN modifies the sex factor DNA and increases the amount of lethal damage by a factor of 7.5.

By employment of the Boyce and Tepper[72] technique described in Section 2.2, it can be shown that TAN has only a small effect on the production of single-strand breaks in viral DNA within the host bacterium [*E. coli* (λ)]. The results presented in Fig. 20 show that TAN increases the number of single-strand breaks by a factor of only about 1.5. Breaks produced in both the presence and the absence of TAN were readily repaired on incubating the irradiated cells. Thus because TAN sensitizes the episomal DNA by a factor of 7.5, and increases the number of single-strand breaks in

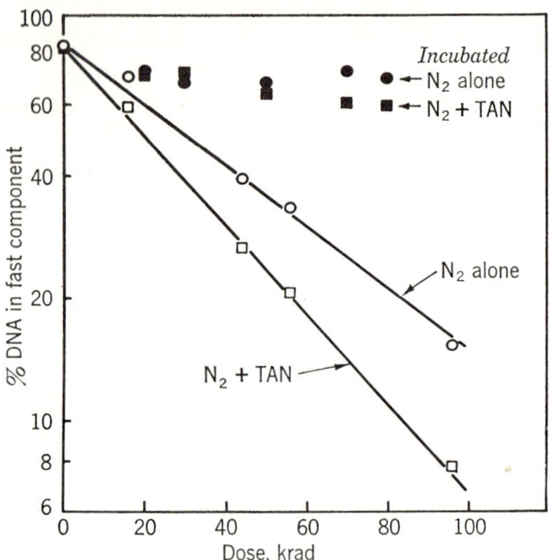

Fig. 20. Percent rapidly sedimenting covalent circular phage λDNA as a function of X-ray dose to anoxic bacteria *E. coli* (λ) superinfected with ^{3}H-phage λ in the presence and absence of 10^{-3} *M* TAN. The data shown by open circles and squares were obtained in experiments in which the irradiated superinfected bacteria were kept at 0°C before lysis. The filled-in circles and squares were obtained in experiments in which the irradiated cells were incubated at 37°C in complete medium for 15 min prior to lysis. Procedures were those of Boyce and Tepper[72] described earlier. (From Emmerson.[159])

the viral λ DNA by only 1.5, it was concluded that the main effect of TAN is to produce damage such as adducts rather than breaks.[159]

Rupp et al.[76] studied the molecular weight of the DNA newly synthesized on DNA templates in cells irradiated under anoxia in the presence of TAN. They found that the newly synthesized DNA was of lower molecular weight than in the unirradiated controls and concluded, by analogy to similar results following ultraviolet irradiation in which pyrimidine dimers are formed,[160] that TAN adducts in DNA interfere with its ability to act as a template for the synthesis of a continuous complementary strand. Because TAN did not significantly affect the molecular weight of the template DNA, these experiments confirmed the fact that this sensitizing agent does not have a large effect on the radiation production of single-strand breaks.

6. MECHANISMS

6.1. Adduct Formation

Of the four types of sensitizing agents considered, all seem to owe at least part of their sensitizing properties to their ability to produce adducts with organic free radicals. It is not known whether adduct formation is the only mechanism by which these four sensitizers work but, if there are other mechanisms involved, we would like to have some idea of their relative importance and, if not, we would like to know if there are other sensitizing agents that operate via a different mechanism.

NEM has been considered in detail in the hope that it is representative of a large group of electron-affinic substances, but it is not yet known whether the others, such as *p*-nitroacetophenone (PNAP),[98, 113] form adducts with DNA radicals. It is striking that most of the electron-affinic sensitizing agents contain carbonyl groups. Perhaps the carbonyl group is able to react directly with an organic radical

$$R\cdot + X\!\!-\!\!\overset{\overset{\displaystyle O}{\|}}{C}\!\!-\!\!Y \;\rightarrow\; X\!\!-\!\!\underset{\underset{\displaystyle R}{|}}{\overset{\overset{\displaystyle O\cdot}{|}}{C}}\!\!-\!\!Y \tag{36}$$

Alternatively, the carbonyl group, which has a large inductive effect, may simply activate another part of the sensitizing molecule.

6.2. Alternatives to Adduct Formation

Several mechanisms alternative to adduct formation which have been proposed to explain the sensitizing properties of electron affinic substances are discussed in Sections 2.1, 3.1, and 3.5. They include the removal of the protecting effect of sulfhydryls,[46, 47, 96] the prolongation of the lifetime of the hydrated electron,[107] and the removal of an electron from an ion pair in or near a target molecule, according to the Adams scheme shown in Eq. (26).[109, 110] These mechanisms are less amenable to direct experimental test than is adduct formation, but some indirect evidence against the first two is presented in Sections 3.1 and 3.5.

6.3. Interference with Repair Processes

In theory, one of the best ways to sensitize a cell to radiation would be to add a substance that would selectively inactivate the repair enzymes. Such a sensitizing agent would be expected to have a very large effect in wild-type cells. No such large effect has yet been observed with any sensitizer thought to work in this way; nevertheless, the approach holds promise for the future. The problem is that even a relatively small cell, such as *E. coli*, contains several thousand genes which code for the synthesis of different enzymes. Of these enzymes, relatively few are concerned with DNA repair and many of the others are vital to the cell. Thus it would be very difficult to inhibit only the repair enzymes and leave the vital enzymes unaffected. Section 2.2 presents two possible examples of experiments in which such selective inhibition may have been partly achieved; in one case by heating the cells to 52°C,[76] and in the other by adding EDTA.[62]

Bridges and Munson[162] found that the extents of sensitization by NEM on anoxic *E. coli* B/r and B_{s-1} are about the same, despite the large genetically controlled differences in absolute sensitivity of these strains. They concluded that the sensitizing properties of NEM are not the result of inactivation of the repair processes. However, even if they had observed a difference, it would not necessarily have indicated a direct interaction between NEM and the repair enzymes. As indicated in Section 5 (Figs. 14 and 15), the sensitization of repair-deficient mutants of *E. coli* K12 by *N*-oxyls is different from that of wild-type cells, but this effect is quite likely to be a result of the ways in which the different repair enzymes *cope* with different types of damage, and not the result of interaction between the *N*-oxyls and the repair enzymes.[147]

Substances that sensitize by reacting with the repair enzymes might be effective when added immediately after irradiation and could thus be readily identified by rapid-mix experiments. It would, however, be necessary to plate out the irradiated bacteria in the presence of the sensitizer, for fresh enzymes could otherwise be synthesized.

Alexander et al.[78] treated irradiated *M. radiodurans* with chloramphenicol, an inhibitor of protein synthesis. They found that this substance strikingly decreases survival. This is an example of sensitization *after* irradiation, but it is not yet clear exactly how it works. Chloramphenicol is currently assumed to operate not on existing enzymes but on the synthesis of enzymes. It is possible that some of the repair enzymes are not constitutive but are inducible by the radiation damage. If such be the case, then the task of finding sensitizers that interfere with induction should be easier than that of finding sensitizers that selectively inactivate specific preexisting enzymes.

Meanwhile, in general, repair processes make the task of radiosensitization more difficult by eliminating, or circumventing, increased damage. Progress in the search for efficient sensitizing agents is therefore closely associated with progress in understanding the fundamental mechanisms of the repair processes.

Repair systems are best understood for the bacteria *E. coli*[7] and *Micrococcus luteus*,[163] in which the genetics and the enzymology, respectively, have been extensively studied. This subject has been reviewed recently[7] and is mentioned only briefly here.

At least two repair processes are believed to promote cell survival by repairing damaged DNA, the *excisional* and the *recombinational* repair processes. Repair of single-strand breaks[61] may be a separate, third, process or it may be part of one of these processes.

The *excisional* repair process is effective in excision from the DNA of ultraviolet photoproducts, such as pyrimidine dimers, although the enzymes involved are believed to act on other distortions in the DNA, such as those caused by mitomycin C.[164] Strains carrying *uvr* mutations are unable to excise pyrimidine dimers and are very sensitive to ultraviolet light, but are only 1.5 to 2 times more sensitive to X-rays in the presence or absence of oxygen than are the wild-type cells. The dimer-excision enzymes participate to only a moderate extent in the repair of X-ray damage. X-Irradiation causes DNA breakdown in cells, presumably reflecting the action of nucleases on the damaged DNA. However, the extent of DNA degradation after irradiation in the presence of oxygen is approximately the same in *uvr* and *uvr*[+] strains.[165] The mechanism of the *excisional* repair process is conceived to involve the recognition of certain types of base damage in the DNA by a specific endonuclease which, in the case of pyrimidine dimers, makes one single cut next to the damaged base. This cut is followed by exonucleolytic degradation of the strand of the DNA that contains the damage and which is unable to pair correctly. In *E. coli*, DNA polymerase I, controlled by the *polA*[+] gene, has the capacity both to remove damaged bases from the 5'-end and to add nucleotides to the 3'-OH terminus.[166] The final sealing is done by polynucleotide ligase which forms a phosphodiester bond between 3'-OH and 5'-phosphate termini held in juxtaposition by hydrogen bonding to the template strand.[167] Single-strand breaks induced in double-stranded DNA by γ rays cannot be repaired directly by polynucleotide ligase.[168] As mentioned in Section 2.2, estimation of the *initial* yield of DNA strand breaks is complicated by the unknown extent to which repair has already occurred when the number of single-strand breaks is determined. Town, Smith, and Kaplan[146] have recently shown that the yield of X-ray-induced single-strand breaks is much higher in strains carrying the mutation *polA*, presumably because of the absence of a rapid repair process in which DNA polymerase I participates. This inability of *polA* strains to carry out rapid repair of single-strand breaks accounts for their increased X-ray sensitivity.[146]

Recombinational repair, controlled by the *rec*[+] genes, appears to share some of the cells' machinery for forming recombinant DNA molecules from two different DNA molecules. In the case of damage produced by ultraviolet-irradiation, *recombinational* repair is thought to occur after DNA replication and may be initiated by gaps produced in the newly synthesized DNA.[160] A possible scheme for such a mechanism is given in Fig. 21. *Recombinational* repair appears to be involved in cell survival after X-irradiation, for the *recA* mutants of *E. coli* are four to five times more sensitive than are wild-type to X-irradiation in the presence or absence of oxygen.[7] The *recA* mutants appear to be deficient or slow in rejoining X-ray-induced single-strand breaks,[169] as measured by the McGrath and Williams technique. However,

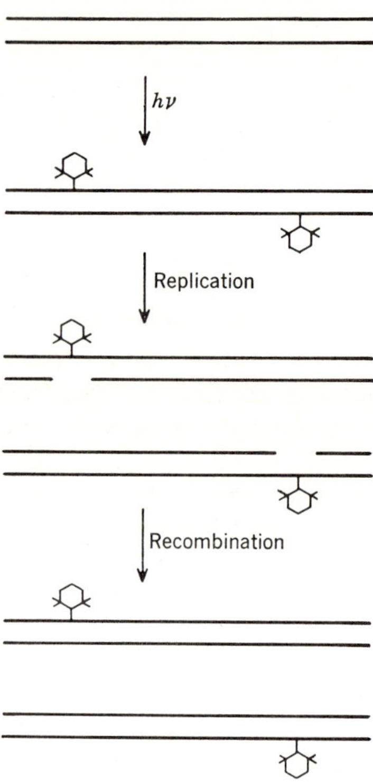

Fig. 21. Scheme showing in general terms how a recombinational repair process could reconstruct good DNA following replication of DNA containing adducts. Replication of templates containing pyrimidine dimers,[160] or TAN adducts[76] has been shown to result in gaps in the newly synthesized DNA. It is possible that the free ends of the DNA strands at these gaps initiate recombination.

repair of single-strand breaks in superinfecting *E. coli* (λ) occurs normally in these mutants.[72]

Repair of X-ray-induced single-strand breaks can be inhibited by quinacrine, a derivative of 9-aminoacridine, in wild-type cells but not in *recA* mutants. This substance appears to inactivate the *recombinational* repair process irreversibly.[122, 141]

There is suggestive evidence that TAN adducts produced in DNA by X-irradiation are acted upon by the *uvr*+-controlled excision enzymes, even in the absence of recombinational repair. This follows from a consideration of the data presented in Table VI, from which it can be seen that the *uvrA recA* double mutant is approximately four times more sensitive to anoxic irradiation in the presence of TAN than is the *recA* single mutant which has the wild-type *uvr*+ gene. Thus the *uvr*+ gene product is responsible, in this

TABLE VI

Sensitization by 10^{-3} M TAN and Oxygen of Bacteria $E.$ $coli$ K12 Carrying Mutations Affecting DNA Repair[a]

Escherichia coli strains	Relevant genotype	X-ray dose in krad at 1 percent surviving fraction				Dose reduction factor in TAN	Dose reduction factor in oxygen
		N_2	N_2 + TAN	O_2	O_2 + TAN		
AB1157	Wild-type	61.2	27.1	20.7	22.5	2.3	3.0
AB1886	*uvrA*	45.1	20.2	19.3	19.3	2.2	2.3
AB2463	*recA*	15.2	4.8	7.1	8.5	3.2	2.1
AB2480	*uvrA recA*	6.3	1.1	5.0	5.1	5.7	1.3

[a] From Rupp et al.[76]

case, for repairing a considerable amount of TAN damage. Damage produced in the presence of oxygen appears to be only slightly repaired by the uvr^+ enzymes, with the 1-percent survival dose decreasing from 7.1 to 5.0 krads. However, damage produced under nitrogen alone seems to be repaired by the uvr^+ enzymes, for the 1-percent survival dose falls from 15.2 to 6.3 in their absence.

Because the *recA uvrA* double mutant is 4.4 times more sensitive to irradiation in the presence of TAN than is the *recA* single mutant, and 18 times more sensitive than is the *uvrA* single mutant, it appears that *recombinational* repair is more efficient than *excisional* repair in overcoming damage caused by TAN–DNA adducts. This statement also applies to damage produced in the DNA on irradiation under anoxia in the absence of TAN.[76]

It is interesting to note the extent to which genetic factors together with an efficient sensitizing agent can increase the sensitivity of a wild-type strain. The 1-percent survival dose for irradiation under anoxia is 61.2 krads for the wild-type strain, AB1157, but only 1.1 for the *recA uvrA* double mutant irradiated in the presence of TAN. Thus the cumulative effect of TAN and two mutations is to increase the sensitivity of the bacteria to X-rays by a factor of almost 60.

The absolute sensitivity of the *recA uvrA* double mutant in the presence of TAN is of particular interest because it places an upper limit on the number of potentially lethal events produced in the DNA, on the assumption of course, that the lethal events are in fact in the DNA. From the $1/e$ dose of approximately 240 rads, it can be calculated that there would be 1.7 × 10^{-6} events per 10^6 daltons per rad in DNA of molecular weight 2.5 × 10^7 daltons; the result corresponds to $G_d = 1.6$ (see Section 2.2 for definition of G_d). This result may be compared with a value of $G_d = 2.5$, which can be

          PETER T. EMMERSON

calculated from the $1/e$ dose of 5.5 krads for the inactivation of the F*lac*[+] episome irradiated in the presence of TAN and introduced by conjugation into a *recA* recipient, as shown in Fig. 19. For this result it is assumed that the molecular weight of the episome is 7.4×10^7 daltons.[170] Although a single intact strand would be effective in the recipient,[171] damage to either strand might inactivate episomal transfer. The F*lac*[+] experiment is of some interest because there is relatively little opportunity for repair of the damage produced by TAN either before or after transfer. In studies such as these, in which the chemical restitution reactions are minimized by sensitizing agents and enzymic repair processes are minimized by mutations, it is quite possible that the estimate of the yield of lethal events approaches the yield of *initial* radiation damage.

6.4. Relationship of Amount of Radiation Damage per Lethal Event to Biological Complexity

Terzi[13] and Kaplan and Moses[12] demonstrated a striking correlation between the radiation sensitivity of various organisms irradiated in the dry state, or in protecting medium, and their biological complexity. Terzi's values for the radiosensitivity, expressed as the lethal efficiency (E) per ion pair produced in the nucleic acids, together with G_d values calculated from the same data are presented in Table VII. To a first approximation, single-strand viruses are inactivated by 1 ionization in their DNA, while the inactivation of double-stranded viruses requires 10, haploid bacteria and yeast

TABLE VII

Comparison of Radiation Sensitivity with Biological Complexity[a]

Organism	E[b]	G_d[c]
RNA and single-stranded DNA viruses	0.64	1.8
Double-stranded DNA viruses	0.062	0.17
Haploid bacteria and yeast	0.013	0.036
Mammalian and avian cells, diploid yeast	0.00069	0.0019

[a] From Terzi.[13]

[b] E is the mean lethal efficiency per ion pair in the DNA.

[c] G_d is the estimated number of lethal events per 100 eV absorbed directly by the DNA (see Section 2.2).

require 50, and mammalian cells and diploid yeast require 1000. As may be expected, the possession of a double-strand genome is an advantage to an organism, because all the known repair processes require a second strand to act as a template. The pronounced radiation resistance of mammalian cells may reflect even more efficient and highly complicated repair processes. The increased resistance of these cells does not appear to be directly associated with their carrying two or more sets of chromosomes, for these data were calculated from the slope of the exponential part of the survival curves. However, it is possible that some of the DNA in mammalian cells is genetically redundant. The value of $G_d = 1.8$ for the inactivation of single-strand viruses is fairly close to the values of $G_d = 1.6$ and 2.5 calculated for the inactivation of the *recA uvrA* double mutant and the F*lac*$^+$ episome, respectively, irradiated in the presence of TAN.

Because the efficiency of inactivation expressed in this way (Table VII) is constant for any one group of organisms, it appears likely that lethality is caused by inactivation of the DNA and that the resistance of the more complex organisms is attributable to their possession of more efficient repair processes. However, the radiation resistance of higher organisms might also be associated with the idea[190, 191] that most of their DNA does not code for proteins but is used for control purposes.

Further support for the association of X-ray sensitivity with DNA repair comes from the observation that radiation-sensitive mutants are often found to owe their sensitivity to an inability to repair DNA damage. For example, the X-ray-sensitive recombination-deficient mutants of *E. coli* break down their DNA after irradiation at rates different from that of the normal strain. In the case of strains carrying the *recB* mutation, the enzyme that is defective is known to be an exonuclease, an enzyme that specifically degrades DNA.[172–174] This case is therefore one example in which X-ray sensitivity can be traced to a defect directly concerned with DNA. In general, if damage to DNA were not to contribute to cell killing, it would be difficult to understand why mutants defective in DNA repair should be sensitive to radiation.

The apparent relationship between the X-ray sensitivity of various organisms and their ability to repair DNA damage emphasizes the overriding importance of DNA damage in cell killing.

7. PREFERENTIAL SENSITIZATION OF ANOXIC CELLS

7.1. Possible Application to Radiation Therapy

The search for substances that might specifically sensitize a tumor during radiation therapy was started many years ago by Mitchell.[175] In order to have any therapeutic advantage, a sensitizing agent would have to sensitize a tumor more than it sensitizes the surrounding healthy tissue. One solution to this problem might be to find sensitizing agents that are accumulated by the tumor. There are some recent indications that certain vitamin-K derivatives fulfill this criterion and experiments are presently being carried out with such compounds which have been labeled with ^{3}H so that the tumor is irradiated internally.[176] The sensitizing agent BU, an analog of thymine, is sometimes preferentially incorporated into the DNA of a tumor relative to that of the adjacent normal tissue. Clinical trials with this sensitizing agent have met with some success.[177, 178]

A certain fraction of the cells of some tumors appear to be anoxic, or at least hypoxic, and therefore less sensitive to radiation. These cells may survive irradiation and serve as foci for regrowth of the tumor. This problem is currently being tackled in several ways. Techniques are being developed in which the patient is treated while breathing oxygen under pressure in an attempt to raise the oxygen concentration in the blood stream and therefore, hopefully, in the normally hypoxic cells of the tumor.[187] However, such treatment can be harmful if the oxygen does not reach the interior of such a tumor. If the tumor remains anoxic and the surrounding tissue is oxygenated, X-ray treatment (in which the OER is about 3) may, contrary to desire, actually result in more tissue kill relative to the tumor. Excessive damage to the surrounding tissue, even under normal conditions of oxygenation, is thus an ever-present danger with X-ray treatment. With irradiation of high LET, the OER is closer to unity. There is thus a good argument for using neutron irradiation instead of X-rays in high-energy radiation therapy.[180]

Another approach, and that relevant to this chapter, has been to search for substances that could preferentially sensitize the anoxic cells.[181, 182] It is at least theoretically possible that a substance could be found that would succeed in penetrating to the anoxic cells despite the failure of oxygen to do so. This hope has been sustained partly because oxygen tends to be used up by the respiration process of the normal tissue surrounding a tumor. A substance that is not metabolized might be more likely to diffuse to the anoxic cells.

The problems involved in finding a suitable preferential sensitizing agent are daunting. For clinical application, an ideal sensitizer would have to be chemically unreactive, nontoxic at concentrations at which it sensitizes significantly, not rapidly metabolized, not pharmacologically active, and water-soluble. NEM fulfills some of these requirements but is generally very toxic to cultured mammalian cells and to animals. Encouraging results have been obtained by Asquith, Willson, and Adams[113] in experiments with another electron-affinic substance, PNAP. This substance is not toxic to cultured mammalian cells (V-79-379-A cells) up to a concentration of

$4 \times 10^{-4}\ M$, at which it sensitizes hypoxic cells with an enhancement ratio of 1.7. PNAP sensitizes the cells in the presence of oxygen by an enhancement ratio of only 1.05.

Another sensitizing agent for anoxic cells which may be relatively free from toxicity is neoarsphenamine.[183] This substance is well tolerated by humans and it preferentially sensitizes anoxic suspensions of mouse leukemia cells (M. Kligerman, personal communication).

A test system for sensitizing agents, which approaches more closely to conditions relevant to radiation therapy, has been developed by Hornsey, Hedges and Bryant.[189] They treat mice with a sensitizing agent, irradiate them while they are temporarily breathing nitrogen, and determine the number of survivors after five days. Death is thought to result primarily from damage to the stem cells in the intestinal tract. Indanetrione was found to sensitize this system by a factor of 1.3,[189] and TAN by a factor of 1.1 (Hornsey, personal communication).

The N-oxyls are well tolerated by bacteria, cultured mammalian cells, and mice. Moreover, their pharmacological activity in various animals has been studied.[184] At rather high doses, di-*tert*-butyl nitroxide and certain other N-oxyls are nerve depressants. However, this effect is also found with molecules of similar structure but without the N-oxyl free electron. No pharmacological activity appears to be associated uniquely with the N-oxyl group

Although N-oxyls are very stable in aqueous solution, there is some doubt as to their stability either in cultured mammalian cells or in animals. Electron spin resonance spectrometry shows that the free spins attributable to the N-oxyl groups are lost over a period of a few hours in high concentrations of Chinese hamster cells.[185] However, Rupp et al.[76] found that the free spins in TAN are stable in, and in fact accumulated by, mouse leukemia cells when in tissue culture medium. The intracellular concentration of TAN is 10 to 100 times greater than would be expected for passive diffusion. Accumulation of the N-oxyls suggests that they may be binding to some particular sites within the mammalian cells. Such specific binding is known to occur from spin-label experiments and may account for the observation of Agnew and Skarsgard[185] that TAN does not sensitize high concentrations of hamster cells to as great an extent as it sensitizes dilute suspensions. A high concentration of N-oxyls relative to the cell concentration may be necessary to saturate the binding sites and ensure an adequate supply for reaction with DNA free radicals.

There are at least two possible explanations for the lack of sensitization in mouse leukemia cells *in vitro*[76] and *in vivo*.[154] On the assumption that TAN sensitizes cells by producing DNA–TAN adducts, one may surmise either that these adducts are not produced or that they are produced but are not lethal to this particular cell, perhaps because they are efficiently repaired. These possibilities can be tested in experiments with ^{3}H-labeled *N*-oxyls. It is possible that failure of *N*-oxyls to sensitize mammalian cells as effectively as they sensitize bacteria is associated with the fact that the DNA is present in higher cells in the form of nucleoprotein. Perhaps the histones wrapped around the DNA prevent the formation of DNA–*N*-oxyl adducts. In this connection, the less sterically hindered *N*-oxyls, such as NPPN[137, 151] and bistrifluoromethyl-*N*-oxyl[158, 179] may have some advantages.

The search for preferential sensitizing agents for anoxic cells using as a test of desirability either sulfhydryl-binding properties or, more recently, electron-affinic properties has produced many interesting sensitizing agents. A third such test, the potential for adduct formation with organic free radicals, might usefully be added. Because higher cells appear to possess very efficient mechanisms for repairing their DNA, it appears that a useful sensitizing agent must necessarily produce a type of damage not easily repaired by the cell.

Acknowledgments

I am very grateful to Dr. Paul Howard-Flanders for introducing me to biology and for many valuable discussions during the early stages of the preparation of this review.

This work was supported by the U.S. Public Health Service, Grant Number CA-06519, and by the Medical Research Council.

Glossary

Action spectrum, a graph of a particular quantitative response as a function of wavelength.

Bacteria, microorganisms with no well-defined cell nucleus. They are usually unicellular and reproduce by division.

Bacteriophage, a virus whose host is a bacterium. Also known simply as phage.

Cancer, uncontrolled growth of cells.

Clone, a mass of cells, usually descended from a single ancestor.

Colony, a mass of cells, usually descended from a single ancestor, growing on a solid surface. The term is usually restricted to bacteria.

Constitutive enzymes, those synthesized in fixed amounts, irrespective of growth conditions.

Cytoplasm, the contents of a cell apart from the nucleus.

Deoxyribonuclease, an enzyme that degrades DNA.

Denaturation, loss of the native configuration of a macromolecule, for example, the breaking of the hydrogen bonds connecting the two strands of DNA in a double helix.

Diploid, having each set of chromosome, except the sex chromosome, represented twice.

Dose reduction factor, ratio of the dose required to produce a given effect under certain conditions to that required to produce the same effect in the presence of a radiosensitizing agent.

DNA polymerase, an enzyme that catalyzes synthesis of DNA from appropriate precursors using a preexisting strand of DNA as template.

Endonuclease, an enzyme that produces breaks in the middle of a nucleic acid strand.

Episome, a genetic element that can exist either free or as part of the cellular chromosome. Examples are the sex factor (F^+) and lysogenic phage DNA.

Excisional repair, a process in which DNA damage is cut out of the DNA and replaced.

Exonuclease, an enzyme that digests DNA from the ends of the strands.

Exponential phase, the phase of rapid growth in which the number of cells doubles in a constant time. Also known as *log phase*.

Gene, a hereditary unit which occupies a fixed chromosomal position and can mutate.

Genetic map, the arrangement of genes on a chromosome. The genetic map of *E. coli* is circular, as is the chromosome.

Genetic recombination, the appearance in the offspring of traits found separately in each of the parents.

Genome, the total genetic complement of an organism.

Haploid, the chromosomal state in which each chromosome is present only once.

Immunity, the state of being resistant to a specific agent. For example, certain strains of *E. coli* are resistant (immune) to infection by phage λ.

Locus, position of a gene on a chromosome.

Lysis, bursting of a cell by destruction of its cell membrane.

Lysogenic bacterium, a bacterium that contains a bacteriophage integrated into its chromosome. By convention, an *E. coli* cell which is lysogenic for λ phage is written "*E. coli* (λ)." Normally phage λ does not grow in such a strain.

Messenger RNA, RNA that carries the genetic message from DNA for use in protein synthesis.

Mutation, an inheritable change in a chromosome. By convention, a modified gene is designated by the exponent "$-$" and the wild-type by "$+$". For convenience, the "$-$" is often omitted. Thus *recA* signifies *recA$^-$*, a mutation in the gene designated *recA$^+$*.

Oxygen enhancement ratio (OER), ratio of the dose required to produce any given effect in the absence of oxygen to the dose required to produce the same effect in the presence of oxygen.

Phage, see bacteriophage.

Plaque, a round clear area in a layer of bacterial cells on an agar plate produced by bacteriophage infection.

Polynucleotide ligase, an enzyme that covalently links DNA backbone chains.

Rapid-mix technique, a technique exploiting short pulses of radiation from an accelerator in which unirradiated samples are mixed with irradiated samples very shortly after irradiation.

Recombinational repair, DNA repair mediated by some of the enzymes responsible for genetic recombination.

Renaturation, return of denatured DNA to its native form.

Replicative form, a double-stranded form of a normally single-stranded phage DNA molecule. It is an intermediate during DNA replication.

Stationary phase, the terminal phase in the growth of a culture of cells in which the number of cells approaches a limiting value.

Template, a macromolecular mold for the synthesis of another macromolecule.

Transformation, genetic modification of a cell induced by the incorporation into that cell of pure DNA from another cell or virus.

Tumor, a contiguous mass of cancer cells.

Wild-type gene, the form of the gene commonly found in nature as distinct from the mutated form.

References

1. B. A. Bridges, *Advances in Radiation Biology*, Vol. 3, L. G. Augenstein, R. Mason, and M. Zelle, Eds., Academic Press, New York, 1969, p. 123.
2. E. L. Powers and A. Tallentire, *Actions Chimiques et Biologiques des Radiations*, Vol. 12, M. Haissinsky, Ed., Masson, Paris, 1968, p. 3.
3. R. B. Setlow and E. C. Pollard, *Molecular Biophysics*, Pergamon Press, Oxford, 1968, p. 16.
4. J. Cairns, *Cold Spring Harbor Symp. Quant. Biol.*, **28,** 43 (1963).
5. R. Beukers, J. Ijlstra, and W. Berends, *Rec. Trav. Chim.*, **79,** 101 (1960).
6. A. Wacker, H. Dellveg, and D. Weinblum, *J. Mol. Biol.*, **3,** 787 (1961).
7. P. Howard-Flanders, *Ann. Rev. Biochem.*, **37,** 175 (1968).
8. W. Szybalski and Z. Lorkiewicz, *Abhandl. Deutsch. Akad. Wiss. Berlin, Kl. Med.*, **1,** 63 (1962).
9. R. H. Haynes, *Radiation Res. Suppl.*, **6,** 1 (1966).
10. P. Howard-Flanders, *Brookhaven Symp. Biol.*, **14,** 18 (1961).
11. H. S. Kaplan, K. C. Smith, and P. A. Tomlin, *Radiation Res.*, **16,** 98 (1962).
12. H. S. Kaplan and L. E. Moses, *Science*, **145,** 21 (1964).
13. M. Terzi, *Nature*, **191,** 461 (1961).
14. R. H. Haynes, *Physical Processes in Radiation Biology*, L. Augenstein, R. Mason, and B. Rosenberg, Eds., Academic Press, New York, 1964, p. 51.
15. H. S. Kaplan and R. Zavarine, *Biochem. Biophys. Res. Commun.*, **8,** 432 (1962).
16. A. D. Hershey and M. Chase, *J. Gen. Physiol.*, **36,** 39 (1952).
17. T. Alper, *Proceedings of the Second Symposium on Microdensitometry*, H. G. Ebert, Ed., Euratom, Brussels, 1969, p. 5.
18. J. N. Mehrishi, in *Radiation Protection and Sensitization*, H. L. Moroson and M. Quintiliani, Eds., Taylor and Francis, London, 1970, p. 265.
19. P. Howard-Flanders and D. Moore, *Radiation Res.*, **9,** 422 (1958).
20. A. Hollaender, G. E. Stapleton, and F. L. Martin, *Nature*, **167,** 103 (1951).
21. H. B. Hewitt and C. W. Willson, *Brit. J. Cancer*, **13,** 675 (1959).
22. J. Read, *British J. Radiol.*, **25,** 89, 154 (1952).
23. P. Howard-Flanders and E. A. Wright, *Nature*, **175,** 428 (1955).
24. H. B. Hewitt and J. Read, *Brit. J. Radiol.*, **23,** 416 (1950).
25. H. E. Ephrussi-Taylor and R. Latarjet, *Biochim. Biophys. Acta*, **16,** 183 (1955).
26. H. Marcovich, *Radiation Res.*, **9,** 149 (1958).
27. P. Howard-Flanders, *Nature*, **186,** 485 (1960).
28. G. Scholes, J. Weiss, and C. M. Wheeler, *Nature*, **178,** 157 (1956).
29. G. Scholes and J. Weiss, *Radiation Res., Suppl.*, **1,** 177 (1959).
30. B. Ekert and R. Monier, *Nature*, **184,** BA58 (1959).
31. G. Scholes, J. F. Ward, and J. Weiss, *J. Mol. Biol.*, **2,** 379 (1960).
32. R. Latarjet, B. Ekert, S. Apelgot, and N. Rebeyrotte, *J. Chim. Phys.*, **58,** 1046 (1961).
33. P. T. Emmerson, *Radiation Res.*, **22,** 187 (1964).
34. G. Scholes, *Radiation Chemistry of Aqueous Systems*, G. Stein, Ed., Interscience, New York, 1968, p. 259.
35. H. Loman and J. Block, *Radiation Res.*, **36,** 1 (1968).
36. E. J. Hart, *J. Am. Chem. Soc.*, **76,** 4198, 4312 (1954).
37. P. T. Emmerson, G. Scholes, D. H. Thomson, J. F. Ward, and J. J. Weiss, *Nature*, **187,** 319 (1960).
38. G. Hems, *Nature*, **196,** 849 (1961).
39. G. Scholes, *Progr. Biophys.*, **13,** 39 (1963).

40. R. E. Cline, and R. M. Fink, *Anal. Chem.*, **28,** 47 (1956).

41. C. Nofre and A. Cier, *Bull. Soc. Chim. France*, 1326 (1966).

42. G. Scholes and R. L. Willson, *Trans. Faraday Soc.*, **63,** 2983 (1967).

43. C. J. Hochanadel and R. Casey, *Radiation Res.*, **25,** 198 (1965).

44. B. Ekert, *Nature*, **194,** 278 (1962).

45. K. C. Smith and J. E. Hays, *Radiation Res.*, **33,** 129 (1968).

46. Z. M. Bacq and P. Alexander, in *The Initial Effects of Ionizing Radiation on Cells*, R. J. C. Harris, Ed., Academic Press, New York, 1960, p. 301.

47. J. P. Lynch and P. Howard-Flanders, *Nature*, **194,** 1247 (1962).

48. R. L. Willson, *Intern. J. Radiation Biol.*, **17,** 349 (1970).

49. G. Scholes, R. L. Willson, and M. Ebert, *Chem. Commun.*, 17 (1969).

50. F. S. Dainton and D. B. Peterson, *Proc. Roy. Soc. (London)*, **A267,** 443 (1962).

51. K. F. Nakken and A. Pihl, *Radiation Res.*, **26,** 519 (1965).

52. W. C. Summers, Ph.D. Thesis, University of Wisconsin (1967).

53. P. A. Cerutti, N. Miller, M. G. Pleiss, F. J. Remsen, and W. J. Ramsay, *Proc. Natl. Acad. Sci., U.S.*, **64,** 731 (1969).

54. P. V. Hariharan and P. A. Cerutti, *Nature, New Biology*, **229,** 247 (1971).

55. J. Holian and W. G. Garrison, *Nature*, **212,** 394 (1966).

56. J. Holian and W. G. Garrison, *J. Phys. Chem.*, **71,** 462 (1964).

57. J. Holian and W. G. Garrison, *Chem. Commun.*, **14,** 676 (1967).

58. W. G. Garrison, in *Current Topics in Radiation Research*, Vol. IV, M. Ebert and A. Howard, Eds., Wiley, New York, 1968, p. 43.

59. G. Hems, *Nature*, **186,** 710 (1960).

60. F. W. Studier, *J. Mol. Biol.*, **11,** 373 (1965).

61. R. A. McGrath and R. W. Williams, *Nature*, **212,** 534 (1966).

62. C. J. Dean, R. W. Serianni, M. G. Ormerod, and P. Alexander, *Nature*, **222,** 1042 (1969).

63. H. S. Kaplan, *Proc. Natl. Acad. Sci., U.S.*, **55,** 1442 (1966).

64. J. T. Lett, I. Caldwell, C. J. Dean, and P. Alexander, *Nature*, **214,** 790 (1967).

65. P. H. M. Lohman, *Mutation Res.*, **6,** 449 (1968).

66. R. M. Humphrey, D. L. Steward, B. A. Sedita, *Mutation Res.*, **6,** 459 (1968).

67. H. Matsudaira, C. Nakagawa, and T. Hishizawa, *Intern. J. Radiation Biology*, **15,** 95 (1969).

68. W. Veatch and S. Okada, *Biophys. J.*, **9,** 330 (1969).

69. M. Furlan and H. Moroson, *Biophys. J.*, A-139 (1969).

70. M. M. Elkind and C. Kamper, *Radiation Res.*, **39,** 518 (1969).

71. T. Terasima and A. Tsuboi, *Biochim. Biophys. Acta*, **174,** 309 (1969).

72. R. P. Boyce and M. Tepper, *Virology*, **34,** 344 (1968).

73. E. T. Young and R. L. Sinsheimer, *J. Mol. Biol.*, **10,** 562 (1964).

74. I. Johansen, W. D. Rupp, and I. Gurvin, submitted for publication.

75. E. C. Pollard and P. K. Weller, *Radiation Res.*, **32,** 417 (1967).

76. W. D. Rupp, E. Zipser, C. von Essen, D. Reno, L. Prosnitz, and P. Howard-Flanders in *Time and Dose Relationships in Radiation Biology as Applied to Radiation Therapy*, V. P. Bond et al., Eds., Natl. Lab. Publ. 50203 (C-57), Upton, N.Y. 1970, p. 1.

77. D. Freifelder, *Radiation Res.*, **29,** 329 (1966).

78. P. Alexander, C. J. Dean, A. R. Lehmann, M. G. Ormerod, P. Feldschreiber, and R. W. Serianni in *Radiation Protection and Sensitization*, H. L. Moroson and M. Quintiliani, Eds., Taylor and Francis, London, 1970, p. 15.

79. D. S. Kapp and K. C. Smith, *Radiation Res.*, **42,** 34 (1970).

80. G. Scholes, P. Shaw, R. L. Willson, and M. Ebert, *Pulse Radiolysis*, M. Ebert et al., Eds., Academic Press, New York, 1965, p. 151.
81. R. Braams, *Pulse Radiolysis*, M. Ebert et al., Eds., Academic Press, New York, 1965, p. 171.
82. W. C. Summers and W. Szybalski, *J. Mol. Biol.*, **26,** 107 (1967).
83. D. Freifelder, *Proc. Natl. Acad. Sci., U.S.*, **54,** 128 (1965).
84. W. C. Summers and W. Szybalski, *J. Mol. Biol.*, **26,** 227 (1967).
85. G. Scholes and J. Weiss, *Exptl. Cell Res. Suppl.*, **2,** 219 (1952).
86. B. Collyns, S. Okada, G. Scholes, J. J. Weiss, and C. M. Wheeler, *Radiation Res.*, **25,** 526 (1965).
87. M. Daniels, G. Scholes, and J. Weiss, *J. Chem. Soc.*, 3771 (1956).
88. J. Weiss, *Progr. Nucleic Acid Res. Mol. Biol.*, **3,** 36 (1964).
89. N. P. Krushinskaya and M. I. Shal'nov, *Radiobiology*, **7,** 36 (1967). (English translation, U.S. Atomic Energy Commission.)
90. D. Freifelder, *J. Mol. Biol.*, **35,** 303 (1968).
91. W. Veatch and S. Okada, *Biophys. J.*, **9,** 330 (1969).
92. S. W. Englander, A. Buzzell, and M. A. Lauffer, *Biochim. Biophys. Acta*, **40,** 385 (1960).
93. B. A. Bridges, *Nature*, **188,** 415 (1960).
94. B. A. Bridges, *J. Gen. Microbiol.*, **26,** 467 (1961).
95. L. Mullenger and M. G. Ormerod, *Intern. J. Radiation Biol.*, **15,** 259 (1969).
96. I. Johansen and P. Howard-Flanders, *Radiation Res.*, **24,** 184 (1965).
97. G. E. Adams, G. S. McNaughton, and B. D. Michael, *Trans. Faraday Soc.*, **64,** 902 (1968).
98. G. E. Adams in *Current Topics in Radiation Research*, Vol. 3, M. Ebert and A. Howard, Eds., Wiley, New York, 1967, p. 35.
99. P. T. Emmerson, *Radiation Res.*, **30,** 841 (1967).
100. W. A. Cramp, *Nature*, **206,** 636 (1965).
101. A. K. Bruce and W. H. Malchman, *Radiation Res.*, **24,** 473 (1965).
102. B. A. Bridges, *Radiation Res.*, **16,** 232 (1962).
103. C. J. Dean, *Brit. J. Radiol.*, **35,** 73 (1962).
104. D. L. Dewey and B. D. Michael, *Biochem., Biophys. Res. Commun.*, **21,** 392 (1968).
105. G. E. Adams, M. S. Cooke, and B. D. Michael, *Nature*, **219,** 1368 (1968).
106. D. L. Dewey, *Progr. Biochem. Pharmacol.*, **1,** 59 (1965).
107. G. E. Adams and D. L. Dewey, *Biochem. Biophys. Res. Commun.*, **12,** 473 (1963).
108. J. Blok, L. H. Luthjens, and A. L. M. Roos, *Radiation Res.*, **30,** 468 (1967).
109. G. E. Adams, *Radiation Chemistry of Aqueous Solutions*, G. Stein, Ed., Interscience, New York, 1968, p. 241.
110. G. E. Adams and M. S. Cooke, *Intern. J. Radiation Biol.*, **15,** 457 (1969).
111. M. J. Ashwood-Smith, D. M. Robinson, J. H. Barnes, and B. A. Bridges, *Nature*, **216,** 137 (1967).
112. M. J. Ashwood-Smith, J. Barnes, J. Huckle, and B. A. Bridges, *Radiation Protection and Sensitization*, H. L. Moroson and M. Quintiliani, Eds., Taylor and Francis, London, 1970, p. 183.
113. J. C. Asquith, R. L. Willson, and G. E. Adams, *4th International Congress of Radiation Research*, 1970, Abstract 27.
114. C. G. Greenstock, G. E. Adams and R. L. Willson, *Radiation Protection and Sensitization*, H. L. Moroson and M. Quintiliani, Eds., Taylor and Francis, London, 1970, p. 65.
115. I. Johansen, J. F. Ward, K. Siegel, and A. Sletten, *Biochem. Biophys. Res. Commun.*, **33,** 949 (1968).

116. R. L. Willson and P. T. Emmerson, in *Radiation Protection and Sensitization*, H. L. Moroson and M. Quintiliani, Eds., Taylor and Francis, London, 1970, p. 73.

117. M. Klimek, *Neoplasma*, **13**, 31 (1966).

118. H. Moroson and M. Furlan, *Intern. J. Radiation Biol.*, **13**, 585 (1968).

119. P. Howard-Flanders, *Nature*, **180**, 1191 (1957).

120. P. Howard-Flanders and P. Jockey, *Radiation Res.*, **13**, 466 (1960).

121. L. A. K. Stavely and C. N. Hinshelwood, *Proc. Roy Soc. (London)* **A154**, 335 (1936).

122. Z. Fuks, K. C. Smith and H. S. Kaplan, *J. Bacteriology*, submitted for publication.

123. W. M. Dale, J. V. Davies, and C. Russell, *Nature*, **189**, 851 (1961).

124. W. M. Dale, J. V. Davies, and C. Russell, *Intern. J. Radiation Biol.*, **4**, 1 (1961).

125. S. Gordon, E. J. Hart, M. S. Matheson, J. Rabani, and J. K. Thomas, *Discuss. Faraday Soc.*, **36**, 193 (1963).

126. J. P. Lynch and P. Howard-Flanders, *Nature*, **194**, 1247 (1962).

127. B. A. Kihlman, *Exptl. Cell Research*, **14**, 639 (1958).

128. L. H. Gray, F. O. Green, and C. A. Hawes, *Nature*, **182**, 952 (1958).

129. T. H. Chang, F. D. Wilson, and W. S. Stone, *Proc. Natl. Acad. Sci., U.S.*, **45**, 1397 (1959).

130. B. Sparrman, L. Ehrenberg, and A. Ehrenberg, *Acta Chem. Scand.*, **13**, 199 (1959).

131. E. L. Powers, B. F. Kaleta, and R. B. Webb, *Radiation Res.*, **7**, 461 (1959).

132. P. Howard-Flanders, *Advan. Biol. Med. Phys.*, **6**, 553 (1958).

133. O. L. Lebedev and S. N. Kazarnovskii, *Tr. po Khim. i Khim. Tekhnol.*, **2**, 649 (1959).

134. A. K. Hoffmann and A. T. Henderson, *J. Am. Chem. Soc.*, **83**, 4671 (1961).

135. R. Briere, H. Lemaire, and A. Rassat, *Tetrahedron Letters*, 1775 (1964).

136. A. R. Forrester and R. H. Thomson, *Nature*, **203**, 74 (1964).

137. R. Dupeyre and A. Rassat, *J. Am. Chem. Soc.*, **88**, 3180 (1966).

138. J. W. Linnett, *J. Amer. Chem. Soc.*, **83**, 2643 (1961).

139. G. Chapelet-Letourneux, H. Lemaire, and A. Rassat, *Bull. Soc. Chim. France*, **11**, 3283 (1965).

140. G. M. Coppinger and J. D. Swallen, *J. Am. Chem. Soc.*, **83**, 4900 (1961).

141. K. C. Smith in *Photophysiology*, A. G. Giese, Ed., Vol. 6, Academic Press, New York, 1971, in press.

142. R. Briere, H. Lemaire, and A. Rassat, *Bull. Soc. Chim. France*, **11**, 3273 (1965).

143. P. T. Emmerson, *Brit. J. Radiol.*, **43**, 429 (1970).

144. P. T. Emmerson and P. Howard-Flanders, *Nature*, **204**, 1005 (1964).

145. P. T. Emmerson and P. Howard-Flanders, *Radiation Res.*, **26**, 54 (1965).

146. C. D. Town, K. C. Smith, and H. S. Kaplan, *Science*, **172**, 851 (1971).

147. P. T. Emmerson, *Radiation Res.*, **36**, 410 (1968).

148. J. H. Barnes, M. J. Ashwood-Smith, and B. A. Bridges, *Intern. J. Radiation Biol.*, **15**, 285 (1969).

149. G. E. Adams, in *Radiation Protection and Sensitization*, H. L. Moroson and M. Quintiliani, Eds., Taylor and Francis, London, 1970, p. 3.

150. H. B. Hewitt and A. S. Walder, *Brit. J. Radiol.*, **42**, 318 (1969).

151. P. T. Emmerson, M. Fielden, and I. Johansen, *Intern. J. Radiation Biol.*, **19**, 229 (1971).

152. L. Parker, L. D. Skarsgard, and P. T. Emmerson, *Radiation Res.*, **38**, 493 (1969).

153. L. Revesz and B. Littbrand, *Radiation Protection and Sensitization*, H. L. Moroson and M. Quintiliani, Eds., Taylor and Francis, London, 1970, p. 217.

154. H. B. Hewitt and E. R. Blake, *Brit. J. Radiol.*, **43**, 91 (1970).

155. J. L. Foster, *Radiation Protection and Sensitization*, H. L. Moroson and M. Quintiliani, Eds., Taylor and Francis, London, 1970, p. 299.

156. P. T. Emmerson and R. L. Willson, *J. Phys. Chem.*, **72**, 3669 (1968).

157. K. F. Nakken, T. Sikkeland, and T. Brustad, *FEBS Letters*, **8,** 33 (1970).

158. W. D. Blackley and R. R. Reinhard, *J. Am. Chem. Soc.*, **87,** 802 (1965).

159. P. T. Emmerson, *Radiation Protection and Sensitization*, H. L. Moroson and M. Quintiliani, Eds., Taylor and Francis, London, 1970, p. 147.

160. W. D. Rupp and P. Howard-Flanders, *J. Mol. Biol.*, **31,** 291 (1968).

161. I. Johansen, *Norwegian Defense Research Establishment Report*, No. 56, 1968.

162. B. A. Bridges and R. J. Munson, *Intern. J. Radiation Biol.*, **13,** 179 (1967).

163. L. Grossman, J. Kaplan, S. Kushner and I. Mahler, *Ann. Ist. Super. Sanita*, **5,** 318 (1969).

164. R. P. Boyce and P. Howard-Flanders, *Z. Vererbungslehre*, **95,** 345 (1964).

165. P. T. Emmerson and P. Howard-Flanders, *Biochem. Biophys. Res. Commun.*, **18,** 24 (1965).

166. R. B. Kelly, M. R. Atkinson, J. A. Huberman, and A. Kornberg, *Nature*, **224,** 495 (1969).

167. B. Olivera and I. R. Lehman, *Proc. Natl. Acad. Sci. U.S.*, **57,** 1426 (1967).

168. D. S. Kapp and K. C. Smith, *Intern. J. Radiation Biol.*, **14,** 567 (1969).

169. D. S. Kapp and K. C. Smith, *J. Bacteriol.*, **103,** 49 (1970).

170. D. Freifelder, *J. Mol. Biol.*, **35,** 95 (1968).

171. W. D. Rupp and G. Ihler, *Cold Spring Harbor Symp. Quant. Biol.*, **33,** 647 (1968).

172. G. Buttin and M. Wright, *Cold Spring Harbor Symp. Quant. Biol.*, **33,** 259 (1968).

173. M. Oishi, *Proc. Natl. Acad. Sci. U.S.*, **64,** 1292 (1969).

174. A. L. Taylor, *Bacteriol. Rev.*, **34,** 155 (1970).

175. J. S. Mitchell and I. Simon-Reuss, *Nature*, **160,** 98 (1947).

176. J. S. Mitchell, in *Radiation Chemistry of Aqueous Solutions*, G. Stein, Ed., Interscience, New York, 1968, p. 287.

177. H. S. Kaplan, *Radiation Protection and Sensitization*, H. L. Moroson and M. Quintiliani, Eds., Taylor and Francis, London, 1970, p. 35.

178. K. Sano, T. Hoshimo, and M. Nagai, *J. Neurosurg.*, **28,** 530 (1968).

179. S. P. Makarov, A. I. Iakubovich, S. S. Dubov, and A. N. Medvedev, *Dokl. Akad. Nauk. SSSR*, **160,** 1319 (1965).

180. E. C. Easson in *Current Topics in Radiation Research*, Vol. 3, M. Ebert and A. Howard, Eds., Wiley, New York, 1967, p. 175.

181. L. G. Lajtha and R. Oliver, *Brit. J. Radiol.*, **35,** 659 (1962).

182. B. A. Bridges, *Brit. J. Radiol.*, **35,** 290 (1962).

183. M. M. Kligerman and L. Schulhof, *Radiation Res.*, **39,** 571 (1969).

184. J. R. Cummings, J. L. Grace, and C. N. Latimer, *J. Pharmacol. Exptl. Therapeutics*, **141,** 349 (1963).

185. D. A. Agnew and L. D. Skarsgard, *Proc. 4th International Congress of Radiation Research, Evian, France, 1970*, Abstract 7.

186. T. Alper, *Nature*, **217,** 862 (1968).

187. I. Churchill-Davidson, *Modern Trends in Radiotherapy*, Butterworths, London, 1967.

188. A. Bopp and U. Hagen, *Biochem. Biophys. Acta*, **209,** 320 (1970).

189. S. Hornsey, M. J. Hedges, and P. E. Bryant, *Intern. J. Radiation Biol.*, **13,** 581 (1968).

190. F. Vogel, *Nature*, **201,** 847 (1964).

191. F. Crick, *Nature*, **234,** 25 (1971).

Author Index

Numbers in parentheses are reference numbers and show that an author's work is referred to although his name is not mentioned in the text. Numbers in *italics* indicate the pages on which the full references appear.

Subject Index

See Tables of Contents for titles of larger content groupings.

Check both formula and word representations for all elements, radicals, ions and compounds. Such check is particularly necessary for cases in which isotopic effects may be expected. Not all species listed in the various tables are necessarily included in the index. In case of doubt, refer also to the appropriate table.

FORMULA REPRESENTATIONS

Note that formula representations are difficult to alphabetize. They require careful search to insure maximum utility.

WORD REPRESENTATIONS

In indexing, the noun, rather than a modifying adjective, is usually taken as the key word. However, commonly employed word pairs, etc. may occasionally be indexed as to the first; e.g., Alkyl chlorides, Michaelis-Menten equation, Pulse radiolysis. The reader is advised, in general to try the first word, as well as the second (or third), in such groupings. Definitions of some unfamiliar biochemical and biological terms may be found on pages 264 and 265.